Engineering Mathematics

Engineering Mathematics

Edited by **Matt Ferrier**

CLANRYE INTERNATIONAL

New Jersey

Published by Clanrye International,
55 Van Reypen Street,
Jersey City, NJ 07306, USA
www.clanryeinternational.com

Engineering Mathematics
Edited by Matt Ferrier

International Standard Book Number: 978-1-63240-211-0 (Hardback)

Printed in the United States of America.

Contents

Preface

Engineering mathematics is a branch that uses mathematical tools, techniques and methodologies to solve issues in engineering and industries. Engineering mathematics mixed with engineering sciences, which includes engineering physics and engineering geology, builds a unique platform which helps in practical and theoretical concept for effective implementation in problem solving. Engineering mathematics mostly studies applied analysis such as differential equations, real and complex analysis, approximation theory, asymptotic, variational, and perturbative methods, representations, numerical analysis, fourier analysis, potential theory; as well as linear algebra and applied probability. Apart from applied analysis, this field of mathematics also includes Newtonian physics, mathematical physics, classical mechanics, and fluid mechanics. The current advancement of numerical computer methods and software developments in engineering mathematics has given birth to computational mathematics, computational engineering, and high performance simulation processes.

This book, which primarily focuses on engineering mathematics, is dedicated to study the effects of mass heat transfer, thermal radiation, spline techniques and viscous dissipation.

Numerical solution of fractional diffusion equation model for freezing in finite media, initiating a mathematical model for prediction of 6-dof motion of planning crafts in regular waves, analytical treatment and convergence of the adomian decomposition method for instability phenomena are relevant model applications in engineering mathematics.

The chapters in this book have been contributed by specialists and mathematicians. I wish to thank them all for their contributions and assistance in the production of this book. Lastly, I would extend gratitude to my family.

Editor

1

Delay-Partitioning Approach to Stability of Linear Discrete-Time Systems with Interval-Like Time-Varying Delay

Priyanka Kokil, V. Krishna Rao Kandanvli, and Haranath Kar

Department of Electronics and Communication Engineering, Motilal Nehru National Institute of Technology Allahabad, Allahabad 211004, India

Correspondence should be addressed to Priyanka Kokil; kokilnit@gmail.com

Academic Editor: J. A. Tenreiro Machado

This paper is concerned with the problem of global asymptotic stability of linear discrete-time systems with interval-like time-varying delay in the state. By utilizing the concept of delay partitioning, a new linear-matrix-inequality-(LMI-) based criterion for the global asymptotic stability of such systems is proposed. The proposed criterion does not involve any free weighting matrices but depends on both the size of delay and partition size. The developed approach is extended to address the problem of global asymptotic stability of state-delayed discrete-time systems with norm-bounded uncertainties. The proposed results are compared with several existing results.

1. Introduction

A source of instability for discrete-time systems is time delay which inevitably exists in various engineering systems. In many applications, time delays are unavoidable and must be taken into account in a realistic system design, for instance, chemical processes, echo cancellation, local loop equalization, multipath propagation in mobile communication, array signal processing, congestion analysis and control in high speed networks, neural networks, and long transmission line in pneumatic systems [1–6]. The stability analysis of time-delay systems has received considerable attention during the last two decades [3–7]. According to dependence of delay, the existing stability criteria are generally classified into two categories: delay-dependent criteria and delay-independent criteria. It is well known that delay-independent stability criteria are usually more conservative than the delay-dependent ones, especially if the size of time delay is small [6–10]. Therefore, much attention has been paid in recent years to the study of delay-dependent stability criteria.

A number of publications relating to the delay-dependent stability of continuous time-delay systems have appeared (see, e.g., [10–17] and the references cited therein). In contrast, less attention has been paid to studying the problem of stability of discrete time-delay systems. Several delay-dependent criteria for the stability of discrete-time systems have appeared [3, 6, 7, 18, 19]. Reference [7] (see [6] also) presents a novel delay-dependent linear-matrix-inequality-(LMI-) based condition for the global asymptotic stability of linear discrete-time systems with interval-like time-varying delay. The criteria proposed in [6, 7] are less conservative with smaller numerical complexity than [19–23]. The delay partitioning approach has been efficiently applied in [24–30] to the stability analysis of systems with time-varying delays. In the context of stability analysis of linear discrete systems with time-varying delay, the delay partitioning concept is first utilized in [30]. The approaches in [29, 30] divide the lower bound of the time-varying delay into a number of partitions. The criteria in [29, 30] are not only delay dependent but also dependent on the partitioning size. Though the approaches in [29, 30] provide less conservative stability results than [19, 20], these approaches would lead to heavier computational burden and more complicated synthesis procedure.

This paper studies the problem of stability analysis of linear discrete-time system with interval-like time-varying delay in the state. In this paper, inspired by the work of [6, 7, 30], an alternative to the approach presented in [29] for the stability analysis of linear discrete-time systems with interval-like time-varying delay in the state is brought out. The proposed method exploits the delay partitioning idea and does not introduce any free weighting matrices. The paper is organized as follows. Section 2 presents a description of the system under consideration. A novel LMI-based criterion for the global asymptotic stability of discrete-time state-delayed systems is proposed in Section 3. The proposed criterion depends on the size of delay as well as partition size. In Section 4, the approach is extended to derive global asymptotic stable conditions for delayed discrete-time systems with norm-bounded uncertainties. Finally, in Section 5, numerical examples are given to illustrate the effectiveness of the presented results.

2. System Description

The following notations are used throughout the paper:

$\mathbf{R}^{p\times q}$: Set of $p \times q$ real matrices,

$\mathbf{R}^{p}$: Set of $p \times 1$ real vectors,

$\mathbf{I}$: Identity matrix of appropriate dimension; the order is specified in subscript as the need arises,

$\mathbf{0}$: Null matrix or null vector of appropriate dimension; the orders are specified in subscripts as the need arises,

$\mathbf{B}^T$: Transpose of the matrix (or vector) $\mathbf{B}$,

$\mathbf{B} > \mathbf{0}$: $\mathbf{B}$ is positive-definite symmetric matrix,

$\mathbf{B} < \mathbf{0}$: $\mathbf{B}$ is negative-definite symmetric matrix.

The system under consideration is given by

$$\mathbf{x}(k+1) = \mathbf{A}\mathbf{x}(k) + \mathbf{A}_1\mathbf{x}(k-d(k)), \tag{1}$$

$$\mathbf{x}(k) = \phi(k), \quad k = -h_2, -h_2+1, \ldots, 0, \tag{2}$$

where $\mathbf{x}(k) \in \mathbf{R}^n$ is the system state vector, $\mathbf{A}$ and $\mathbf{A}_1$ are constant matrices with appropriate dimensions, $d(k)$ is a positive integer representing interval-like time-varying delay satisfying

$$1 \le h_1 \le d(k) \le h_2, \tag{3}$$

where h_1 and h_2 are known positive integers representing the lower and upper delay bounds, respectively, and $\phi(k)$ is an initial value at time k. Let the lower bound of the delay h_1 be divided into m number of partitions such that

$$h_1 = \tau m, \tag{4}$$

where τ is an integer representing partition size.

3. Proposed Criterion

In this section, inspired by the work of [6, 7, 30], an LMI-based criterion for the global asymptotic stability of system (1)–(4) is established.

The main result may be stated as follows.

Theorem 1. *For given positive integers τ, m, and h_2, the system in (1)–(4) is asymptotically stable if there exist real matrices* $\mathbf{P} = \mathbf{P}^T > \mathbf{0}$, $\mathbf{Q}_i = \mathbf{Q}_i^T > \mathbf{0}$ $(i = 1, 2, 3)$, *and* $\mathbf{Z}_i = \mathbf{Z}_i^T > \mathbf{0}$ $(i = 1, 2)$ *such that*

$$\begin{aligned}\mathbf{\Psi}_1 &= \mathbf{\Psi} - \left[\mathbf{0}_{n\times mn}\ \mathbf{I}_n\ -\mathbf{I}_n\ \mathbf{0}_n\right]^T \mathbf{Z}_2 \\ &\quad \times \left[\mathbf{0}_{n\times mn}\ \mathbf{I}_n\ -\mathbf{I}_n\ \mathbf{0}_n\right] < \mathbf{0}, \\ \mathbf{\Psi}_2 &= \mathbf{\Psi} - \left[\mathbf{0}_{n\times(m+1)n}\ \mathbf{I}_n\ -\mathbf{I}_n\right]^T \mathbf{Z}_2 \\ &\quad \times \left[\mathbf{0}_{n\times(m+1)n}\ \mathbf{I}_n\ -\mathbf{I}_n\right] < \mathbf{0},\end{aligned} \tag{5}$$

where

$$\begin{aligned}\mathbf{\Psi} &= \mathbf{\Lambda}_1^T\mathbf{\Phi}_1\mathbf{\Lambda}_1 + \mathbf{\Lambda}_2^T\mathbf{\Phi}_2\mathbf{\Lambda}_2 + \mathbf{\Lambda}_3^T\mathbf{\Phi}_3\mathbf{\Lambda}_3 - \mathbf{\Lambda}_4^T\mathbf{\Phi}_4\mathbf{\Lambda}_4 + \mathbf{\Lambda}_1^T\mathbf{\Phi}_5\mathbf{\Lambda}_3 \\ &\quad + \mathbf{\Lambda}_3^T\mathbf{\Phi}_5^T\mathbf{\Lambda}_1 - \mathbf{\Lambda}_5^T\mathbf{Z}_1\mathbf{\Lambda}_5 + \mathbf{\Lambda}_1^T\mathbf{Z}_1\mathbf{\Lambda}_5 + \mathbf{\Lambda}_5^T\mathbf{Z}_1\mathbf{\Lambda}_1 - \mathbf{\Lambda}_6^T\mathbf{Z}_2\mathbf{\Lambda}_6 \\ &\quad + \mathbf{\Lambda}_3^T\mathbf{Z}_2\mathbf{\Lambda}_4 + \mathbf{\Lambda}_4^T\mathbf{Z}_2\mathbf{\Lambda}_3 + \mathbf{\Lambda}_6^T\mathbf{Z}_2\mathbf{\Lambda}_3 + \mathbf{\Lambda}_3^T\mathbf{Z}_2\mathbf{\Lambda}_6,\end{aligned} \tag{6}$$

$$\mathbf{\Lambda}_1 = \left[\mathbf{I}_n\ \mathbf{0}_{n\times(m+2)n}\right], \qquad \mathbf{\Lambda}_2 = \begin{bmatrix} \mathbf{I}_{mn} & \mathbf{0}_{mn\times 3n} & \\ \mathbf{0}_{mn\times n} & \mathbf{I}_{mn} & \mathbf{0}_{mn\times 2n}\end{bmatrix}, \tag{7}$$

$$\mathbf{\Lambda}_3 = \left[\mathbf{0}_{n\times(m+1)n}\ \mathbf{I}_n\ \mathbf{0}_n\right], \qquad \mathbf{\Lambda}_4 = \left[\mathbf{0}_{n\times(m+2)n}\ \mathbf{I}_n\right], \tag{8}$$

$$\mathbf{\Lambda}_5 = \left[\mathbf{0}_n\ \mathbf{I}_n\ \mathbf{0}_{n\times(m+1)n}\right], \qquad \mathbf{\Lambda}_6 = \left[\mathbf{0}_{n\times mn}\ \mathbf{I}_n\ \mathbf{0}_{n\times 2n}\right], \tag{9}$$

$$\begin{aligned}\mathbf{\Phi}_1 &= \mathbf{A}^T\mathbf{P}\mathbf{A} - \mathbf{P} + \mathbf{Q}_2 + (h_2 - \tau m + 1)\,\mathbf{Q}_3 - \mathbf{Z}_1 \\ &\quad + (\mathbf{A} - \mathbf{I}_n)^T\left(\tau^2\mathbf{Z}_1 + (h_2 - \tau m)^2\mathbf{Z}_2\right)(\mathbf{A} - \mathbf{I}_n),\end{aligned} \tag{10}$$

$$\mathbf{\Phi}_2 = \begin{bmatrix}\mathbf{Q}_1 & \mathbf{0} \\ \mathbf{0} & -\mathbf{Q}_1\end{bmatrix}, \tag{11}$$

$$\mathbf{\Phi}_3 = \mathbf{A}_1^T\left(\mathbf{P} + \tau^2\mathbf{Z}_1 + (h_2 - \tau m)^2\mathbf{Z}_2\right)\mathbf{A}_1 - \mathbf{Q}_3 - 2\mathbf{Z}_2, \tag{12}$$

$$\mathbf{\Phi}_4 = \mathbf{Q}_2 + \mathbf{Z}_2, \tag{13}$$

$$\mathbf{\Phi}_5 = \mathbf{A}^T\mathbf{P}\mathbf{A}_1 + (\mathbf{A} - \mathbf{I}_n)^T\left(\tau^2\mathbf{Z}_1 + (h_2 - \tau m)^2\mathbf{Z}_2\right)\mathbf{A}_1. \tag{14}$$

Proof. Consider a Lyapunov function [29, 30]

$$\begin{aligned}V(k) &= \mathbf{x}^T(k)\,\mathbf{P}\mathbf{x}(k) + \sum_{i=k-\tau}^{k-1}\mathbf{Y}^T(i)\,\mathbf{Q}_1\mathbf{Y}(i) \\ &\quad + \sum_{i=k-h_2}^{k-1}\mathbf{x}^T(i)\,\mathbf{Q}_2\mathbf{x}(i) + \sum_{j=-h_2}^{-\tau m}\sum_{i=k+j}^{k-1}\mathbf{x}^T(i)\,\mathbf{Q}_3\mathbf{x}(i) \\ &\quad + \sum_{j=-\tau}^{-1}\sum_{i=k+j}^{k-1}\tau\Delta\mathbf{x}^T(i)\,\mathbf{Z}_1\Delta\mathbf{x}(i) \\ &\quad + \sum_{j=-h_2}^{-\tau m-1}\sum_{i=k+j}^{k-1}(h_2 - \tau m)\,\Delta\mathbf{x}^T(i)\,\mathbf{Z}_2\Delta\mathbf{x}(i),\end{aligned} \tag{15}$$

where

$$\mathbf{\Upsilon}(i) = \left[\mathbf{x}^T(i)\ \ \mathbf{x}^T(i-\tau)\ \ \cdots\ \ \mathbf{x}^T(i-(m-1)\tau)\right]^T, \quad (16)$$

$$\Delta\mathbf{x}(i) = \mathbf{x}(i+1) - \mathbf{x}(i). \quad (17)$$

Taking the forward difference of (15) along the solution of the system (1)-(2), we have

$$\begin{aligned}
&\Delta V(k)\\
&= V(k+1) - V(k)\\
&= \left[\mathbf{A}\mathbf{x}(k) + \mathbf{A}_1\mathbf{x}(k-d(k))\right]^T \mathbf{P}\left[\mathbf{A}\mathbf{x}(k) + \mathbf{A}_1\mathbf{x}(k-d(k))\right]\\
&\quad - \mathbf{x}^T(k)\mathbf{P}\mathbf{x}(k) + \mathbf{\Upsilon}^T(k)\mathbf{Q}_1\mathbf{\Upsilon}(k) - \mathbf{\Upsilon}^T(k-\tau)\mathbf{Q}_1\mathbf{\Upsilon}(k-\tau)\\
&\quad + \mathbf{x}^T(k)\mathbf{Q}_2\mathbf{x}(k) - \mathbf{x}^T(k-h_2)\mathbf{Q}_2\mathbf{x}(k-h_2)\\
&\quad + (h_2 - \tau m + 1)\mathbf{x}^T(k)\mathbf{Q}_3\mathbf{x}(k) + \Delta\mathbf{x}^T(k)\tau^2\mathbf{Z}_1\Delta\mathbf{x}(k)\\
&\quad + \Delta\mathbf{x}^T(k)(h_2 - \tau m)^2\mathbf{Z}_2\Delta\mathbf{x}(k) - \sum_{i=k-h_2}^{k-\tau m}\mathbf{x}^T(i)\mathbf{Q}_3\mathbf{x}(i)\\
&\quad - \sum_{i=k-\tau}^{k-1}\tau\Delta\mathbf{x}^T(i)\mathbf{Z}_1\Delta\mathbf{x}(i)\\
&\quad - \sum_{i=k-h_2}^{k-\tau m-1}(h_2 - \tau m)\Delta\mathbf{x}^T(i)\mathbf{Z}_2\Delta\mathbf{x}(i).
\end{aligned} \quad (18)$$

Using Lemma 1 in [22], we obtain [6, 7]

$$\begin{aligned}
&-\sum_{i=k-\tau}^{k-1}\tau\Delta\mathbf{x}^T(i)\mathbf{Z}_1\Delta\mathbf{x}(i)\\
&\quad \le -(\mathbf{x}(k) - \mathbf{x}(k-\tau))^T\mathbf{Z}_1(\mathbf{x}(k) - \mathbf{x}(k-\tau)).
\end{aligned} \quad (19)$$

Note that

$$-\sum_{i=k-h_2}^{k-\tau m}\mathbf{x}^T(i)\mathbf{Q}_3\mathbf{x}(i) \le -\mathbf{x}^T(k-d(k))\mathbf{Q}_3\mathbf{x}(k\quad d(k)). \quad (20)$$

Define

$$\boldsymbol{\zeta}(k) = \left[\mathbf{\Upsilon}^T(k)\ \ \mathbf{x}^T(k-\tau m)\ \ \mathbf{x}^T(k-d(k))\ \ \mathbf{x}^T(k-h_2)\right]^T. \quad (21)$$

It follows from (1) and (17) that

$$\Delta\mathbf{x}(k) = (\mathbf{A} - \mathbf{I}_n)\mathbf{x}(k) + \mathbf{A}_1\mathbf{x}(k-d(k)). \quad (22)$$

Now, we have the following relation [6, 7]:

$$\begin{aligned}
&-\sum_{i=k-h_2}^{k-\tau m-1}(h_2 - \tau m)\Delta\mathbf{x}^T(i)\mathbf{Z}_2\Delta\mathbf{x}(i)\\
&\quad = -\sum_{i=k-h_2}^{k-d(k)-1}(h_2 - \tau m)\Delta\mathbf{x}^T(i)\mathbf{Z}_2\Delta\mathbf{x}(i)\\
&\qquad - \sum_{i=k-d(k)}^{k-\tau m-1}(h_2 - \tau m)\Delta\mathbf{x}^T(i)\mathbf{Z}_2\Delta\mathbf{x}(i)\\
&\quad = -\sum_{i=k-h_2}^{k-d(k)-1}(h_2 - d(k))\Delta\mathbf{x}^T(i)\mathbf{Z}_2\Delta\mathbf{x}(i)\\
&\qquad - \sum_{i=k-h_2}^{k-d(k)-1}(d(k) - \tau m)\Delta\mathbf{x}^T(i)\mathbf{Z}_2\Delta\mathbf{x}(i)\\
&\qquad - \sum_{i=k-d(k)}^{k-\tau m-1}(h_2 - d(k))\Delta\mathbf{x}^T(i)\mathbf{Z}_2\Delta\mathbf{x}(i)\\
&\qquad - \sum_{i=k-d(k)}^{k-\tau m-1}(d(k) - \tau m)\Delta\mathbf{x}^T(i)\mathbf{Z}_2\Delta\mathbf{x}(i).
\end{aligned} \quad (23)$$

Letting $\beta = (d(k) - \tau m)/(h_2 - \tau m)$, one obtains [6]

$$\begin{aligned}
&-\sum_{i=k-h_2}^{k-d(k)-1}(d(k) - \tau m)\Delta\mathbf{x}^T(i)\mathbf{Z}_2\Delta\mathbf{x}(i)\\
&\quad = -\beta\sum_{i=k-h_2}^{k-d(k)-1}(h_2 - \tau m)\Delta\mathbf{x}^T(i)\mathbf{Z}_2\Delta\mathbf{x}(i)\\
&\quad \le -\beta\sum_{i=k-h_2}^{k-d(k)-1}(h_2 - d(k))\ \Delta\mathbf{x}^T(i)\mathbf{Z}_2\Delta\mathbf{x}(i),\\
&-\sum_{i=k-d(k)}^{k-\tau m-1}(h_2 - d(k))\Delta\mathbf{x}^T(i)\mathbf{Z}_2\Delta\mathbf{x}(i)\\
&\quad = -(1-\beta)\sum_{i=k-d(k)}^{k-\tau m-1}(h_2 - \tau m)\Delta\mathbf{x}^T(i)\mathbf{Z}_2\Delta\mathbf{x}(i)\\
&\quad \le -(1-\beta)\sum_{i=k-d(k)}^{k-\tau m-1}(d(k) - \tau m)\Delta\mathbf{x}^T(i)\mathbf{Z}_2\Delta\mathbf{x}(i).
\end{aligned} \quad (24)$$

Applying Lemma 1 in [22] and using (23)-(24), we have [6, 7]

$$\begin{aligned}
&-\sum_{i=k-h_2}^{k-\tau m-1}(h_2 - \tau m)\Delta\mathbf{x}^T(i)\mathbf{Z}_2\Delta\mathbf{x}(i)\\
&\quad \le -(\mathbf{x}(k-d(k)) - \mathbf{x}(k-h_2))^T\\
&\qquad \times\mathbf{Z}_2(\mathbf{x}(k-d(k)) - \mathbf{x}(k-h_2))
\end{aligned}$$

$$
\begin{aligned}
&- (\mathbf{x}(k-\tau m) - \mathbf{x}(k-d(k)))^T \\
&\times \mathbf{Z}_2 (\mathbf{x}(k-\tau m) - \mathbf{x}(k-d(k))) \\
&- \beta(\mathbf{x}(k-d(k)) - \mathbf{x}(k-h_2))^T \\
&\times \mathbf{Z}_2 (\mathbf{x}(k-d(k)) - \mathbf{x}(k-h_2)) \\
&- (1-\beta)(\mathbf{x}(k-\tau m) - \mathbf{x}(k-d(k)))^T \\
&\times \mathbf{Z}_2 (\mathbf{x}(k-\tau m) - \mathbf{x}(k-d(k))).
\end{aligned} \tag{25}
$$

Employing (18)–(22) and (25), we have the following inequality:

$$\Delta V(k) \le \boldsymbol{\zeta}^T(k)\left[(1-\beta)\boldsymbol{\Psi}_1 + \beta\boldsymbol{\Psi}_2\right]\boldsymbol{\zeta}(k). \tag{26}$$

Since $0 \le \beta \le 1$, $\Delta V(k) < 0$ for all nonzero $\boldsymbol{\zeta}(k)$ if (5) hold true. This completes the proof of Theorem 1. □

Remark 2. To prove Theorem 1, a methodology similar to [6, 7] has been adopted.

Remark 3. It may be noted that the inequalities (5) are LMIs and can be effectively solved by using MATLAB LMI toolbox [31, 32].

Remark 4. For a given τ and m, the allowable maximum value of h_2 for guaranteeing the global asymptotic stability of system (1)–(4) can be obtained by iteratively solving (5).

Remark 5. For $m = 1$, after some algebraic manipulations, it can be shown that Theorem 1 is equivalent to Theorem 1 in [7]. The parameters m and τ (subject to (4)) in Theorem 1 represent some additional degrees of freedom, that is, in comparison with Theorem 1 in [7] which is free of these parameters.

Remark 6. A comparison of the number of the decision variables involved in several recent stability results is summarized in Table 1. It may be observed that the size of complexity in [6, 7, 18, 19, 23] is only related to state dimension n, whereas the complexity of [29, Theorem 3.1], [30, Theorem 2], and Theorem 1 depends on both n and m.

The total number of scalar decision variables of Theorem 1 is $D_1 = (n/2)[n(m^2+5)+(m+5)]$, and the total row size of the LMIs is $L_1 = n(3m+11)$. The numerical complexity of Theorem 1 is proportional to $L_1 D_1^3$ [14]. The total number of scalar decision variables of Theorem 2 in [30] is $D_2 = (n/2)[n(3m^2+18m+41)+(3m+11)]$, the total row size of the LMIs is $L_2 = n(7m+31)$, and the numerical complexity is proportional to $L_2 D_2^3$. Therefore, Theorem 1 has much smaller numerical complexity than Theorem 2 in [30]. Since no free weighting matrix has been introduced in the presented method, Theorem 1 involves less number of decision variables as compared to Theorem 3.1 in [29] (see Table 1). Thus, in general, Theorem 1 is numerically less complex than Theorem 3.1 in [29].

Table 1: Comparison of the number of decision variables involved in various methods.

Methods	Number of the decision variables
Theorem 1 in [19]	$(23n^2+5n)/2$
Theorem 3 in [19]	$(67n^2+9n)/2$
Theorem 1 in [23]	$9n^2+3n$
Theorem 1 in [18]	$13n^2+5n$
Proposition 2 in [6]	$3n^2+3n$
Theorem 1 in [7]	$3n^2+3n$
Theorem 2 in [30]	$(n/2)[n(3m^2+18m+41)+(3m+11)]$
Theorem 3.1 in [29]	$(n/2)[n(m^2+4m+19)+(m+5)]$
Theorem 1	$(n/2)\left[n(m^2+5)+(m+5)\right]$

Remark 7. Using similar steps as in the proof of Proposition 8 in [30], it is easy to establish that the conservatism of the stability result obtained via Theorem 1 is nonincreasing as the number of partitions increases.

4. Extensions to Uncertain State-Delayed Discrete-Time Systems

In this section, we extend the previously discussed approach to derive global asymptotic stable conditions for delayed discrete-time systems with norm-bounded uncertainties.

Consider the system given by

$$\mathbf{x}(k+1) = (\mathbf{A}+\Delta\mathbf{A})\,\mathbf{x}(k) + (\mathbf{A}_1+\Delta\mathbf{A}_1)\,\mathbf{x}(k-d(k)). \tag{27}$$

The parameter uncertainties $\Delta\mathbf{A}$ and $\Delta\mathbf{A}_1$ are assumed to be norm-bounded and of the following form:

$$\left[\Delta\mathbf{A}\ \ \Delta\mathbf{A}_1\right] = \mathbf{H}\mathbf{F}\left[\mathbf{E}_0\ \ \mathbf{E}_1\right], \tag{28}$$

where $\mathbf{H} \in \mathbf{R}^{n\times q}$ and $\mathbf{E}_i \in \mathbf{R}^{p\times n}$ $(i = 0, 1)$ are known constant matrices and $\mathbf{F} \in \mathbf{R}^{q\times p}$ is an unknown matrix which satisfies

$$\mathbf{F}^T\mathbf{F} \le \mathbf{I}. \tag{29}$$

The parameter uncertainties $\Delta\mathbf{A}$ and $\Delta\mathbf{A}_1$ are said be admissible if both (28) and (29) are satisfied.

The following lemma is needed in the proof of our next result.

Lemma 8 (see [5, 25]). *Let* $\boldsymbol{\Sigma}$, $\boldsymbol{\Gamma}$, $\mathbf{F}$, *and* $\mathbf{M}$ *be the real matrices of appropriate dimensions with* $\mathbf{M}$ *satisfying* $\mathbf{M} = \mathbf{M}^T$; *then*

$$\mathbf{M} + \boldsymbol{\Sigma}\mathbf{F}\boldsymbol{\Gamma} + \boldsymbol{\Gamma}^T\mathbf{F}^T\boldsymbol{\Sigma}^T < \mathbf{0}, \tag{30}$$

for all $\mathbf{F}^T\mathbf{F} \le \mathbf{I}$, *if and only if there exists a scalar* $\varepsilon > 0$ *such that*

$$\mathbf{M} + \varepsilon^{-1}\boldsymbol{\Sigma}\boldsymbol{\Sigma}^T + \varepsilon\boldsymbol{\Gamma}^T\boldsymbol{\Gamma} < \mathbf{0}. \tag{31}$$

Theorem 9. *For given scalars* h_1 *and* h_2, *system described by (27)–(29) and (2)–(4) is globally asymptotically stable for all the admissible uncertainties, if there exist real matrices*

TABLE 2: Comparison of the maximum values of h_2 for given h_1.

Methods	$h_1 = 4$	$h_1 = 6$	$h_1 = 10$	$h_1 = 12$	$h_1 = 20$
Theorem 1 in [20]	8	9	12	13	20
Theorem 1 in [19]	13	14	15	16	22
Proposition 2 in [6]	15	16	18	19	25
Theorem 1 in [7]	15	16	18	19	25
Theorem 1	15 ($m = 1, \tau = 4$) 15 ($m = 2, \tau = 2$)	16 ($m = 1, \tau = 6$) 16 ($m = 2, \tau = 3$)	18 ($m = 1, \tau = 10$) 18 ($m = 2, \tau = 5$)	19 ($m = 1, \tau = 12$) 20 ($m = 2, \tau = 6$)	25 ($m = 1, \tau = 20$) 26 ($m = 2, \tau = 10$)

TABLE 3: Comparison of upper bound $\overline{\alpha}$ for given h_1 and h_2.

Methods	$2 \le d(k) \le 7$	$3 \le d(k) \le 9$	$5 \le d(k) \le 10$	$6 \le d(k) \le 12$	$10 \le d(k) \le 15$	$20 \le d(k) \le 26$
Theorem 5 in [20]	0.0830	Infeasible	Infeasible	Infeasible	Infeasible	Infeasible
Corollary 2 in [19]	0.1901	0.1457	0.1313	0.0906	0.0655	Infeasible
Theorem 4 in [7]	0.1920	0.1548	0.1425	0.1146	0.1023	Infeasible
Theorem 9	0.1927 ($m = 2, \tau = 1$)	0.1560 ($m = 3, \tau = 1$)	0.1476 ($m = 5, \tau = 1$)	0.1194 ($m = 2, \tau = 3$)	0.1080 ($m = 2, \tau = 5$)	0.0834 ($m = 2, \tau = 10$) 0.0837 ($m = 5, \tau = 4$)

$\mathbf{P} = \mathbf{P}^T > \mathbf{0}$, $\mathbf{Q}_i = \mathbf{Q}_i^T > \mathbf{0}$ $(i = 1,2,3)$, *and* $\mathbf{Z}_i = \mathbf{Z}_i^T > \mathbf{0}$ $(i = 1,2)$ *and a scalar* $\varepsilon > 0$ *such that the following LMIs hold:*

$$\begin{bmatrix} \overline{\Psi}_1 + \varepsilon\overline{\mathbf{E}}^T\overline{\mathbf{E}} & \Delta_1^T\mathbf{P} & \Delta_2^T\overline{\mathbf{Z}} & \mathbf{0}_{(m+3)n\times q} \\ \mathbf{P}\Delta_1 & -\mathbf{P} & \mathbf{0}_n & \mathbf{PH} \\ \overline{\mathbf{Z}}\Delta_2 & \mathbf{0}_n & -\overline{\mathbf{Z}} & \overline{\mathbf{Z}}\mathbf{H} \\ \mathbf{0}_{q\times(m+3)n} & \mathbf{H}^T\mathbf{P} & \mathbf{H}^T\overline{\mathbf{Z}} & -\varepsilon\mathbf{I}_q \end{bmatrix} < 0, \quad (32)$$

$$\begin{bmatrix} \overline{\Psi}_2 + \varepsilon\overline{\mathbf{E}}^T\overline{\mathbf{E}} & \Delta_1^T\mathbf{P} & \Delta_2^T\overline{\mathbf{Z}} & \mathbf{0}_{(m+3)n\times q} \\ \mathbf{P}\Delta_1 & -\mathbf{P} & \mathbf{0}_n & \mathbf{PH} \\ \overline{\mathbf{Z}}\,\Delta_2 & \mathbf{0}_n & -\overline{\mathbf{Z}} & \overline{\mathbf{Z}}\mathbf{H} \\ \mathbf{0}_{q\times(m+3)n} & \mathbf{H}^T\mathbf{P} & \mathbf{H}^T\overline{\mathbf{Z}} & -\varepsilon\mathbf{I}_q \end{bmatrix} < 0, \quad (33)$$

where

$$\overline{\Psi}_1 = \widetilde{\Psi} - \Lambda_7^T\mathbf{Z}_2\Lambda_7, \qquad \overline{\Psi}_2 = \widetilde{\Psi} - \Lambda_8^T\mathbf{Z}_2\Lambda_8, \quad (34)$$

$$\begin{aligned} \widetilde{\Psi} = {} & \Lambda_1^T\widetilde{\Phi}_1\Lambda_1 + \Lambda_2^T\widetilde{\Phi}_2\Lambda_2 + \Lambda_3^T\widetilde{\Phi}_3\Lambda_3 - \Lambda_4^T\widetilde{\Phi}_4\Lambda_4 - \Lambda_5^T\mathbf{Z}_1\Lambda_5 \\ & + \Lambda_1^T\mathbf{Z}_1\Lambda_5 + \Lambda_5^T\mathbf{Z}_1\Lambda_1 - \Lambda_6^T\mathbf{Z}_2\Lambda_6 + \Lambda_3^T\mathbf{Z}_2\Lambda_4 + \Lambda_4^T\mathbf{Z}_2\Lambda_3 \\ & + \Lambda_6^T\mathbf{Z}_2\Lambda_3 + \Lambda_3^T\mathbf{Z}_2\Lambda_6, \end{aligned} \quad (35)$$

$$\begin{aligned} \Lambda_7 &= [\mathbf{0}_{n\times mn} \ \ \mathbf{I}_n \ \ -\mathbf{I}_n \ \ \mathbf{0}_n], \\ \Lambda_8 &= [\mathbf{0}_{n\times(m+1)n} \ \ \mathbf{I}_n \ \ -\mathbf{I}_n], \end{aligned} \quad (36)$$

$$\widetilde{\Phi}_1 = -\mathbf{P} + \mathbf{Q}_2 + (h_2 - \tau m + 1)\,\mathbf{Q}_3 - \mathbf{Z}_1,$$

$$\widetilde{\Phi}_2 = \begin{bmatrix} \mathbf{Q}_1 & \mathbf{0} \\ \mathbf{0} & -\mathbf{Q}_1 \end{bmatrix}, \quad (37)$$

$$\widetilde{\Phi}_3 = -\mathbf{Q}_3 - 2\mathbf{Z}_2, \qquad \widetilde{\Phi}_4 = \mathbf{Q}_2 + \mathbf{Z}_2, \quad (38)$$

$$\overline{\mathbf{Z}} = \tau^2\mathbf{Z}_1 + (h_2 - \tau m)\,\mathbf{Z}_2, \quad (39)$$

$$\begin{aligned} \Delta_1 &= [\mathbf{A} \ \ \mathbf{0}_{n\times mn} \ \ \mathbf{A}_1 \ \ \mathbf{0}_n], \\ \Delta_2 &= [\mathbf{A} - \mathbf{I}_n \ \ \mathbf{0}_{n\times mn} \ \ \mathbf{A}_1 \ \ \mathbf{0}_n], \\ \overline{\mathbf{E}} &= [\mathbf{E}_0 \ \ \mathbf{0}_{p\times mn} \ \ \mathbf{E}_1 \ \ \mathbf{0}_{p\times n}], \end{aligned} \quad (40)$$

and Λ_i $(i = 1,\ldots,6)$ *in* (35) *are given by* (7)–(9).

Proof. Applying Theorem 1, the sufficient conditions for the global asymptotic stability of system described by (27)–(29) and (2)–(4) are obtained as follows:

$$\left[\begin{array}{cccc|cc} & & & & \mathbf{A}^T\mathbf{P} & (\mathbf{A}-\mathbf{I}_n)^T\overline{\mathbf{Z}} \\ & \overline{\Psi}_1 & & & \mathbf{0}_{mn\times n} & \mathbf{0}_{mn\times n} \\ & & & & \mathbf{A}_1^T\mathbf{P} & \mathbf{A}_1^T\overline{\mathbf{Z}} \\ & & & & \mathbf{0}_n & \mathbf{0}_n \\ \hline \mathbf{PA} & \mathbf{0}_{n\times mn} & \mathbf{PA}_1 & \mathbf{0}_n & -\mathbf{P} & \mathbf{0}_n \\ \overline{\mathbf{Z}}(\mathbf{A}-\mathbf{I}_n) & \mathbf{0}_{n\times mn} & \overline{\mathbf{Z}}\mathbf{A}_1 & \mathbf{0}_n & \mathbf{0}_n & -\overline{\mathbf{Z}} \end{array}\right] + \begin{bmatrix} \mathbf{0}_{(m+3)n\times q} \\ \mathbf{PH} \\ \overline{\mathbf{Z}}\mathbf{H} \end{bmatrix} \mathbf{F}\,[\mathbf{E}_0 \ \ \mathbf{0}_{p\times mn} \ \ \mathbf{E}_1 \ \ \mathbf{0}_{p\times 3n}]$$

$$+\begin{bmatrix} \mathbf{E}_0^T \\ \mathbf{0}_{mn\times p} \\ \mathbf{E}_1^T \\ \mathbf{0}_{3n\times p} \end{bmatrix} \mathbf{F}^T \begin{bmatrix} \mathbf{0}_{q\times(m+3)n} & \mathbf{H}^T\mathbf{P} & \mathbf{H}^T\overline{\mathbf{Z}} \end{bmatrix} < 0, \tag{41}$$

$$\left[\begin{array}{cccc|cc} & & & & \mathbf{A}^T\mathbf{P} & (\mathbf{A}-\mathbf{I}_n)^T\overline{\mathbf{Z}} \\ & \overline{\Psi}_2 & & & \mathbf{0}_{mn\times n} & \mathbf{0}_{mn\times n} \\ & & & & \mathbf{A}_1^T\mathbf{P} & \mathbf{A}_1^T\overline{\mathbf{Z}} \\ & & & & \mathbf{0}_n & \mathbf{0}_n \\ \hline \mathbf{PA} & \mathbf{0}_{n\times mn} & \mathbf{PA}_1 & \mathbf{0}_n & -\mathbf{P} & \mathbf{0}_n \\ \overline{\mathbf{Z}}(\mathbf{A}-\mathbf{I}_n) & \mathbf{0}_{n\times mn} & \overline{\mathbf{Z}}\mathbf{A}_1 & \mathbf{0}_n & \mathbf{0}_n & -\overline{\mathbf{Z}} \end{array}\right] + \begin{bmatrix} \mathbf{0}_{(m+3)n\times q} \\ \mathbf{PH} \\ \overline{\mathbf{Z}}\mathbf{H} \end{bmatrix} \mathbf{F}\begin{bmatrix} \mathbf{E}_0 & \mathbf{0}_{p\times mn} & \mathbf{E}_1 & \mathbf{0}_{p\times 3n} \end{bmatrix} + \begin{bmatrix} \mathbf{E}_0^T \\ \mathbf{0}_{mn\times p} \\ \mathbf{E}_1^T \\ \mathbf{0}_{3n\times p} \end{bmatrix} \mathbf{F}^T \begin{bmatrix} \mathbf{0}_{q\times(m+3)n} & \mathbf{H}^T\mathbf{P} & \mathbf{H}^T\overline{\mathbf{Z}} \end{bmatrix} < 0. \tag{42}$$

By Lemma 8 and Schur complement, it can be shown that (41) and (42) are equivalent to (32) and (33), respectively. This completes the proof of Theorem 9. □

Remark 10. With $m = 1$, Theorem 9 reduces to an equivalent form of Theorem 4 in [7]. The parameters m and τ (subject to (4)) in Theorem 9 may be treated as additional degrees of freedom, that is, in comparison with Theorem 4 in [7] which is free of such parameters.

5. Numerical Examples

To demonstrate the applicability of the presented results and compare them with previous results, we now consider the following examples.

Example 1 (see [6, 7, 19, 20, 29]). Consider the delayed discrete-time system (1) with the following parameters:

$$\mathbf{A} = \begin{bmatrix} 0.8 & 0 \\ 0.05 & 0.9 \end{bmatrix}, \qquad \mathbf{A}_1 = \begin{bmatrix} -0.1 & 0 \\ -0.2 & -0.1 \end{bmatrix}. \tag{43}$$

For given τ and m, the admissible maximum value of h_2 guaranteeing the asymptotically stability of the delayed discrete-time system can be obtained by solving the LMIs (5). For a given h_1, a comparison of the maximum values of h_2 calculated by different methods is summarized in Table 2, from which one can easily see that the stability conditions obtained in this paper are less conservative than those in [7, 19, 20].

A comparison of [29, Table 2] and Table 2 reveals that Theorem 3.1 in [29] may provide a larger upper delay bound h_2 than Theorem 1. However, as discussed in Remark 6, Theorem 1 involves less number of decision variables as compared to Theorem 3.1 in [29].

Example 2 (see [7, 19]). Consider the following delayed discrete-time system with parameter uncertainty:

$$\mathbf{x}(k+1) = \begin{bmatrix} 0.8+\alpha(k) & 0 \\ 0 & 0.9 \end{bmatrix}\mathbf{x}(k) + \begin{bmatrix} -0.1 & 0 \\ -0.1 & -0.1 \end{bmatrix}\mathbf{x}(k-d(k)), \tag{44}$$

where $\mathbf{x}(k) = [x_1(k)\ \ x_2(k)]^T$ and $|\alpha(k)| \le \overline{\alpha}$. System (44) can be represented in the form of (27) with

$$\mathbf{A} = \begin{bmatrix} 0.8 & 0 \\ 0 & 0.9 \end{bmatrix}, \qquad \mathbf{A}_1 = \begin{bmatrix} -0.1 & 0 \\ -0.1 & -0.1 \end{bmatrix}, \qquad \mathbf{H} = \begin{bmatrix} \overline{\alpha} \\ 0 \end{bmatrix},$$
$$\mathbf{E}_0 = [0\ \ 1], \qquad \mathbf{E}_1 = [0\ \ 0], \qquad \mathbf{F} = \frac{\alpha(k)}{\overline{\alpha}}. \tag{45}$$

For given h_1 and h_2, we wish to find $\overline{\alpha}$ such that the uncertain system (44) is asymptotically stable for any $|\alpha(k)| \le \overline{\alpha}$. Table 3 presents a comparison of the values of $\overline{\alpha}$ obtained by various methods. From Table 3, it is clear that Theorem 9 can provide less conservative results than [7, 19, 20]. In the situation where $20 \le d(k) \le 26$, the stability criteria in [7, 19, 20] are invalid whereas the value of $\overline{\alpha}$ obtained by Theorem 9 is 0.0837 for $m = 5$, $\tau = 4$.

6. Conclusion

In this paper, we have considered the problem of global asymptotic stability of linear discrete-time systems with interval-like time-varying delay in the state. By utilizing the concept of delay partitioning, an LMI-based criterion for the global asymptotic stability of such systems has been established. The proposed criterion depends on both the size of delay and partition size. With the help of numerical examples, it has been illustrated that the proposed stability condition may provide less conservative result than most of the existing results [6, 7, 19, 20] due to the delay partitioning, and it may become even less conservative when the partitioning goes finer. Since no free weighting matrix has been introduced in the presented method, Theorem 1 involves less number of decision variables as compared to Theorem 3.1 in [29]. However, a comparison of [29, Table 2] and Table 2 reveals that Theorem 3.1 in [29] may provide a larger upper delay bound h_2 than Theorem 1. The proposed criterion is numerically less complex than [30]. The method has been extended to deal with the problem of global asymptotic stability of state-delayed discrete-time systems with norm-bounded uncertainties.

References

[1] J. K. Hale, *Theory of Functional Differential Equations*, Springer, New York, NY, USA, 1977.

[2] J. Richard, "Time-delay systems: an overview of some recent advances and open problems," *Automatica*, vol. 39, no. 10, pp. 1667–1694, 2003.

[3] H. Shao, "New delay-dependent stability criteria for systems with interval delay," *Automatica*, vol. 45, no. 3, pp. 744–749, 2009.

[4] V. K. R. Kandanvli and H. Kar, "Robust stability of discrete-time state-delayed systems employing generalized overflow nonlinearities," *Nonlinear Analysis: Theo y, Methods and Applications*, vol. 69, no. 9, pp. 2780–2787, 2008.

[5] K. F. Chen and I. K. Fong, "Stability analysis and output-feedback stabilisation of discrete-time systems with an interval time-varying state delay," *IET Control Theo y and Applications*, vol. 4, no. 4, pp. 563–572, 2010.

[6] H. Shao and Q. Han, "New stability criteria for linear discrete-time systems with interval-like time-varying delays," *IEEE Transactions on Automatic Control*, vol. 56, no. 3, pp. 619–625, 2011.

[7] H. Huang and G. Feng, "Improved approach to delay-dependent stability analysis of discrete-time systems with time-varying delay," *IET Control Theo y and Applications*, vol. 4, no. 10, pp. 2152–2159, 2010.

[8] M. S. Mahmoud, F. M. Al-Sunni, and Y. Shi, "Switched discrete-time delay systems: delay-dependent analysis and synthesis," *Circuits, Systems, and Signal Processing*, vol. 28, no. 5, pp. 735–761, 2009.

[9] M. S. Mahmoud and Y. Xia, "Robust stability and stabilization of a class of nonlinear switched discrete-time systems with time-varying delays," *Journal of Optimization Theory and Applications*, vol. 143, no. 2, pp. 329–355, 2009.

[10] S. Xu and J. Lam, "On equivalence and efficiency of certain stability criteria for time-delay systems," *IEEE Transactions on Automatic Control*, vol. 52, no. 1, pp. 95–101, 2007.

[11] E. Fridman and U. Shaked, "A descriptor system approach to H_∞ control of linear time-delay systems," *IEEE Transactions on Automatic Control*, vol. 47, no. 2, pp. 253–270, 2002.

[12] Q. Han, "On stability of linear neutral systems with mixed time delays: a discretized Lyapunov functional approach," *Automatica*, vol. 41, no. 7, pp. 1209–1218, 2005.

[13] Q. Han, "A discrete delay decomposition approach to stability of linear retarded and neutral systems," *Automatica*, vol. 45, no. 2, pp. 517–524, 2009.

[14] Q. Han, "Improved stability criteria and controller design for linear neutral systems," *Automatica*, vol. 45, no. 8, pp. 1948–1952, 2009.

[15] Q. Han and K. Gu, "Stability of linear systems with time-varying delay: a generalized discretized lyapunov functional approach," *Asian Journal of Control*, vol. 3, no. 3, pp. 170–180, 2001.

[16] X. Jiang and Q. Han, "On H_∞ control for linear systems with interval time-varying delay," *Automatica*, vol. 41, no. 12, pp. 2099–2106, 2005.

[17] M. Wu, Y. He, J. She, and G. Liu, "Delay-dependent criteria for robust stability of time-varying delay systems," *Automatica*, vol. 40, no. 8, pp. 1435–1439, 2004.

[18] Y. He, M. Wu, G. Liu, and J. She, "Output feedback stabilization for a discrete-time system with a time-varying delay," *IEEE Transactions on Automatic Control*, vol. 53, no. 10, pp. 2372–2377, 2008.

[19] H. Gao and T. Chen, "New results on stability of discrete-time systems with time-varying state delay," *IEEE Transactions on Automatic Control*, vol. 52, no. 2, pp. 328–334, 2007.

[20] H. Gao, J. Lam, C. Wang, and Y. Wang, "Delay-dependent output-feedback stabilisation of discrete-time systems with time varying state delay," *IEE Proceedings—Control Theo y and Applications*, vol. 151, no. 6, pp. 691–698, 2004.

[21] E. Fridman and U. Shaked, "Stability and guaranteed cost control of uncertain discrete delay systems," *International Journal of Control*, vol. 78, no. 4, pp. 235–246, 2005.

[22] X. Jiang, Q. Han, and X. Yu, "Stability criteria for linear discrete-time systems with interval-like time-varying delay," in *Proceedings of the American Control Conference (ACC '05)*, pp. 2817–2822, Portland, Ore, USA, June 2005.

[23] B. Zhang, S. Xu, and Y. Zou, "Improved stability criterion and its applications in delayed controller design for discrete-time systems," *Automatica*, vol. 44, no. 11, pp. 2963–2967, 2008.

[24] H. Huang and G. Feng, "State estimation of recurrent neural networks with time-varying delay: a novel delay partition approach," *Neurocomputing*, vol. 74, no. 5, pp. 792–796, 2011.

[25] B. Du, J. Lam, Z. Shu, and Z. Wang, "A delay-partitioning projection approach to stability analysis of continuous systems with multiple delay components," *IET Control Theo y and Applications*, vol. 3, no. 4, pp. 383–390, 2009.

[26] Y. Zhao, H. Gao, J. Lam, and B. Du, "Stability and stabilization of delayed T-S fuzzy systems: a delay partitioning approach," *IEEE Transactions on Fuzzy Systems*, vol. 17, no. 4, pp. 750–762, 2009.

[27] J. Liu, B. Yao, and Z. Gu, "Delay-dependent H_∞ filtering for Markovian jump time-delay systems: a piecewise analysis method," *Circuits, Systems, and Signal Processing*, vol. 30, no. 6, pp. 1253–1273, 2011.

[28] L. Wu, J. Lam, X. Yao, and J. Xiong, "Robust guaranteed cost control of discrete-time networked control systems," *Optimal Control Applications and Methods*, vol. 32, no. 1, pp. 95–112, 2011.

[29] P. Kokil, H. Kar, and V. K. R. Kandanvli, "Stability analysis of linear discrete-time systems with interval delay: a delay-partitioning approach," *ISRN Applied Mathematics*, vol. 2011, Article ID 624127, 10 pages, 2011.

[30] X. Meng, J. Lam, B. Du, and H. Gao, "A delay-partitioning approach to the stability analysis of discrete-time systems," *Automatica*, vol. 46, no. 3, pp. 610–614, 2010.

[31] S. Boyd, L. El Ghaoui, E. Feron, and V. Balakrishnan, *Linear Matrix Inequalities in System and Control Theo y*, SIAM, Philadelphia, Pa, USA, 1994.

[32] P. Gahinet, A. Nemirovski, A. J. Laub, and M. Chilali, *LMI Control Toolbox for Use with Matlab*, MATH Works, Natick, Mass, USA, 1995.

Influence of Hall Current and Thermal Radiation on MHD Convective Heat and Mass Transfer in a Rotating Porous Channel with Chemical Reaction

Dulal Pal[1] and Babulal Talukdar[2]

[1] *Department of Mathematics, Visva-Bharati University, Santiniketan, West Bengal 731235, India*
[2] *Department of Mathematics, Gobindapur High School, Kalabagh, Murshidabad, West Bengal 742213, India*

Correspondence should be addressed to Dulal Pal; dulalp123@rediffmail.com

Academic Editor: Song Cen

A theoretical study is carried out to obtain an analytic solution of heat and mass transfer in a vertical porous channel with rotation and Hall current. A constant suction and injection is applied to the two insulating porous plates. A strong magnetic field is applied in the transverse direction. The entire system rotates with uniform angular velocity Ω about the axis normal to the plates. The governing equations are solved by perturbation technique to obtain the analytical results for velocity, temperature, and concentration fields and shear stresses. The steady and unsteady resultant velocities along with the phase differences for various values of physical parameters are discussed in detail. The effects of rotation, buoyancy force, magnetic field, thermal radiation, and heat generation parameters on resultant velocity, temperature, and concentration fields are analyzed.

1. Introduction

Free convection in channel flow has many important applications in designing ventilating and heating of buildings, cooling of electronic components of a nuclear reactor, bed thermal storage, and heat sink in the turbine blades. Convective flows driven by temperature difference of the bounding walls of channels are important in industrial applications. El-Hakiem [1] studied the unsteady MHD oscillatory flow on free convection radiation through a porous medium with a vertical infinite surface that absorbs the fluid with a constant velocity. Jaiswal and Soundalgekar [2] analyzed the effects of suction with oscillating temperature on a flow past an infinite porous plate. Singh et al. [3] studied the unsteady free convective flow in a porous medium bounded by an infinite vertical porous plate in the presence of rotation. Pal and Shivakumara [4] studied the mixed convection heat transfer from a vertical plate in a porous medium.

Hydromagnetic convection with heat transfer in a rotating medium has important applications in MHD generators and accelerators design, geophysics, and nuclear power reactors. MHD free convection and mass transfer flows in a rotating system have diverse applications. The effects of Hall currents cannot be neglected as the conducting fluid when it is an ionized gas, and applied field strength is strong then the electron cyclotron frequency $\omega = eB/m$ (where e, B, and m denote the electron charge, the applied magnetic field, and mass of an electron, resp.) exceeds the collision frequency so that the electron makes cyclotron orbit between the collisions which will divert in a direction perpendicular to the magnetic and electric fields directions. Thus, if an electric field is applied perpendicular to the magnetic field then whole current will not pass along the electric field. This phenomena of flow of the electric current across an electric field with magnetic field is known as Hall effect, and accordingly this current is known as Hall current [5]. Thus, it is essential to analyze the effects of Hall currents in many industrial problems. Gupta [6] has studied the influence of Hall current on steady MHD flow in a viscous fluid. Jana et al. [7] analyzed the hall effect in steady flow past an infinite porous flat plate. Makinde and Mhone [8] studied hydromagnetic oscillatory flow through a channel having porous medium. Zhang and Wang [9] analyzed the effect of magnetic field in a power-law fluid over a vertical stretching sheet. Hameed and Nadeem [10] analyzed unsteady hydromagnetic flow of a non-Newtonian fluid over a porous plate. Makinde et al. [11] examined the effect of magnetic field in a rotating porous medium cylindrical annulus. Sibanda and Makinde [12] analyzed effects of magnetic fields on heat transfer on a rotating disk in a porous medium with Ohmic heating and viscous dissipation. Pop and Watanabe [13] analyzed convective flow of a conducting fluid in the presence of magnetic field and Hall current. Saha et al. [14] studied Hall current effect on MHD natural convection flow from vertical flat plate. Recently, Pal et al. [15] examined the influence of Hall current and chemical reaction on oscillatory mixed convection radiation of a micropolar fluid in a rotating system.

Radiation effects on free convection flow have become very important due to its applications in space technology, processes having high temperature, and design of pertinent equipments. Moreover, heat and mass transfer with thermal radiation on convective flows is very important due its significant role in the surface heat transfer. Recent developments in gas cooled nuclear reactors, nuclear power plants, gas turbines, space vehicles, and hypersonic flights have attracted research in this field. The unsteady convective flow in a moving plate with thermal radiation were examined by Cogley et al. [16] and Mansour [17]. The combined effects of radiation and buoyancy force past a vertical plate were analyzed by Hossain and Takhar [18]. Hossain et al. [19] analyzed the influence of thermal radiation on convective flows over a porous vertical plate. Seddeek [20] explained the importance of thermal radiation and variable viscosity on unsteady forced convection with an align magnetic field. Muthucumaraswamy and Senthil [21] studied the effects of thermal radiation on heat and mass transfer over a moving vertical plate. Pal [22] investigated convective heat and mass transfer in a stagnation-point flow towards a stretching sheet with thermal radiation. Aydin and Kaya [23] justified the effects of thermal radiation on mixed convection flow over a permeable vertical plate with magnetic field. Mohamed [24] studied unsteady MHD flow over a vertical moving porous plate with heat generation and Soret effect. Chauhan and Rastogi [25] analyzed the effects of thermal radiation, porosity, and suction on unsteady convective hydromagnetic vertical rotating channel. Ibrahim and Makinde [26] investigated radiation effect on chemically reaction MHD boundary layer flow of heat and mass transfer past a porous vertical flat plate. Pal and Mondal [27] studied the effects of thermal radiation on MHD Darcy-Forchheimer convective flow past a stretching sheet in a porous medium. Palani and Kim [28] analyzed the effect of thermal radiation on convection flow past a vertical cone with surface heat flux. Recently, Mahmoud and Waheed [29] examined thermal radiation on flow over an infinite flat plate with slip velocity.

The study of heat and mass transfer due to chemical reaction is also very importance because of its occurrence in most of the branches of science and technology. The processes involving mass transfer effects are important in chemical processing equipments which are designed to draw high value products from cheaper raw materials with the involvement of chemical reaction. In many industrial processes, the species undergo some kind of chemical reaction with the ambient fluid which may affect the flow behaviour and the production quality of final products. Aboeldahab and Elbarbary [30] examined heat and mass transfer over a vertical plate in the presence of magnetic field and Hall effect. Abo-Eldahab and El Aziz [31] investigated the Hall current and Joule heating effects on electrically conducting fluid past a semi-infinite plate with strong magnetic field and heat generation/absorption. Kandasamy et al. [32] discussed the effects of chemical reaction and magnetic field on heat and mass transfer over a vertical stretching surface. Muthucumaraswamy and Janakiraman [33] analyzed the effects of mass transfer over a vertical oscillating plate with chemical reaction. Sharma and Singh [34] have analyzed the unsteady MHD free convection flow and heat transfer over a vertical porous plate in the presence of internal heat generation and variable suction. Sudheer Babu and Satya Narayan [35] examined chemical reaction and thermal radiation effects on MHD convective flow in a porous medium in the presence of suction. Makinde and Chinyoka [36] studied the effects of magnetic field on MHD Couette flow of a third-grade fluid with chemical reaction. Recently, Pal and Talukdar [37] investigated the influence of chemical reaction and Joule heating on unsteady convective viscous dissipating fluid over a vertical plate in porous media with thermal radiation and magnetic field.

The objective of the present study is to analyze the effects of Hall current, thermal radiation, and first-order chemical reaction on the oscillatory convective flow and mass transfer with suction injection in a rotating vertical porous channel. The present results are compared with those of Singh and Kumar [38], and a very good agreement is found.

2. Problem Formulation

We consider unidirectional oscillatory free convective flow of a viscous incompressible and electrically conducting fluid between two insulating infinite vertical permeable plates separated by a distance d. A constant injection velocity w_0 is applied at the stationary plate $z^* = 0$. Also, a constant suction velocity w_0 is applied at the plate $z^* = d$, which oscillates in its own plane with a velocity $U^*(t^*)$ about a nonzero constant mean velocity U_0. The channel rotates as a rigid body with angular velocity Ω^* about the z^*-axis perpendicular to the planes of the plates. A strong transverse magnetic field of uniform strength H_0 is applied along the axis of rotation by neglecting induced electric and magnetic fields. The fluid is assumed to be a gray, emitting, and absorbing, but nonscattering medium. The radiative heat flux term can be simplified by using the Rosseland approximation. It is also assumed that the chemically reactive species undergo first-order irreversible chemical reaction.

The solenoidal relation for the magnetic field $\nabla \cdot \vec{H} = 0$, where $\vec{H} = (H_x^*, H_y^*, H_z^*)$ gives $H_z^* = H_0$ (constant) everywhere in the flow field, which gives $\vec{H} = (0, 0, H_0)$. If (J_x^*, J_y^*, J_z^*) are the component of electric current density $\vec{J}$, then the equation of conservation of electric charge $\nabla \cdot \vec{J} = 0$ gives $J_z^* = \text{constant}$. This constant is zero, that is, $J_z^* = 0$

everywhere in the flow since the plate is electrically non-conducting. The generalized Ohm's law, in the absence of the electric field [39], is of the form

$$\vec{J} + \frac{\omega_e \tau_e}{H_0}\left(\vec{J} \times \vec{H}\right) = \sigma\left(\mu_e \vec{V} \times \vec{H} + \frac{1}{e n_e}\nabla p_e\right), \quad (1)$$

where $\vec{V}$, σ, μ_e, ω_e, τ_e, e, n_e, and p_e are the velocity, the electrical conductivity, the magnetic permeability, the cyclotron frequency, the electron collision time, the electric charge, the number density of the electron, and the electron pressure, respectively. Under the usual assumption, the electron pressure (for a weakly ionized gas), the thermoelectric pressure, and ion slip are negligible, so we have from the Ohm's law

$$J_x^* + \omega_e \tau_e J_y^* = \sigma \mu_e H_0 v^*$$
$$J_y^* - \omega_e \tau_e J_x^* = -\sigma \mu_e H_0 u^*, \quad (2)$$

from which we obtain that

$$J_x^* = \frac{\sigma \mu_e H_0 \left(m u^* + v^*\right)}{1 + m^2}, \qquad J_y^* = \frac{\sigma \mu_e H_0 \left(m v^* - u^*\right)}{1 + m^2}. \quad (3)$$

Since the plates are infinite in extent, all the physical quantities except the pressure depend only on z^* and t^*. The physical configuration of the problem is shown in Figure 1. A Cartesian coordinate system is assumed, and z^*-axis is taken normal to the plates, while x^*- and y^*-axes are in the upward and perpendicular directions on the plate $z^* = 0$ (origin), respectively. The velocity components u^*, v^*, w^* are in the x^*-, y^*-, z^*-directions, respectively. The governing equations in the rotating system in presence of Hall current, thermal radiation, and chemical reaction are given by the following equations:

$$\frac{\partial w^*}{\partial z^*} = 0 \Longrightarrow w^* = w_0, \quad (4)$$

$$\frac{\partial u^*}{\partial t^*} + w_0 \frac{\partial u^*}{\partial z^*} - 2\Omega^* v^* = -\frac{1}{\rho}\frac{\partial P^*}{\partial x^*} + \nu \frac{\partial^2 u^*}{\partial z^{*2}} + g_0 \beta \left(T^* - T_d\right) + g_0 \beta^* \left(C^* - C_d\right) + \frac{H_0}{\rho} J_y^*, \quad (5)$$

$$\frac{\partial v^*}{\partial t^*} + w_0 \frac{\partial v^*}{\partial z^*} + 2\Omega^* u^* = -\frac{1}{\rho}\frac{\partial P^*}{\partial y^*} + \nu \frac{\partial^2 v^*}{\partial z^{*2}} - \frac{H_0}{\rho} J_x^*, \quad (6)$$

$$\frac{\partial T^*}{\partial t^*} + w_0 \frac{\partial T^*}{\partial z^*} = \frac{\kappa}{\rho c_p}\frac{\partial^2 T^*}{\partial z^{*2}} - \frac{Q_0}{\rho c_p}\left(T^* - T_d\right) - \frac{1}{\rho c_p}\frac{\partial q_r^*}{\partial z^*}, \quad (7)$$

$$\frac{\partial C^*}{\partial t^*} + w_0 \frac{\partial C^*}{\partial z^*} = D_m \frac{\partial^2 C^*}{\partial z^{*2}} - k_1 \left(C^* - C_d\right), \quad (8)$$

where $m(= \omega_e \tau_e)$ is the Hall parameter, β and β^* are the coefficients of thermal and solutal expansion, c_p is the specific heat at constant pressure, ρ is the density of the fluid, ν is the kinematics viscosity, κ is the fluid thermal conductivity, g_0 is the acceleration of gravity, Q_0 is the additional heat source, q_r^* is the radiative heat flux, D_m is the molecular diffusivity, k_1 is the chemical reaction rate constant. The radiative heat flux is given by $q_r^* = -(4\sigma^*/3k^*)(\partial T^{*4}/\partial z^*)$, in which σ^* and k^* are the Stefan-Boltzmann constant and the mean absorption coefficient, respectively.

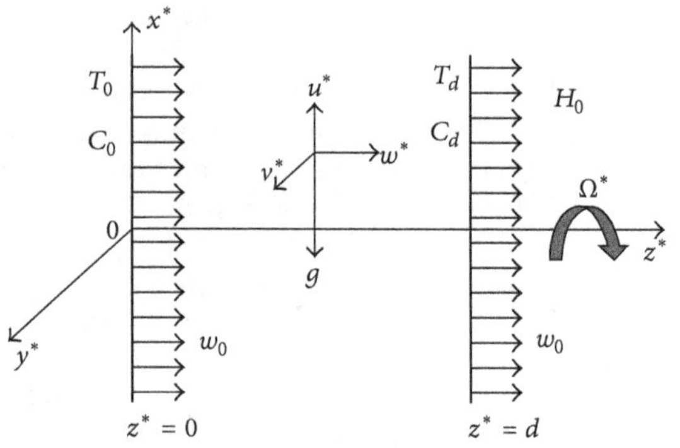

FIGURE 1: Physical configuration of the problem.

The initial and boundary conditions as suggested by the physics of the problem are

$$u^* = v^* = 0, \qquad T^* = T_0 + \epsilon\left(T_0 - T_d\right)\cos\omega^* t^*,$$
$$C^* = C_0 + \epsilon\left(C_0 - C_d\right)\cos\omega^* t^* \quad \text{at } z^* = 0$$
$$u^* = U^*\left(t^*\right) = U_0\left(1 + \epsilon\cos\omega^* t^*\right), \quad v^* = 0, \quad (9)$$
$$T^* = T_d, \quad C^* = C_d \quad \text{at } z^* = d,$$

where ϵ is a small constant.

We now introduce the dimensionless variables as follows:

$$\eta = \frac{z^*}{d}, \qquad u = \frac{u^*}{U_0}, \qquad v = \frac{v^*}{U_0}, \qquad t = \frac{\omega^*}{t^*},$$
$$\omega = \frac{\omega^* d^2}{\nu}, \qquad \Omega = \frac{\Omega^* d^2}{\nu}, \qquad \lambda = \frac{w_0 d}{\nu}, \quad (10)$$
$$\theta = \frac{T^* - T_d}{T_0 - T_d}, \qquad \phi = \frac{C^* - C_d}{C_0 - C_d}.$$

After combining (5) and (6) and taking $q = u + iv$, then (5)–(8) reduce to

$$\omega\frac{\partial q}{\partial t} + \lambda\frac{\partial q}{\partial \eta} = \frac{\partial^2 q}{\partial \eta^2} + w\frac{\partial U}{\partial t} - 2i\Omega\left(q - U\right) - \frac{M^2\left(1 + im\right)}{1 + m^2}\left(q - U\right) + \lambda^2\left(\mathrm{Gr}\,\theta + \mathrm{Gm}\,\phi\right),$$

$$\omega\frac{\partial \theta}{\partial t} + \lambda\frac{\partial \theta}{\partial \eta} = \frac{1}{\mathrm{Pr}}\left(1 + \frac{4}{3R}\right)\frac{\partial^2 q}{\partial \eta^2} - \frac{Q_H}{\mathrm{Pr}}\theta,$$

$$\omega\frac{\partial \phi}{\partial t} + \lambda\frac{\partial \phi}{\partial \eta} = \frac{1}{\mathrm{Sc}}\frac{\partial^2 \phi}{\partial \eta^2} - \xi\phi, \quad (11)$$

where $\mathrm{Gr} = g_0\beta\nu(T0_w - T_d)/U_0 w_0^2$ is the modified thermal Grashof number, $\mathrm{Gm} = g_0\beta^*\nu(C_0 - C_d)/U_0 w_0^2$ is the modified solutal Grashof number, $\mathrm{Pr} = \nu\rho c_p/\kappa$ is the Prandtl number, $M = H_0 d\sqrt{\sigma/\mu}$ is the Hartmann number, $Q_H = Q_0 d^2/\kappa$ is the heat source parameter, $R = \kappa k^*/4\sigma^* T_d$ is the radiation parameter, $\mathrm{Sc} = \nu/Dm$ is the Schmidt number, and $\xi = k_1 d^2/\nu$ is the reaction parameter.

The boundary conditions (9) can be expressed in complex form as

$$q = 0, \quad \theta = 1 + \frac{\epsilon}{2}\left(e^{it} + e^{-it}\right), \quad \phi = 1 + \frac{\epsilon}{2}\left(e^{it} + e^{-it}\right) \quad \text{at } \eta = 0,$$
$$q = U(t) = 1 + \frac{\epsilon}{2}\left(e^{it} + e^{-it}\right), \quad \theta = 0, \quad \phi = 0 \quad \text{at } \eta = 1. \tag{12}$$

3. Method of Solution

The set of partial differential equations (11) cannot be solved in closed form. So it is solved analytically after these equations are reduced to a set of ordinary differential equations in dimensionless form. We assume that

$$\mathfrak{R}(\eta, t) = \mathfrak{R}_0(\eta) + \frac{\epsilon}{2}\left(\mathfrak{R}_1(\eta) e^{it} + \mathfrak{R}_2(\eta) e^{-it}\right), \tag{13}$$

where $\mathfrak{R}$ stands for q or θ or ϕ, and $\epsilon \ll 1$ which is a perturbation parameter. The method of solution is applicable for small perturbation.

Substituting (13) into (11) and comparing the harmonic and nonharmonic terms, we obtain the following ordinary differential equations:

$$\begin{aligned}
&q_0'' - \lambda q_0' - S q_0 = -S - \lambda^2(\mathrm{Gr}\,\theta_0 + \mathrm{Gm}\,\phi_0),\\
&q_1'' - \lambda q_1' - (S + i\omega) q_1 = -(S + i\omega) - \lambda^2(\mathrm{Gr}\,\theta_1 + \mathrm{Gm}\,\phi_1),\\
&q_2'' - \lambda q_2' - (S - i\omega) q_1 = -(S - i\omega) - \lambda^2(\mathrm{Gr}\,\theta_2 + \mathrm{Gm}\,\phi_2),\\
&\theta_0'' - \frac{3\lambda \mathrm{Pr} R}{3R+4}\theta_0' - \frac{3RQ_H}{3R+4}\theta_0 = 0,\\
&\theta_1'' - \frac{3\lambda \mathrm{Pr} R}{3R+4}\theta_1' - \frac{3R}{3R+4}(i\omega \mathrm{Pr} + Q_H)\theta_1 = 0,\\
&\theta_2'' - \frac{3\lambda \mathrm{Pr} R}{3R+4}\theta_2' + \frac{3R}{3R+4}(i\omega \mathrm{Pr} - Q_H)\theta_2 = 0,\\
&\phi_0'' - \mathrm{Sc}\lambda\phi_0' - \mathrm{Sc}\xi\phi_0 = 0,\\
&\phi_1'' - \mathrm{Sc}\lambda\phi_1' - \mathrm{Sc}(i\omega + \xi)\phi_1 = 0,\\
&\phi_2'' - \mathrm{Sc}\lambda\phi_2' + \mathrm{Sc}(i\omega - \xi)\phi_2 = 0,
\end{aligned} \tag{14}$$

where $S = (M^2(1 + im)/(1 + m^2)) + 2i\Omega$ and dashes denote the derivatives w.r.t. η.

The transformed boundary conditions are

$$\begin{aligned}
&q_0 = 0, \quad q_1 = 0, \quad q_2 = 0,\\
&\theta_0 = 1, \quad \theta_1 = 1, \quad \theta_2 = 1,\\
&\phi_0 = 1, \quad \phi_1 = 1, \quad \phi_2 = 1 \quad \text{at } \eta = 0,\\
&q_0 = 1, \quad q_1 = 1, \quad q_2 = 1,\\
&\theta_0 = 0, \quad \theta_1 = 0, \quad \theta_2 = 0,\\
&\phi_0 = 0, \quad \phi_1 = 0, \quad \phi_2 = 0 \quad \text{at } \eta = 1.
\end{aligned} \tag{15}$$

The solutions of (14) under the boundary conditions (15) are

$$\begin{aligned}
q_0 = {}& 1 - e^{h_{14}\eta} + A_1\left(e^{h_{14}\eta} - e^{h_8\eta}\right) - A_2\left(e^{h_{14}\eta} - e^{h_7\eta}\right)\\
&+ A_3\left(e^{h_{14}\eta} - e^{h_2\eta}\right) - A_4\left(e^{h_{14}\eta} - e^{h_1\eta}\right) + \frac{e^{h_{13}\eta} - e^{h_{14}\eta}}{e^{h_{13}} - e^{h_{14}}}\\
&\times\left[e^{h_{14}} - A_1\left(e^{h_{14}} - e^{h_8}\right) + A_2\left(e^{h_{14}} - e^{h_7}\right)\right.\\
&\left. - A_3\left(e^{h_{14}} - e^{h_2}\right) + A_4\left(e^{h_{14}} - e^{h_1}\right)\right],
\end{aligned} \tag{16}$$

$$\begin{aligned}
q_1 = {}& 1 - e^{h_{16}\eta} + A_5\left(e^{h_{16}\eta} - e^{h_{10}\eta}\right) - A_6\left(e^{h_{16}\eta} - e^{h_9\eta}\right)\\
&+ A_7\left(e^{h_{16}\eta} - e^{h_4\eta}\right) - A_8\left(e^{h_{16}\eta} - e^{h_3\eta}\right) + \frac{e^{h_{15}\eta} - e^{h_{16}\eta}}{e^{h_{15}} - e^{h_{16}}}\\
&\times\left[e^{h_{16}} - A_5\left(e^{h_{16}} - e^{h_{10}}\right) + A_6\left(e^{h_{16}} - e^{h_9}\right)\right.\\
&\left. - A_7\left(e^{h_{16}} - e^{h_4}\right) + A_8\left(e^{h_{16}} - e^{h_3}\right)\right],
\end{aligned} \tag{17}$$

$$\begin{aligned}
q_2 = {}& 1 - e^{h_{18}\eta} + A_9\left(e^{h_{18}\eta} - e^{h_{12}\eta}\right) - A_{10}\left(e^{h_{18}\eta} - e^{h_{11}\eta}\right)\\
&+ A_{11}\left(e^{h_{18}\eta} - e^{h_6\eta}\right) - A_{12}\left(e^{h_{18}\eta} - e^{h_5\eta}\right) + \frac{e^{h_{17}\eta} - e^{h_{18}\eta}}{e^{h_{17}} - e^{h_{18}}}\\
&\times\left[e^{h_{18}} - A_9\left(e^{h_{18}} - e^{h_{12}}\right) + A_{10}\left(e^{h_{18}} - e^{h_{11}}\right)\right.\\
&\left. - A_{11}\left(e^{h_{18}} - e^{h_6}\right) + A_{12}\left(e^{h_{18}} - e^{h_5}\right)\right],
\end{aligned} \tag{18}$$

$$\theta_0 = \frac{e^{h_7 + h_8\eta} - e^{h_8 + h_7\eta}}{e^{h_7} - e^{h_8}}, \tag{19}$$

$$\theta_1 = \frac{e^{h_9 + h_{10}\eta} - e^{h_{10} + h_9\eta}}{e^{h_9} - e^{h_{10}}}, \tag{20}$$

$$\theta_2 = \frac{e^{h_{11} + h_{12}\eta} - e^{h_{12} + h_{11}\eta}}{e^{h_{11}} - e^{h_{12}}}, \tag{21}$$

$$\phi_0 = \frac{e^{h_1 + h_2\eta} - e^{h_2 + h_1\eta}}{e^{h_1} - e^{h_2}}, \tag{22}$$

$$\phi_1 = \frac{e^{h_3+h_4\eta} - e^{h_4+h_3\eta}}{e^{h_3} - e^{h_4}}, \tag{23}$$

$$\phi_2 = \frac{e^{h_5+h_6\eta} - e^{h_6+h_5\eta}}{e^{h_5} - e^{h_6}}. \tag{24}$$

4. Amplitude and Phase Difference due to Steady and Unsteady Flow

Equation (16) corresponds to the steady part, which gives u_0 as the primary and v_0 as secondary velocity components. The amplitude (resultant velocity) and phase difference due to these primary and secondary velocities for the steady flow are given by

$$R_0 = \sqrt{u_0^2 + v_0^2}, \qquad \alpha_0 = \tan^{-1}\left(\frac{v_0}{u_0}\right), \tag{25}$$

where $u_0(\eta) + iv_0(\eta) = q_0(\eta)$.

Equations (17) and (18) together give the unsteady part of the flow. Thus, unsteady primary and secondary velocity components $u_1(\eta)$ and $v_1(\eta)$, respectively, for the fluctuating flow can be obtained from the following:

$$\begin{aligned} u_1(\eta, t) &= [\text{Real}\, q_1(\eta) + \text{Real}\, q_2(\eta)] \cos t \\ &\quad - [\text{Im}\, q_1(\eta) - \text{Im}\, q_2(\eta)] \sin t, \\ v_1(\eta, t) &= [\text{Real}\, q_1(\eta) - \text{Real}\, q_2(\eta)] \sin t \\ &\quad + [\text{Im}\, q_1(\eta) + \text{Im}\, q_2(\eta)] \cos t. \end{aligned} \tag{26}$$

The amplitude (resultant velocity) and the phase difference of the unsteady flow are given by

$$R_v = \sqrt{u_1^2 + v_1^2}, \qquad \alpha_1 = \tan^{-1}\left(\frac{v_1}{u_1}\right), \tag{27}$$

where $u_1(\eta) + iv_1(\eta) = q_1(\eta)e^{it} + q_2(\eta)e^{-it}$.

The amplitude (resultant velocity) and the phase difference

$$R_n = \sqrt{u^2 + v^2}, \qquad \alpha = \tan^{-1}\left(\frac{v}{u}\right), \tag{28}$$

where u = Real part of q and v = Imaginary part of q.

5. Amplitude and Phase Difference of Shear Stresses due to Steady and Unsteady Flow at the Plate

The amplitude and phase difference of shear stresses at the stationary plate ($\eta = 0$) for the steady flow can be obtained as

$$\tau_{0r} = \sqrt{\tau_{0x}^2 + \tau_{0y}^2}, \qquad \beta_0 = \tan^{-1}\left(\frac{\tau_{0y}}{\tau_{0x}}\right), \tag{29}$$

where

$$\begin{aligned} \left(\frac{\partial q_0}{\partial \eta}\right)_{\eta=0} &= \tau_{0x} + i\tau_{0y} = -h_{14} + A_1(h_{14} - h_8) - A_2(h_{14} - h_7) \\ &\quad + A_3(h_{14} - h_2) - A_4(h_{14} - h_1) + \frac{h_{13} - h_{14}}{e^{h_{13}} - e^{h_{14}}} \\ &\quad \times \left[e^{h_{14}} - A_1\left(e^{h_{14}} - e^{h_8}\right) + A_2\left(e^{h_{14}} - e^{h_7}\right)\right. \\ &\quad \left. - A_3\left(e^{h_{14}} - e^{h_2}\right) + A_4\left(e^{h_{14}} - e^{h_1}\right)\right]. \end{aligned} \tag{30}$$

For the unsteady part of flow, the amplitude and phase difference of shear stresses at the stationary plate ($\eta = 0$) can be obtained as

$$\tau_{1r} = \sqrt{\tau_{1x}^2 + \tau_{1y}^2}, \qquad \beta_1 = \tan^{-1}\left(\frac{\tau_{1y}}{\tau_{1x}}\right), \tag{31}$$

where

$$\tau_{1x} + i\tau_{1y} = \left(\frac{\partial q_1}{\partial \eta}\right)_{\eta=0} e^{it} + \left(\frac{\partial q_2}{\partial \eta}\right)_{\eta=0} e^{-it},$$

$$\begin{aligned} \left(\frac{\partial q_1}{\partial \eta}\right)_{\eta=0} &= -h_{16} + A_5(h_{16} - h_{10}) - A_6(h_{16} - h_9) \\ &\quad + A_7(h_{16} - h_4) - A_8(h_{16} - h_3) + \frac{h_{15} - h_{16}}{e^{h_{15}} - e^{h_{16}}} \\ &\quad \times \left[e^{h_{16}} - A_5\left(e^{h_{16}} - e^{h_{10}}\right) + A_6\left(e^{h_{16}} - e^{h_9}\right)\right. \\ &\quad \left. - A_7\left(e^{h_{16}} - e^{h_4}\right) + A_8\left(e^{h_{16}} - e^{h_3}\right)\right], \\ \left(\frac{\partial q_2}{\partial \eta}\right)_{\eta=0} &= -h_{18} + A_9(h_{18} - h_{12}) - A_{10}(h_{18} - h_{11}) \\ &\quad + A_{11}(h_{16} - h_6) - A_{12}(h_{18} - h_5) + \frac{h_{17} - h_{18}}{e^{h_{17}} - e^{h_{18}}} \\ &\quad \times \left[e^{h_{18}} - A_9\left(e^{h_{18}} - e^{h_{12}}\right) + A_{10}\left(e^{h_{18}} - e^{h_{11}}\right)\right. \\ &\quad \left. - A_{11}\left(e^{h_{18}} - e^{h_6}\right) + A_{12}\left(e^{h_{18}} - e^{h_5}\right)\right]. \end{aligned} \tag{32}$$

The amplitude and phase difference of shear stresses at the stationary plate ($\eta = 0$) for the flow can be obtained as

$$\tau = \left(\frac{\partial q}{\partial \eta}\right)_{\eta=0} = \sqrt{\tau_x^2 + \tau_y^2}, \qquad \beta_2 = \tan^{-1}\left(\frac{\tau_y}{\tau_x}\right), \tag{33}$$

where τ_x = Real part of $(\partial q/\partial \eta)_{\eta=0}$ and τ_y = Imaginary part of $(\partial q/\partial \eta)_{\eta=0}$.

The Nusselt number

$$\text{Nu} = -\left(1 + \frac{4}{3R}\right)\left(\frac{\partial \theta}{\partial \eta}\right)_{\eta=0} = N_x + iN_y. \tag{34}$$

The rate of heat transfer (i.e., heat flux) at the plate in terms of amplitude and phase is given by

$$\Theta = \sqrt{N_x^2 + N_y^2}, \qquad \gamma = \tan^{-1}\left(\frac{N_y}{N_x}\right). \tag{35}$$

The Sherwood number

$$\mathrm{Sh} = \left(\frac{\partial \phi}{\partial \eta}\right)_{\eta=0} = M_x + iM_y. \quad (36)$$

The rate of mass transfer (i.e., mass flux) at the plate in terms of amplitude and phase is given by

$$\Phi = \sqrt{M_x^2 + M_y^2}, \qquad \delta = \tan^{-1}\left(\frac{M_y}{M_x}\right). \quad (37)$$

6. Results and Discussion

The system of ordinary differential equations (14) with boundary conditions (15) is solved analytically by employing the perturbation technique. The solutions are obtained for the steady and unsteady velocity fields from (16)–(18), temperature fields from (19)–(21), and concentration fields are given by (22)–(24). The effects of various parameters on the thermal, mass, and hydrodynamic behaviors of buoyancy-induced flow in a rotating vertical channel are studied. The results are presented graphically and in tabular form. Temperature of the heated wall (left wall) at $z^* = 0$ is a function of time as given in the boundary conditions, and the cooled wall at $z^* = d$ is maintained at a constant temperature. Further, it is assumed that the temperature difference is small enough so that the density changes of the fluid in the system will be small. When the injection/suction parameter λ is positive, fluid is injected through the hot wall into the channel and sucked out through the cold wall. The numerical results of the amplitude of the shear stresses and the phase difference of the shear stresses at the stationary plate ($z^* = 0$) for the steady and unsteady flow are presented in Table 1. The effect of various physical parameters on flow, heat, concentration fields, skin-friction, Nusselt number, and Sherwood number are presented graphically in Figures 2–16.

Table 1 shows a comparative study of the present results of amplitude and phase difference of shear stresses for the steady flow with those of Singh and Kumar [38]. It is seen from this table that the present results coincide very well with those of Singh and Kumar [38]. This confirms that the present analytical solutions are correct and accurate. Further, it is observed from this table that the effects of increasing the value of thermal Grashof number Gr, magnetic field M, and injection/suction parameter λ are to increase amplitude and decrease the phase difference of shear stresses for the steady flow, whereas reverse effect is found by increasing the Hall parameter, m. The effects of increasing the angular velocity Ω are to increase both amplitude and phase difference of shear stresses, whereas reverse effects are seen by increasing the values of the Prandtl number. The computed results τ_{0r}, τ_{1r}, β_0, β_1 for the present problem are provided in Table 2 for various values of Gm, R, Q_H, Sc, ξ, and m. It is seen from this table that the values of τ_{0r} and τ_{1r} increase whereas the values of β_0 and β_1 decrease with the increase of solutal Grashof number Gm, but the effects are reversed with an increase in the Hall parameter m, that is, the value of β_0 and β_1 are increased whereas there is decrease in the values of τ_{0r} and τ_{1r}. Also, it is found that the values of τ_{0r}, $\beta_0\beta_1$ increase with

TABLE 1: Comparison results for the resultant velocity or amplitude and the phase difference of the unsteady flow with Singh and Kumar [38] for different values of Gr, M, m, λ, Ω, Pr, and Gm = 0.0, $R = \infty$, $\xi = 0.0$, Sc $\rightarrow$ 0.0 (in present problem).

Gr	M	m	λ	Ω	Pr	Singh and Kumar [38]		Present results	
						τ_{0r}	β_0	τ_{0r}	β_0
5	2	1	0.5	10	0.71	4.5847	0.7279	4.5847	0.7279
10	2	1	0.5	10	0.71	4.6515	0.6815	4.6515	0.6815
5	4	1	0.5	10	0.71	5.2797	0.6395	5.2797	0.6395
5	2	3	0.5	10	0.71	4.4862	0.7627	4.4862	0.7627
5	2	1	1.0	10	0.71	4.6162	0.6229	4.6162	0.6229
5	2	1	0.5	20	0.71	6.3323	0.7614	6.3323	0.7614
5	2	1	0.5	40	0.71	8.8924	0.7783	8.8924	0.7783
5	2	1	0.5	80	0.71	12.558	0.7857	12.558	0.7857
5	2	1	0.5	10	7.0	4.5726	0.7198	4.5726	0.7198

TABLE 2: Values of τ_{0r}, β_0, τ_{1r}, and β_1 for the reference values of Gr = 5.0, $M = 2.0$, $\lambda = 0.5$, $\Omega = 10.0$, Pr = 0.71, and $\omega = 5.0$ at $t = \pi/4$.

Gm	R	Q_H	Sc	ξ	m	τ_{0r}	β_0	τ_{1r}	β_1
5.0	1.0	5.0	0.15	0.1	1.0	4.6540	0.6877	6.4123	0.7980
10.0	1.0	5.0	0.15	0.1	1.0	4.7299	0.6437	6.4542	0.7485
5.0	5.0	5.0	0.15	0.1	1.0	4.6543	0.6900	6.4094	0.8011
5.0	1.0	10.0	0.15	0.1	1.0	4.6545	0.6908	6.4219	0.8010
5.0	1.0	5.0	0.60	0.1	1.0	4.6534	0.6869	6.3805	0.8006
5.0	1.0	5.0	0.15	1.0	1.0	4.6541	0.6880	6.4132	0.7983
5.0	1.0	5.0	0.15	1.0	3.0	4.5464	0.7197	6.3064	0.8366

an increase in the radiation parameter R. It is noted that the values of both τ_{0r} and β_0 increase due to an increase in the heat source parameter Q_H and chemical reaction parameter ξ, whereas the effects are reversed with the increase in the Schmidt number, that is, the values of τ_{0r}, $\tau_{1r}\beta_1$ decrease with an increase in the Schmidt number. Also, it is found that the value of β_1 decreases with an increase in the heat source parameter Q_H, chemical reaction parameter ξ and Schmidt number Sc.

The profiles for resultant velocity R_n for the flow are shown in Figures 2–6 for suction/injection parameter λ and for small and large values of rotation parameter Ω, ϵ, and η, respectively. From Figure 2, it is observed that the increase in the suction parameter λ leads to an increase of R_n within the stationary plates. Similar trend of R_n profiles is seen by increasing the rotation parameter Ω, that is, resultant velocity profiles increase with increase in the rotation parameter Ω (small values) as shown in Figure 3. However, the opposite effect occurs near the right wall for large values of Ω as shown in Figure 4. This effect is due to the rotation effects being more dominant near the walls, so when Ω reaches high values, the secondary velocity component v decreases with increase in Ω while approaching to the right plate. From Figure 5, it is observed that the increase in the ϵ leads to an increase of R_n within the stationary plates. From Figure 6, it is seen that the resultant velocity profiles increases with increase in η; also it is observed that the velocity oscillates with increasing time. The phase difference α for the flow is shown graphically in

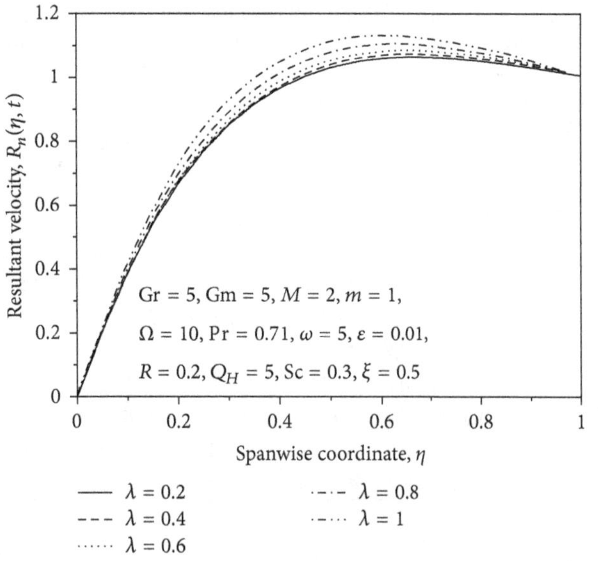

FIGURE 2: Resultant velocity R_n due to u and v versus η at $t = \pi/4$.

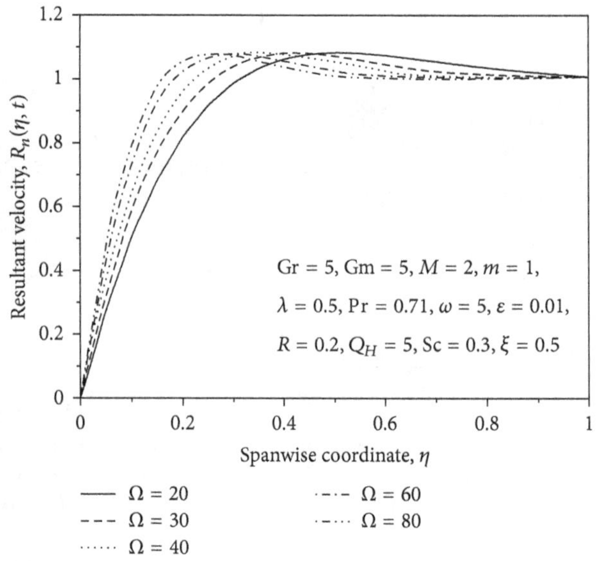

FIGURE 4: Resultant velocity R_n due to u and v versus η for large values of Ω at $t = \pi/4$.

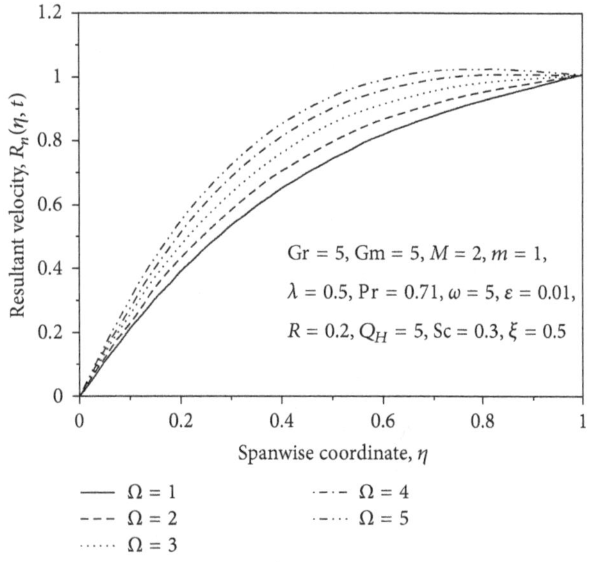

FIGURE 3: Resultant velocity R_n due to u and v versus η for small values of Ω at $t = \pi/4$.

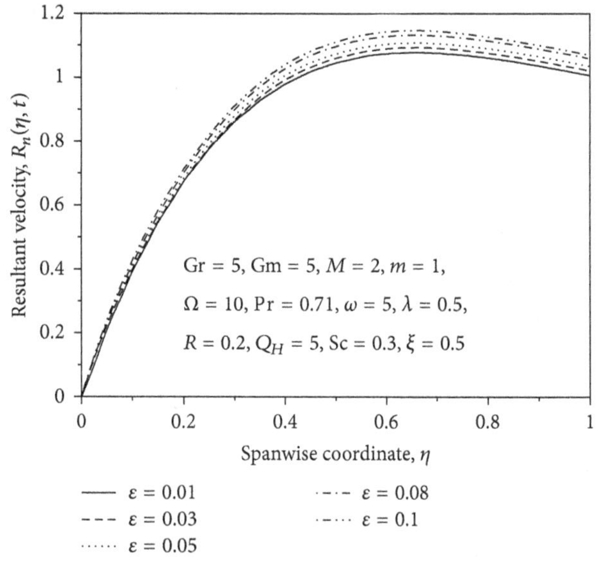

FIGURE 5: Resultant velocity R_n due to u and v versus η for different values of ϵ at $t = \pi/4$.

Figure 7 for various values of rotation parameter Ω. From this figure, it is observed that the phase angle α decreases with an increase in rotation parameter. Figure 8 shows the variation of α against η for different values of thermal Grashof number Gr, solutal Grashof number Gm, Hartmann number M, and Hall parameter m. From this figure it is found that the values of α decrease with an increase in the value of Gr, Gm and M, whereas reverse trend is seen on the values of α by increasing the value of the Hall parameter m. The phase difference α for the flow is shown graphically in Figure 9 for various positive values of suction/injection parameter λ. The figure shows that the phase angle α decreases with the increase of suction parameter.

The effect of reaction rate parameter ξ on the species concentration profiles for generative chemical reaction is shown in Figure 10. It is noticed for the graph that there is a marked effect of increasing the value of the chemical reaction rate parameter ξ on concentration distribution ϕ in the boundary layer. It is observed that increasing the value of the chemical reaction parameter ξ decreases the concentration of species in the boundary layer; this is due to the fact that destructive chemical reduces the solutal boundary layer thickness and increases the mass transfer. Opposite trend is seen in the case when Schmidt number is increased as noted in Figure 11. It may also be observed from this figure that the

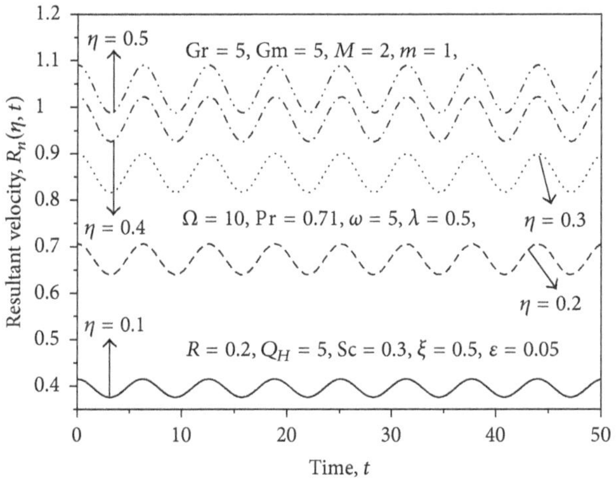

FIGURE 6: Resultant velocity R_n due to u and v versus t for different values of η.

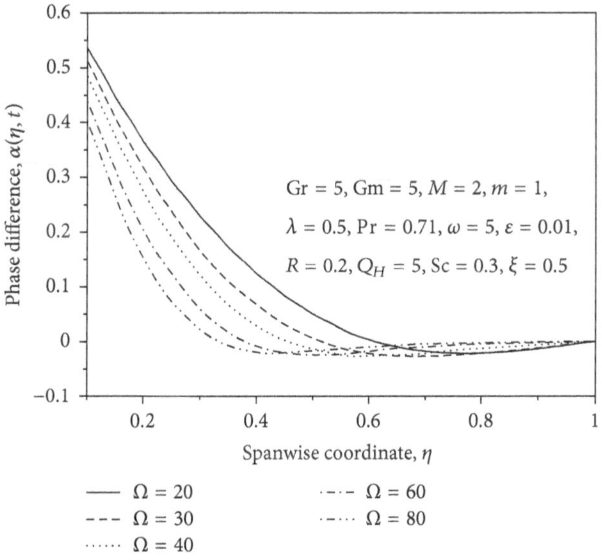

FIGURE 7: Phase angle α due to u and v versus η at $t = \pi/4$.

effect of Schmidt number Sc is to increase the concentration distribution in the solutal boundary layer.

Figure 12 has been plotted to depict the variation of temperature profiles against η for different values of heat absorption parameter Q_H by fixing other physical parameters. From this graph, we observe that temperature θ decreases with increase in the heat absorption parameter Q_H because when heat is absorbed, the buoyancy force decreases the temperature profile. Figure 13 represents graph of temperature distribution with η for different values of radiation parameter. From this figure, we note that the initial temperature $\theta = 1.0$ decreases to zero satisfying the boundary condition at $\eta = 1.0$. Further, it is observed from this figure that increase in the radiation parameter decreases the temperature distribution in the thermal boundary layer due to decrease in the thickness of the thermal boundary layer with thermal radiation parameter R. This is because large

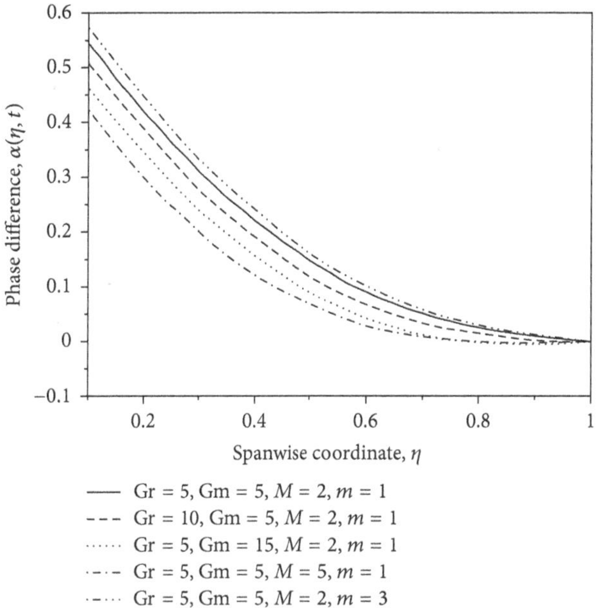

FIGURE 8: Phase angle α due to u and v versus η for $\lambda = 0.5$, $\Omega = 10$, $Sc = 0.3$, $Pr = 0.71$, $R = 0.2$, $Q_H = 5.0$, $\epsilon = 0.01$, and $\xi = 0.5$ at $t = \pi/4$.

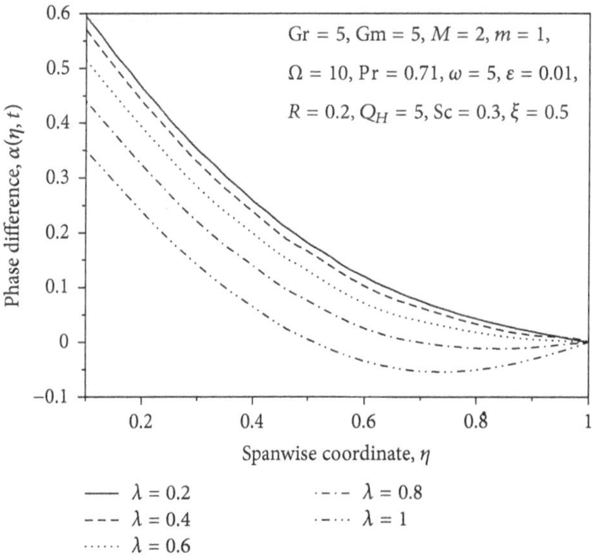

FIGURE 9: Phase angle α due to u and v versus η for $\omega = 5.0$ at $t = \pi/4$.

values of radiation parameter correspond to an increase in dominance of conduction over radiation, thereby decreasing the buoyancy force and the temperature in the thermal boundary layer.

Figures 14–16 show the amplitude of skin-friction, Nusselt number, and Sherwood number against frequency parameter ω for different values of Gr, Q_H and ξ, respectively. From Figure 14, it is observed that the skin friction increases with increasing the values of Gr. Also, the skin friction decreases slowly with increasing the value of ω. The amplitude of

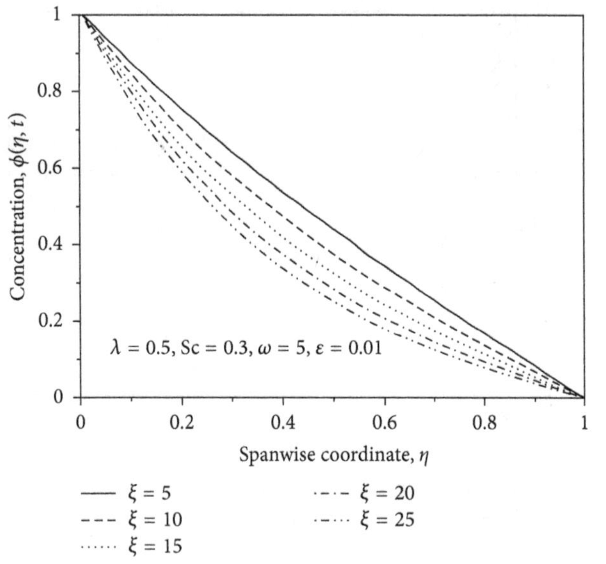

FIGURE 10: Concentration profiles against η for different values of ξ at $t = \pi/4$.

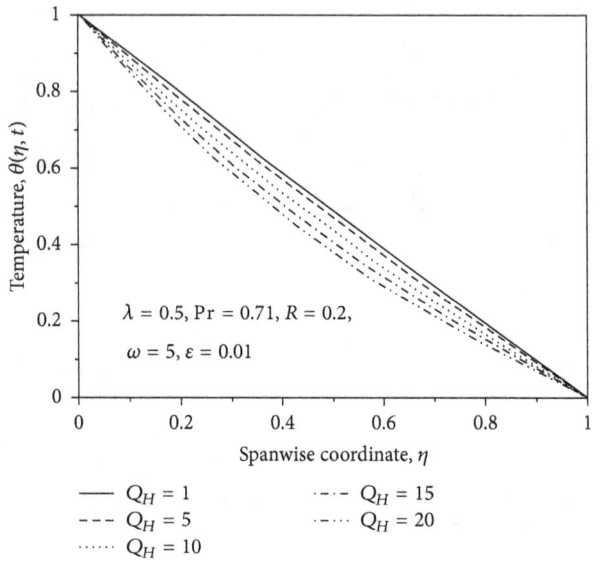

FIGURE 12: Temperature profiles against η for different values of Q_H at $t = \pi/4$.

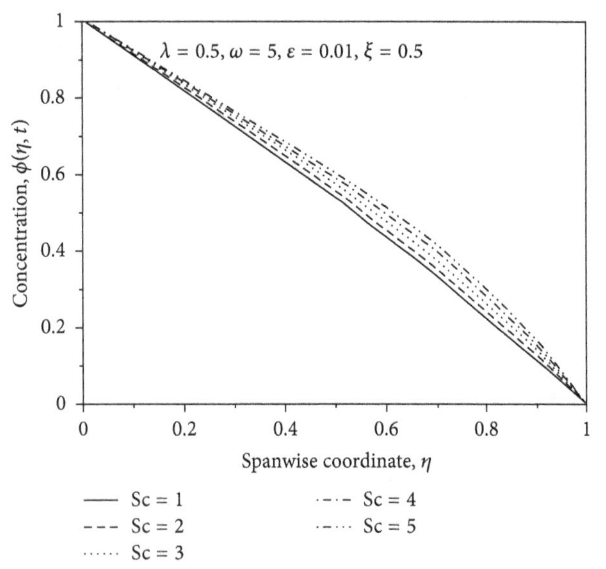

FIGURE 11: Concentration profiles against η for different values of Sc at $t = \pi/4$.

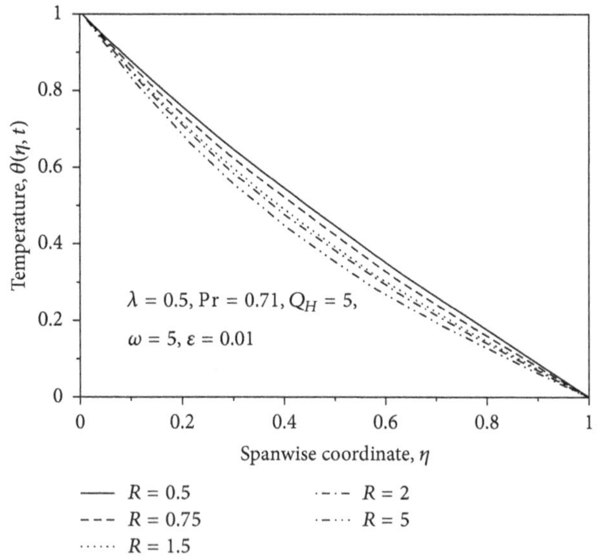

FIGURE 13: Temperature profiles against η for different values of R at $t = \pi/4$.

Nusselt number decreases with increasing the value of heat source parameter Q_H which is shown in Figure 15. Figure 16 shows the variation of Sherwood number with ξ and ω. From this figure, it is seen that the Sherwood number decreases with increasing the values of chemical reaction parameter ξ, and the opposite trend is seen with increasing the values of ω.

7. Conclusions

The influence of hall current and chemical reaction on unsteady MHD heat and mass transfer of an oscillatory convective flow in a rotating vertical porous channel with thermal radiation and injection is studied analytically. Computed results are presented to exhibit their dependence on the important physical parameters. We conclude the following from the numerical results.

(i) An increase in Q_H leads to an increase in τ_{0r}, β_0, τ_{1r} and decrease in β_1.

(ii) An increase in radiation parameter R and chemical reaction parameter ξ leads to increase in τ_{0r}, β_0, and β_1 but decrease in τ_{1r}.

(iii) An increase in Gr, Gm, M, Ω, and λ leads to decrease in α_0 and α_1, whereas reverse effect is seen by increasing Hall parameter m.

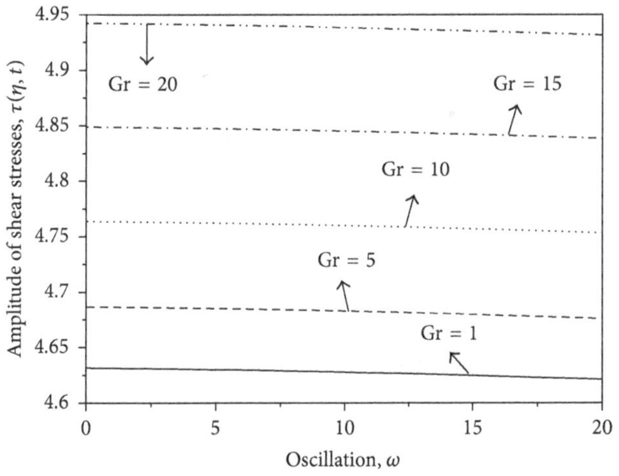

FIGURE 14: Skin friction coefficient against ω for different values of Gr with Gm $= 5.0$, $M = 2.0$, $m = 1.0$, $\Omega = 10.0$, Pr $= 0.71$, $R = 0.2$, $Q_H = 5.0$, Sc $= 0.3$, $\xi = 0.5$, $\epsilon = 0.01$ at $t = \pi/4$.

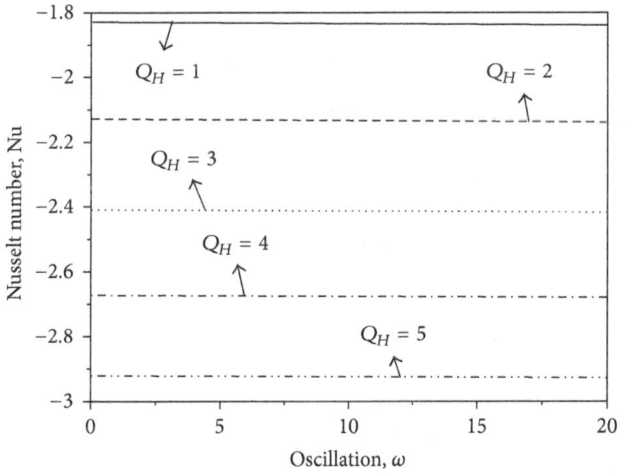

FIGURE 15: Nusselt number against ω for different values of Q_H with $\lambda = 0.5$, Pr $= 0.71$, $R = 2.0$, $\epsilon = 0.01$ at $t = \pi/4$.

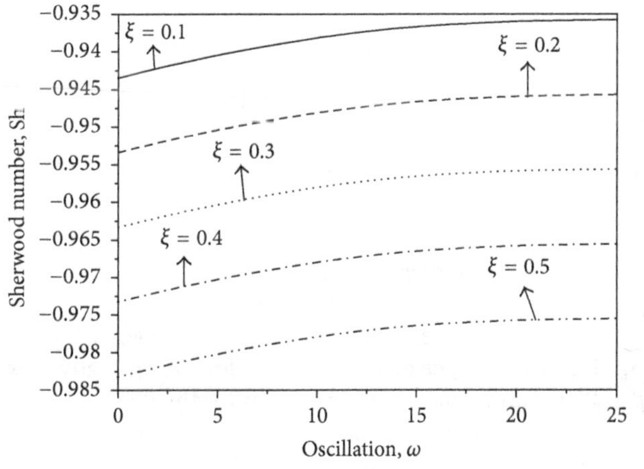

FIGURE 16: Sherwood number against ω for different values of ξ with $\lambda = 0.5$, $\epsilon = 0.01$, Sc $= 0.3$ at $t = \pi/4$.

(iv) The amplitude R_n increases with the increase of λ and Ω.

(v) The value of R_n decreases with the increase in chemical reaction parameter ξ and oscillation parameter ω.

(vi) The skin friction increases with increase in thermal Grashof number Gr.

Appendix

Consider the following:

$$R_1 = \frac{3R}{3R+4}, \qquad R_2 = R_1 \mathrm{Pr},$$

$$h_1 = \frac{\mathrm{Sc}\,\lambda + \sqrt{(\mathrm{Sc}\,\lambda)^2 + 4\xi \mathrm{Sc}}}{2},$$

$$h_2 = \frac{\mathrm{Sc}\,\lambda - \sqrt{(\mathrm{Sc}\,\lambda)^2 + 4\xi \mathrm{Sc}}}{2},$$

$$h_3 = \frac{\mathrm{Sc}\,\lambda + \sqrt{(\mathrm{Sc}\,\lambda)^2 + 4\,(i\omega + \xi)\,\mathrm{Sc}}}{2},$$

$$h_4 = \frac{\mathrm{Sc}\,\lambda - \sqrt{(\mathrm{Sc}\,\lambda)^2 + 4\,(i\omega + \xi)\,\mathrm{Sc}}}{2},$$

$$h_5 = \frac{\mathrm{Sc}\,\lambda + \sqrt{(\mathrm{Sc}\,\lambda)^2 - 4\,(i\omega - \xi)\,\mathrm{Sc}}}{2},$$

$$h_6 = \frac{\mathrm{Sc}\,\lambda - \sqrt{(\mathrm{Sc}\,\lambda)^2 - 4\,(i\omega - \xi)\,\mathrm{Sc}}}{2},$$

$$h_7 = \frac{R_2\lambda + \sqrt{(R_2\lambda)^2 + 4R_1 Q_H}}{2},$$

$$h_8 = \frac{R_2\lambda - \sqrt{(R_2\lambda)^2 + 4R_1 Q_H}}{2},$$

$$h_9 = \frac{R_2\lambda + \sqrt{(R_2\lambda)^2 + 4R_1\left(i\omega \mathrm{Pr} + Q_H\right)}}{2},$$

$$h_{10} = \frac{R_2\lambda - \sqrt{(R_2\lambda)^2 + 4R_1\left(i\omega \mathrm{Pr} + Q_H\right)}}{2},$$

$$h_{11} = \frac{R_2\lambda + \sqrt{(R_2\lambda)^2 - 4R_1\left(i\omega \mathrm{Pr} - Q_H\right)}}{2},$$

$$h_{12} = \frac{R_2\lambda - \sqrt{(R_2\lambda)^2 - 4R_1\left(i\omega \mathrm{Pr} + Q_H\right)}}{2},$$

$$h_{13} = \frac{\lambda + \sqrt{\lambda^2 + 4S}}{2}, \qquad h_{14} = \frac{\lambda - \sqrt{\lambda^2 + 4S}}{2},$$

$$h_{15} = \frac{\lambda + \sqrt{\lambda^2 + 4(S + i\omega)}}{2},$$

$$h_{16} = \frac{\lambda - \sqrt{\lambda^2 + 4(S + i\omega)}}{2},$$

$$h_{17} = \frac{\lambda + \sqrt{\lambda^2 + 4(S - i\omega)}}{2},$$

$$h_{18} = \frac{\lambda - \sqrt{\lambda^2 + 4(S - i\omega)}}{2},$$

$$A_1 = \frac{\lambda^2 \mathrm{Gr}\, e^{h_7}}{\left(e^{h_7} - e^{h_8}\right)\left[h_8^2 - \lambda h_8 - S\right]},$$

$$A_2 = \frac{\lambda^2 \mathrm{Gr}\, e^{h_8}}{\left(e^{h_7} - e^{h_8}\right)\left[h_7^2 - \lambda h_7 - S\right]},$$

$$A_3 = \frac{\lambda^2 \mathrm{Gr}\, e^{h_1}}{\left(e^{h_1} - e^{h_2}\right)\left[h_2^2 - \lambda h_2 - S\right]},$$

$$A_4 = \frac{\lambda^2 \mathrm{Gr}\, e^{h_2}}{\left(e^{h_1} - e^{h_2}\right)\left[h_1^2 - \lambda h_1 - S\right]},$$

$$A_5 = \frac{\lambda^2 \mathrm{Gr}\, e^{h_9}}{\left(e^{h_9} - e^{h_{10}}\right)\left[h_{10}^2 - \lambda h_{10} - (S + i\omega)\right]},$$

$$A_6 = \frac{\lambda^2 \mathrm{Gr}\, e^{h_{10}}}{\left(e^{h_9} - e^{h_{10}}\right)\left[h_9^2 - \lambda h_9 - (S + i\omega)\right]},$$

$$A_7 = \frac{\lambda^2 \mathrm{Gm}\, e^{h_3}}{\left(e^{h_3} - e^{h_4}\right)\left[h_4^2 - \lambda h_4 - (S + i\omega)\right]},$$

$$A_8 = \frac{\lambda^2 \mathrm{Gm}\, e^{h_4}}{\left(e^{h_3} - e^{h_4}\right)\left[h_3^2 - \lambda h_3 - (S + i\omega)\right]},$$

$$A_9 = \frac{\lambda^2 \mathrm{Gr}\, e^{h_{11}}}{\left(e^{h_{11}} - e^{h_{12}}\right)\left[h_{12}^2 - \lambda h_{12} - (S - i\omega)\right]},$$

$$A_{10} = \frac{\lambda^2 \mathrm{Gr}\, e^{h_{12}}}{\left(e^{h_{11}} - e^{h_{12}}\right)\left[h_{11}^2 - \lambda h_{11} - (S - i\omega)\right]},$$

$$A_{11} = \frac{\lambda^2 \mathrm{Gm}\, e^{h_5}}{\left(e^{h_5} - e^{h_6}\right)\left[h_6^2 - \lambda h_6 - (S - i\omega)\right]},$$

$$A_{12} = \frac{\lambda^2 \mathrm{Gm}\, e^{h_6}}{\left(e^{h_5} - e^{h_6}\right)\left[h_5^2 - \lambda h_5 - (S - i\omega)\right]}. \qquad \text{(A.1)}$$

Nomenclature

C^*: Dimensional concentration
C_0: Concentration at the left plate
C_d: Concentration at the right plate
c_p: Specific heat at constant pressure
d: Distance of the plates
D_m: Chemical molecular diffusivity
e: Electric charge
g_0: Acceleration due to gravity
Gm: Modified Grashof number for mass transfer
Gr: Modified Grashof number for heat transfer
$\vec{H}$: Magnetic field
H_0: Magnetic field of uniform strength
H_x: x-component of magnetic field
$\vec{J}$: Current density
J_x: x-component of current density
k^*: Mean absorption coefficient
k_1: Chemical reaction rate constant
m: Hall parameter
M: Hartmann number
Nu: Nusselt number
n_e: Number density of the electron
P^*: Dimensional pressure
p_e: Electron pressure
Pr: Prandtl number
q_r^*: Radiative heat flux
Q_0: Dimensional heat source
Q_H: Heat source parameter
R: Radiation parameter
R_0: Amplitude for steady flow
R_n: Resultant velocity
R_v: Amplitude for unsteady flow
Sc: Schmidt number
Sh: Sherwood number
t^*: Dimensional time
T^*: Dimensional temperature
T_0: Temperature at the left wall
T_d: Temperature at the right wall
U_0: Nonzero constant mean velocity
u_0: Primary velocity component for steady flow
u_1: Primary velocity component for unsteady flow
$\vec{V}$: Electron velocity
v_0: Secondary velocity component for steady flow
v_1: Secondary velocity component for unsteady flow
u^*, v^*, w^*: Velocity components are in the x^*-, y^*-, z^*-directions, respectively
w_0: Dimensional injection/suction velocity.

Greek Symbols

α_0: Phase difference for steady flow
α_1: Phase difference for unsteady flow
α: Phase difference of the flow
β: Coefficient of thermal expansion
β^*: Coefficient of solutal expansion
β_0: Phase difference of shear stresses for the steady flow
β_1: Phase difference of shear stresses for the unsteady flow
β_2: Phase difference of shear stresses for the flow
δ: Phase difference of mass flux
ϵ: Small positive constant
η: Dimensionless distance
γ: Phase difference of heat flux

κ: Fluid thermal conductivity
λ: Injection/suction parameter
μ: Dynamic viscosity
μ_e: Magnetic permeability
ν: Kinematic viscosity
ω: Oscillation parameter
Ω^*: Dimensional angular velocity
Ω: Angular velocity
ω_e: Cyclotron frequency
Φ: Amplitude of mass flux
ϕ: Nondimensional concentration
ρ: Density
σ: Electric conductivity
σ^*: Stefan-Boltzmann constant
τ: Amplitude of shear stresses for the flow
τ_{0r}: Amplitude of shear stresses for the steady flow
τ_{1r}: Amplitude of shear stresses for the unsteady flow
τ_e: Electron collision time
θ: Non-dimensional temperature
Θ: Amplitude of heat flux.

Acknowledgment

One of the authors (Dulal Pal) is grateful to the University Grants Commission (UGC), New Delhi, for providing financial support under SAP-DRS (Phase-II) Grant.

References

[1] M. A. El-Hakiem, "MHD oscillatory flow on free convection-radiation through a porous medium with constant suction velocity," *Journal of Magnetism and Magnetic Materials*, vol. 220, no. 2-3, pp. 271–276, 2000.

[2] B. S. Jaiswal and V. M. Soundalgekar, "Oscillating plate temperature effects on a flow past an infinite vertical porous plate with constant suction and embedded in a porous medium," *Heat and Mass Transfer*, vol. 37, no. 2-3, pp. 125–131, 2001.

[3] K. D. Singh, M. G. Gorla, and H. Raj, "A periodic solution of oscillatory couette flow through porous medium in rotating system," *Indian Journal of Pure and Applied Mathematics*, vol. 36, no. 3, pp. 151–159, 2005.

[4] D. Pal and I. S. Shivakumara, "Mixed convection heat transfer from a vertical heated plate embedded in a sparsely packed porous medium," *International Journal of Applied Mechanics and Engineering*, vol. 11, no. 4, pp. 929–939, 2006.

[5] T. G. Cowling, *Magnetohydrodynamics*, Interscience Publishers, New York, NY, USA, 1957.

[6] A. S. Gupta, "Hydromagnetic flow past a porous flat plate with hall effects," *Acta Mechanica*, vol. 22, no. 3-4, pp. 281–287, 1975.

[7] R. N. Jana, A. S. Gupta, and N. Datta, "Hall effects on the hydro magnetic flow past an infinite porous flat plate," *Journal of the Physical Society of Japan*, vol. 43, no. 5, pp. 1767–1772, 1977.

[8] O. D. Makinde and P. Y. Mhone, "Heat transfer to MHD oscillatory flow in a channel filled with porous medium," *Romanian Journal of Physics*, vol. 50, pp. 931–938, 2005.

[9] Z. Zhang and J. Wang, "On the similarity solutions of magnetohydrodynamic flows of power-law fluids over a stretching sheet," *Journal of Mathematical Analysis and Applications*, vol. 330, no. 1, pp. 207–220, 2007.

[10] M. Hameed and S. Nadeem, "Unsteady MHD flow of a non-Newtonian fluid on a porous plate," *Journal of Mathematical Analysis and Applications*, vol. 325, no. 1, pp. 724–733, 2007.

[11] O. D. Makinde, O. A. Beg, and H. S. Takhar, "Magnetohydrodynamic viscous flow in a rotating porous medium cylindrical annalus with on applied radial magnetic field," *International Journal of Applied Mathematics and Mechanics*, vol. 5, pp. 68–81, 2009.

[12] P. Sibanda and O. D. Makinde, "On steady MHD flow and heat transfer past a rotating disk in a porous medium with ohmic heating and viscous dissipation," *International Journal of Numerical Methods for Heat and Fluid Flow*, vol. 20, no. 3, pp. 269–285, 2010.

[13] I. Pop and T. Watanabe, "Hall effects on magnetohydrodynamic free convection about a semi-infinite vertical flat plate," *International Journal of Engineering Science*, vol. 32, no. 12, pp. 1903–1911, 1994.

[14] L. K. Saha, S. Siddiqa, and M. A. Hossain, "Effect of Hall current on MHD natural convection flow from vertical permeable flat plate with uniform surface heat flux," *Applied Mathematics and Mechanics (English Edition)*, vol. 32, no. 9, pp. 1127–1146, 2011.

[15] D. Pal, B. Talukdar, I. S. Shivakumara, and K. Vajravelu, "Effects of Hall current and chemical reaction on oscillatory mixed convection-radiation of a micropolar fluid in a rotating system," *Chemical Engineering Communications*, vol. 199, pp. 943–965, 2012.

[16] A. C. Cogley, W. C. Vincenti, and S. E. Gilles, "Differential approximation for radiation transfer in a nongray gas near equilibrium," *American Institute of Aeronautics and Astronautics Journal*, vol. 6, pp. 551–555, 1968.

[17] M. A. Mansour, "Radiative and free-convection effects on the oscillatory flow past a vertical plate," *Astrophysics and Space Science*, vol. 166, no. 2, pp. 269–275, 1990.

[18] M. A. Hossain and H. S. Takhar, "Radiation effect on mixed convection along a vertical plate with uniform surface temperature," *Heat and Mass Transfer*, vol. 31, no. 4, pp. 243–248, 1996.

[19] M. A. Hossain, M. A. Alim, and D. A. S. Rees, "The effect of radiation on free convection from a porous vertical plate," *International Journal of Heat and Mass Transfer*, vol. 42, no. 1, pp. 181–191, 1999.

[20] M. A. Seddeek, "Effects of radiation and variable viscosity on a MHD free convection flow past a semi-infinite flat plate with an aligned magnetic field in the case of unsteady flow," *International Journal of Heat and Mass Transfer*, vol. 45, no. 4, pp. 931–935, 2002.

[21] R. Muthucumaraswamy and G. K. Senthil, "Studied the effect of heat and mass transfer on moving vertical plate in the presence of thermal radiation," *Journal of Theoretical And Applied Mechanics*, vol. 31, no. 1, pp. 35–46, 2004.

[22] D. Pal, "Heat and mass transfer in stagnation-point flow towards a stretching surface in the presence of buoyancy force and thermal radiation," *Meccanica*, vol. 44, no. 2, pp. 145–158, 2009.

[23] O. Aydin and A. Kaya, "Radiation effect on MHD mixed convection flow about a permeable vertical plate," *Heat and Mass Transfer*, vol. 45, no. 2, pp. 239–246, 2008.

[24] R. A. Mohamed, "Double-diffusive convection-radiation interaction on unsteady MHD flow over a vertical moving porous plate with heat generation and Soret effects," *Applied Mathematical Sciences*, vol. 3, no. 13–16, pp. 629–651, 2009.

[25] D. S. Chauhan and P. Rastogi, "Radiation effects on natural convection MHD flow in a rotating vertical porous channel partially filled with a porous medium," *Applied Mathematical Sciences*, vol. 4, no. 13–16, pp. 643–655, 2010.

[26] S. Y. Ibrahim and O. D. Makinde, "Radiation effect on chemically reacting magnetohydrodynamics (MHD) boundary layer flow of heat and mass transfer through a porous vertical flat plate," *International Journal of Physical Sciences*, vol. 6, no. 6, pp. 1508–1516, 2011.

[27] D. Pal and H. Mondal, "The influence of thermal radiation on hydromagnetic darcy-forchheimer mixed convection flow past a stretching sheet embedded in a porous medium," *Meccanica*, vol. 46, no. 4, pp. 739–753, 2011.

[28] G. Palani and K. Y. Kim, "Influence of magnetic field and thermal radiation by natural convection past vertical cone subjected to variable surface heat flux," *Applied Mathematics and Mechanics (English Edition)*, vol. 33, pp. 605–620, 2012.

[29] M. A. A. Mahmoud and S. E. Waheed, "Variable fluid properties and ther-28 mal radiation effects on flow and heat transfer in micropolar fluid film past moving permeable infinite flat plate with slip velocity," *Applied Mathematics and Mechanics (English Edition)*, vol. 33, pp. 663–678, 2012.

[30] E. M. Aboeldahab and E. M. E. Elbarbary, "Hall current effect on magnetohydrodynamic free-convection flow past a semi-infinite vertical plate with mass transfer," *International Journal of Engineering Science*, vol. 39, no. 14, pp. 1641–1652, 2001.

[31] E. M. Abo-Eldahab and M. A. El Aziz, "Viscous dissipation and Joule heating effects on MHD-free convection from a vertical plate with power-law variation in surface temperature in the presence of Hall and ion-slip currents," *Applied Mathematical Modelling*, vol. 29, no. 6, pp. 579–595, 2005.

[32] R. Kandasamy, K. Periasamy, and K. K. Sivagnana Prabhu, "Chemical reaction, heat and mass transfer on MHD flow over a vertical stretching surface with heat source and thermal stratification effects," *International Journal of Heat and Mass Transfer*, vol. 48, no. 21-22, pp. 4557–4561, 2005.

[33] R. Muthucumaraswamy and B. Janakiraman, "Mass transfer effects on isothermal vertical oscillating plate in the presence of chemical reaction," *International Journal of Applied Mathematics and Mechanics*, vol. 4, no. 1, pp. 66–74, 2008.

[34] P. R. Sharma and K. D. Singh, "Unsteady MHD free convective flow and heat transfer along a vertical porous plate with variable suction and internal heat generation," *International Journal of Applied Mathematics and Mechanics*, vol. 4, no. 5, pp. 1–8, 2009.

[35] M. Sudheer Babu and P. V. Satya Narayan, "Effects of the chemical reaction and radiation absorption on free convection flow through porous medium with variable suction in the presence of uniform magnetic field," *Journal of Heat and Mass Transfer*, vol. 3, pp. 219–234, 2009.

[36] O. D. Makinde and T. Chinyoka, "Numerical study of unsteady hydromagnetic Generalized Couette flow of a reactive third-grade fluid with asymmetric convective cooling," *Computers and Mathematics with Applications*, vol. 61, no. 4, pp. 1167–1179, 2011.

[37] D. Pal and B. Talukdar, "Combined effects of Joule heating and chemical reaction on unsteady magnetohydrodynamic mixed convection of a viscous dissipating fluid over a vertical plate in porous media with thermal radiation," *Mathematical and Computer Modelling*, vol. 54, no. 11-12, pp. 3016–3036, 2011.

[38] K. D. Singh and R. Kumar, "Combined effects of hall current and rotation on free convection MHD flow in a porous channel," *Indian Journal of Pure and Applied Physics*, vol. 47, no. 9, pp. 617–623, 2009.

[39] R. C. Meyer, "On reducing aerodynamic heat transfer rates by magnetohydrodynamic techniques," *Journal of the Aerospace Sciences*, vol. 25, p. 561, 1958.

The Effect of Heat Transfer on MHD Marangoni Boundary Layer Flow Past a Flat Plate in Nanofluid

D. R. V. S. R. K. Sastry,[1] A. S. N. Murti,[2] and T. Poorna Kantha[2]

[1] *Department of Mathematics, Aditya Engineering College, Surampalem, Ardhra Pradesh 533437, India*
[2] *Deptartment of Engineering Mathematics, GITAM University, Visakhapatnam, Ardhra Pradesh 530023, India*

Correspondence should be addressed to D. R. V. S. R. K. Sastry; sastry_dev@yahoo.co.in

Academic Editor: Yurong Liu

The problem of heat transfer on the Marangoni convection boundary layer flow in an electrically conducting nanofluid is studied. Similarity transformations are used to transform the set of governing partial differential equations of the flow into a set of nonlinear ordinary differential equations. Numerical solutions of the similarity equations are then solved through the MATLAB "bvp4c" function. Different nanoparticles like Cu, Al_2O_3, and TiO_2 are taken into consideration with water as base fluid. The velocity and temperature profiles are shown in graphs. Also the effects of the Prandtl number and solid volume fraction on heat transfer are discussed.

1. Introduction

The convection induced by the variations of the surface tension gradients is known as the Marangoni convection. This convection has received great consideration in view of its application in the fields of welding and crystal growth. Also this convection is necessary to stabilize the soap films and drying silicon wafers. During the study of the existence of the steady dissipative layers which occur along the liquid-liquid or liquid-gas interfaces, Napolitano [1] first called the boundary layer as the Marangoni boundary layer. Many researchers such as Okano et al. [2], Christopher and Wang [3], Pop et al. [4] and Magyari and Chamkha [5] have investigated the Marangoni convection in various geometries. Al-Mudhaf and Chamkha [6] obtained the similarity solution for the MHD thermosolutal Marangoni convection over a flat surface in the presence of heat generation or absorption with fluid suction and injection. Chen [7] investigated the flow and the heat transfer characteristics on the forced convection in a power law liquid film under an applied Marangoni convection over a stretching sheet. In recent years, the study on convective transport of nanofluids has become one of the popular topics of interest. Nanotechnology takes important part for the development of high performance, compact, and cost-effective liquid cooling systems. Moreover, nanofluids have effective applications in many industries such as electronics, transportation, biomedical, and many more. Nanotechnology has been an ongoing topic of discussion in public health as some of the researchers claimed that nanoparticles could present possible dangers in health and environment. Jang and Choi [8] have introduced nanosized particle in a base fluid, which is also termed nanofluid, for the first time. Arifin et al. [9] have examined the influence of nanoparticles on the Marangoni boundary layer flow using a model proposed by Tiwari and Das [10]. An extended work was done by Buongiorno [11], Daungthongsuk and Wongwises [12], Trisaksri and Wongwises [13], Wang and Mujumdar [14], and Kakaç and Pramuanjaroenkij [15]. Recently Hamid et al. [16] studied the radiation effects on the Marangoni boundary layer flow past a flat plate in nanofluid. In the present paper, we study a numerical solution of MHD heat transfer problem in nanofluid with nanoparticles Cu, Al_2O_3, and TiO_2. We also observed the effects of the Prandtl number and solid volume fraction on the Nusselt number. The results are shown graphically.

2. Mathematical Formulation

Consider a steady two-dimensional Marangoni boundary layer flow past a permeable flat plate in a water-based nanofluid containing different types of nano particles like Cu (Copper), Al_2O_3 (Aluminium Oxide), and TiO_2 (Titanium dioxide). Assume that the fluid is incompressible and the flow is laminar. Also it is assumed that the base fluid and the particles are in thermal equilibrium and no slip occurs between them. The thermophysical properties of nanoparticles are given in the Table 1. Further, we consider a Cartesian coordinate system (x, y), where x and y are the coordinates measured along the plate and normal to it, respectively, and the flow takes place at $y \geq 0$. Assume that the temperature of the plate is $T_w(x)$ and that of the ambient fluid is T_∞.

We further assume that the surface tension σ is to vary linearly with temperature as

$$\sigma = \sigma_0 \left[1 - \gamma (T - T_\infty) \right], \tag{1}$$

where σ_0 is the surface tension at the interface and γ is the rate of change of surface tension with temperature (a positive fluid property). It is also assumed that a uniform magnetic field, H_0 is imposed in the direction normal to the surface (Figure 1). Then, the steady state boundary layer equations for a nanofluid in the Cartesian coordinates are given by

$$\frac{\partial u}{\partial x} + \frac{\partial v}{\partial y} = 0, \tag{2}$$

$$u\frac{\partial u}{\partial x} + v\frac{\partial u}{\partial y} = \frac{\mu_{\text{nf}}}{\rho_{\text{nf}}}\frac{\partial^2 u}{\partial y^2} - \frac{\sigma}{\rho_{\text{nf}}} H_0^2 u, \tag{3}$$

$$u\frac{\partial T}{\partial x} + v\frac{\partial T}{\partial y} = \alpha_{\text{nf}}\frac{\partial^2 T}{\partial y^2}, \tag{4}$$

together with the boundary conditions

$$\begin{gathered} v = 0, \quad T = T_\infty + ax^2, \\ \mu_{\text{nf}}\frac{\partial u}{\partial y} = \frac{\partial \sigma}{\partial T}\frac{\partial T}{\partial x} \quad \text{at } y = 0, \\ u = 0, \quad T = T_\infty \quad \text{as } y \longrightarrow \infty. \end{gathered} \tag{5}$$

Here u and v are the components of velocity along the x- and y-axes, respectively. T is the temperature, α_{nf} is the thermal diffusivity, ρ_{nf} is the effective density, k_{nf} is the effective thermal conductivity, and μ_{nf} is the effective viscosity of the nanofluid. Moreover, a is the coefficient of temperature gradient. Consider the following:

$$\begin{gathered} \alpha_{\text{nf}} = \frac{k_{\text{nf}}}{(\rho C_p)_{\text{nf}}}, \\ \rho_{\text{nf}} = (1 - \phi)\rho_f + \phi\rho_s, \\ \mu_{\text{nf}} = \frac{\mu_f}{(1 - \phi)^{2.5}}, \\ (\rho C_p)_{\text{nf}} = (1 - \phi)(\rho C_p)_f + \phi(\rho C_p)_s, \\ \frac{k_{\text{nf}}}{k_f} = \frac{(k_s + 2k_f) - 2\phi(k_f - k_s)}{(k_s + 2k_f) + \phi(k_f - k_s)}, \\ a = \frac{\Delta T}{L^2}, \end{gathered} \tag{6}$$

Table 1: Thermophysical properties of pure water and nanoparticles (Oztop and Abu-Nada [17]).

Physical property	Pure water	Cu	Al_2O_3	TiO_2
ρ (kg/m^3)	997.1	8933	3970	4250
C_p (J/kg K)	4179	385	765	686.2
k (W/m K)	0.613	401	40	8.9538

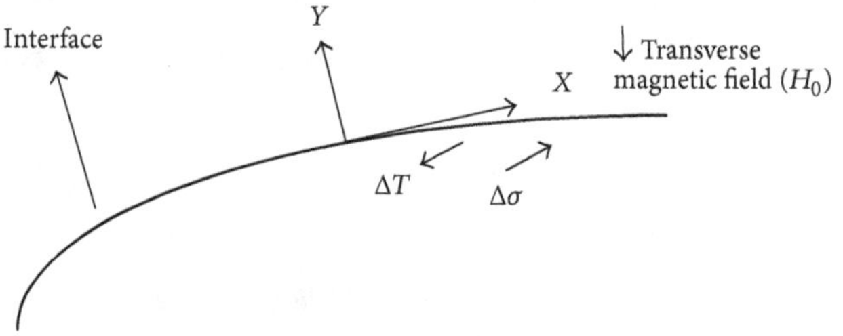

Figure 1: Schematic diagram of the problem.

where ϕ is the solid volume fraction of the nanofluid, ρ_f is the reference density of the fluid fraction, ρ_s is the reference density of the solid fraction, μ_f is the viscosity of the fluid fraction, k_f is the thermal conductivity of the fluid, k_s is the thermal conductivity of the solid, and $(\rho C_p)_{\text{nf}}$ is the heat capacity of the nanofluid. L is the length of the surface, and ΔT is the constant characteristic temperature.

A similarity solution of (2)–(5) is obtained by introducing an independent variable η and a dependent variable f in terms of the stream function ψ as

$$\psi = C_1 x f(\eta), \qquad \eta = C_2 y, \tag{7}$$

with $u = \partial\psi/\partial y$ and $v = -\partial\psi/\partial x$.

The constants C_1 and C_2 are given by

$$C_1 = \left(\frac{\sigma_0 \gamma a \mu_f}{\rho_f^2} \right)^{1/3}, \qquad C_2 = \left(\frac{\sigma_0 \gamma a \rho_f}{\mu_f^2} \right)^{1/3}. \tag{8}$$

Further, the dimensionless temperature θ is given by

$$\theta(\eta) = \frac{T - T_\infty}{ax^2}. \tag{9}$$

Substituting (6), (7), (8), and (9) into (3) and (4), we obtain a set of nonlinear ordinary differential equations:

$$f''' = (1-\phi)^{2.5}\left[(1-\phi)+\phi\frac{\rho_s}{\rho_f}\right]\left(f'^2 - ff''\right) + M(P_r)^{1/3}(k_f)^{1/3}(Cp_f)^{-1/3}(\rho_f)^{-2/3}f', \quad (10)$$

$$\theta'' = \frac{\left[(1-\phi)+\phi\left(\rho_s Cp_s/\rho_f Cp_f\right)\right]}{k_{nf}/k_f}\left(2f'\theta - f\theta'\right)P_r,$$

and the boundary conditions become

$$f(0)=0, \quad \theta(0)=1, \quad \frac{1}{(1-\phi)^{2.5}}f''(0)=-2,$$
$$f'(\infty)=0, \quad \theta(\infty)=0, \quad (11)$$

where the magnetic field parameter $M = \sigma^{1/3}H_0^2/(\gamma a)^{2/3}$.

Also one can define the surface velocity and the local Nusselt number, respectively, as

$$u_w(x) = \sqrt[3]{\frac{(\sigma_0\gamma a)^2}{\rho_f\mu_f}}xf'(0), \quad (12)$$

$$\mathrm{Nu}_x = \frac{xq_w(x)}{k_f\left[T(x,0)-T(x,\infty)\right]}, \quad (13)$$

where $q_w(x)$ is the heat flux from the surface of the plate and is given by

$$q_w(x) = -k_{nf}\left(\frac{\partial T}{\partial y}\right)_{y=0}. \quad (14)$$

Using the above nondimension quantities, one can obtain the local Nusselt number as

$$\mathrm{Nu}_x = -\frac{k_{nf}}{k_f}C_2\theta'(0). \quad (15)$$

Based on the average temperature difference between the temperature of the surface and the ambient fluid temperature we define

$$\mathrm{Nu}_L = -\frac{k_{nf}}{k_f}\left(\frac{\mathrm{Ma}_L}{\mathrm{Pr}}\right)^{1/3}\theta'(0), \quad (16)$$

where Ma_L is the Marangoni based on L and is defined as

$$\mathrm{Ma}_L = \frac{(\partial\sigma/\partial T)(\Delta T)L}{\alpha_f\mu_f}. \quad (17)$$

3. Results and Discussion

Numerical solutions were obtained for the effect of the Prandtl number and solid volume fraction on the Marangoni heat transfer in a nanofluid. In this paper, we considered three different nanoparticles whose thermophysical properties were given in Table 1. The nonlinear ordinary differential equations (10) subject to the boundary conditions (11) were solved numerically using the MATLAB "bvp4c" routine. We considered the range of nanoparticles volume fraction ϕ as $0 \le \emptyset \le 0.3$ and the Prandtl number Pr as $2 \le \mathrm{Pr} \le 8$ (for the base fluid (water) Pr = 6.2). The influences of the magnetic field parameter (M), the nanoparticles volume fraction (ϕ) on velocity and, temperature and also the influcnce of the Prandtl number (Pr) and solid volume fraction (ϕ) on the Nusselt number are presented in graphs.

Figures 2, 3, and 4 display the velocity profiles, and Figures 5, 6, and 7 display the temperature profiles of Cu-water, Al_2O_3-water, and TiO_2-water, respectively, for different values of magnetic field parameter M. It is observed from the figures that the velocity in the boundary layer decreases and temperature increases as the Magnetic field parameter increases; this is due to the resistive force, called the Lorentz force, which is produced by the induced magnetic field within the boundary layer.

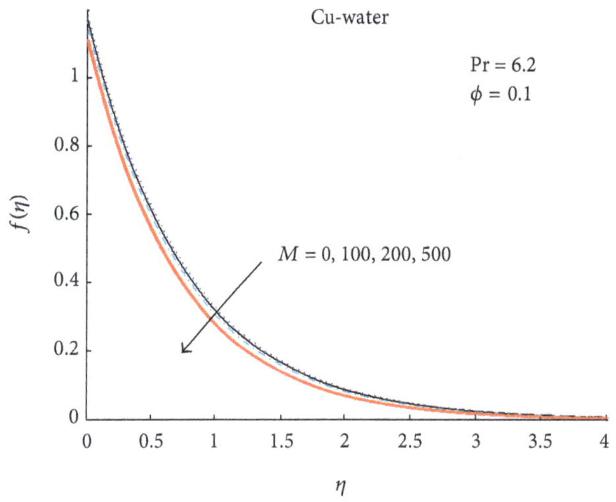

FIGURE 2: Velocity profile for Cu nanoparticles for various M.

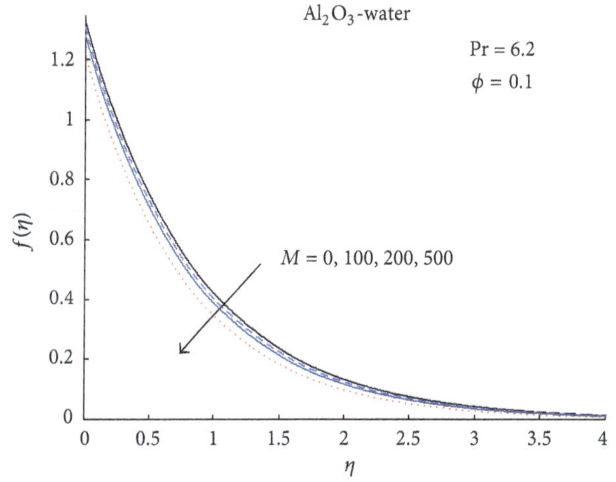

FIGURE 3: Velocity profile for Al_2O_3 nanoparticles for various M.

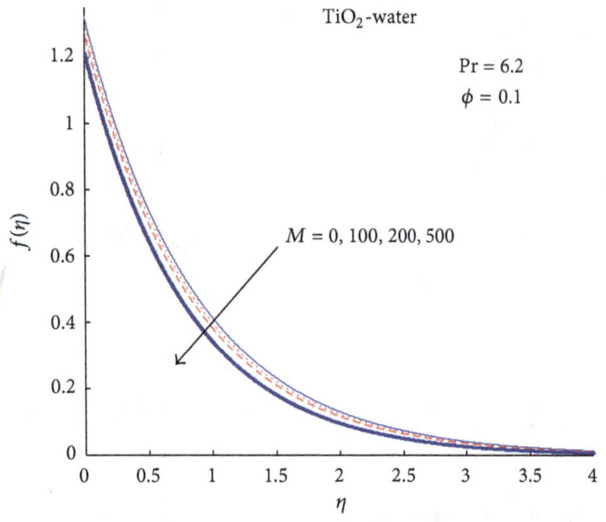

FIGURE 4: Velocity profile for TiO_2 nanoparticles for various M.

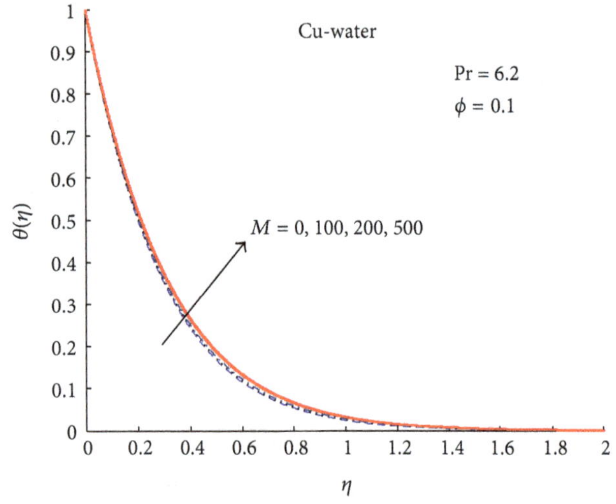

FIGURE 5: Temperature profile for Cu nanoparticles for various M.

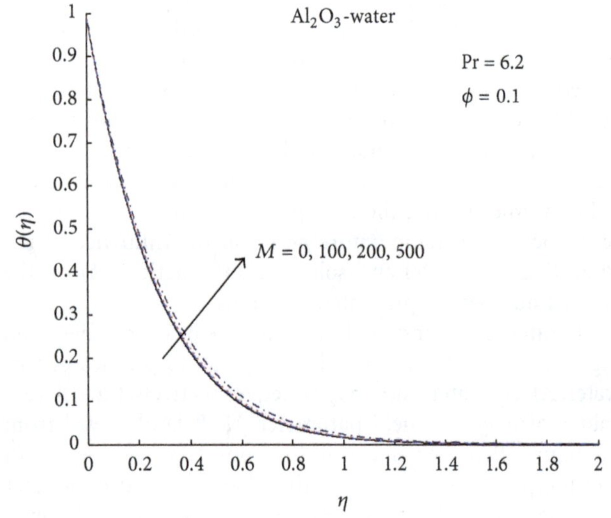

FIGURE 6: Temperature profile for Al_2O_3 nanoparticles for various M.

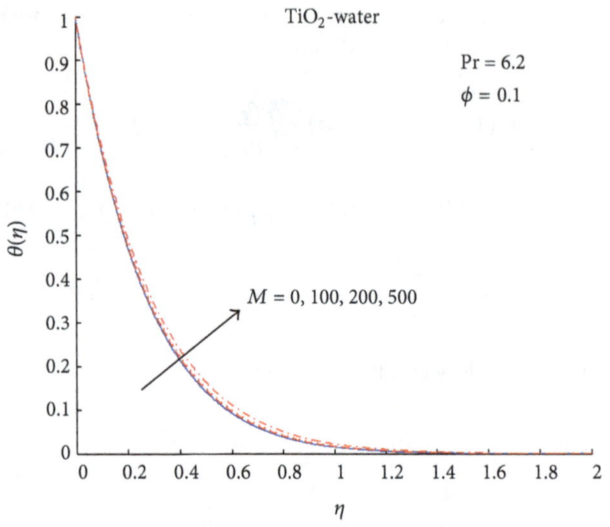

FIGURE 7: Temperature profile for TiO_2 nanoparticles for various M.

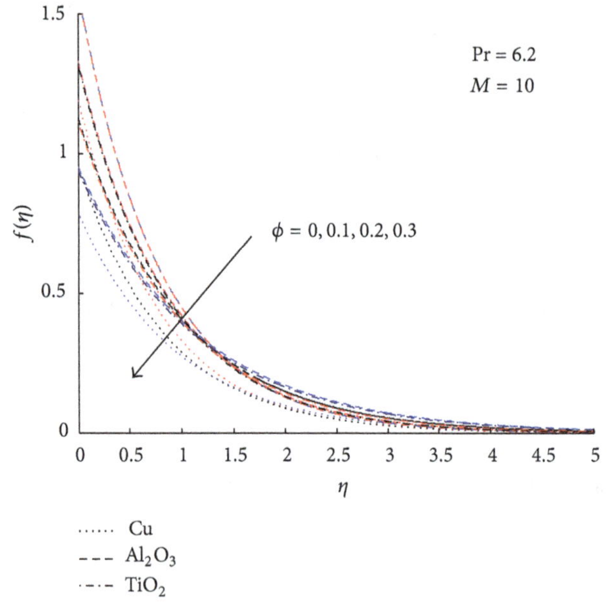

FIGURE 8: Velocity profile for different ϕ.

Figure 8 depicts the influence of volume fraction on the velocity profile of the nanofluid particles. It is observed near the wall that velocity decreases with an increase in the volume fraction ϕ. Also it is observed that the velocity of TiO_2 nanoparticles is higher than that of Cu nanoparticles. From Figure 9, it is clear that an increase in the value of volume fraction enhances the temperature profile, and Cu nanoparticles exhibit more temperature than that of the other nanoparticles. It is also known from Figure 10 that temperature decreases with an increase in the Prandtl number. This is because of a decrease in thermal diffusivity with an increase in the Prandtl number (Pr).

Figures 11 and 12 depict the influence of the Prandtl number and volume fraction on heat transfer, respectively. It is observed that the Nusselt number increases with an

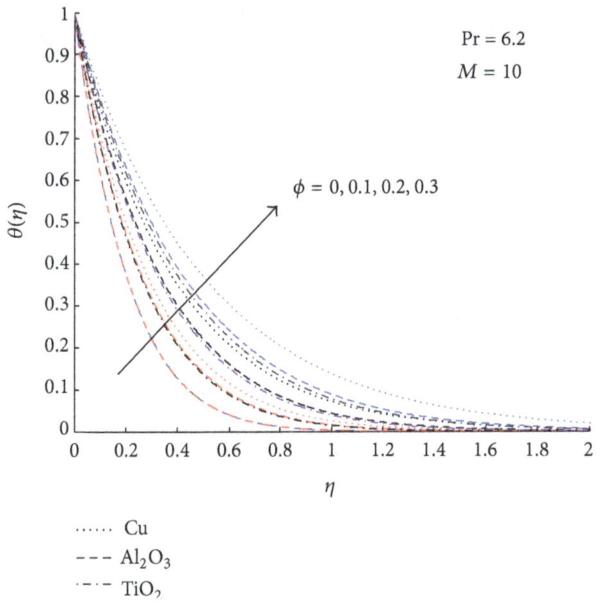

FIGURE 9: Temperature profile for different ϕ.

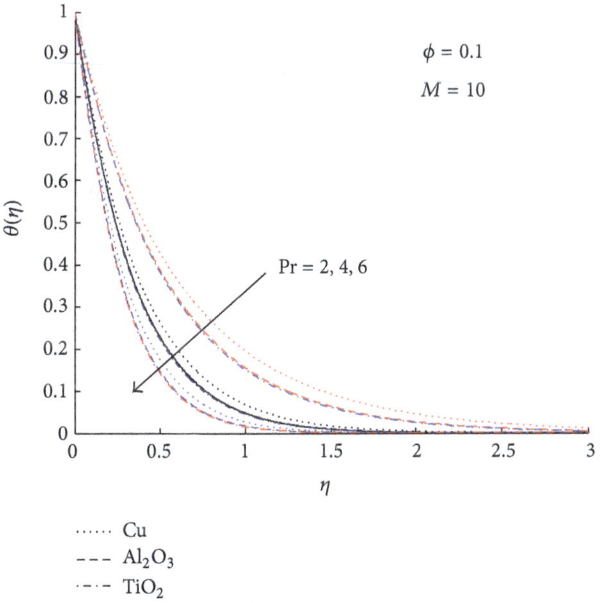

FIGURE 10: Temperature profile for different Pr.

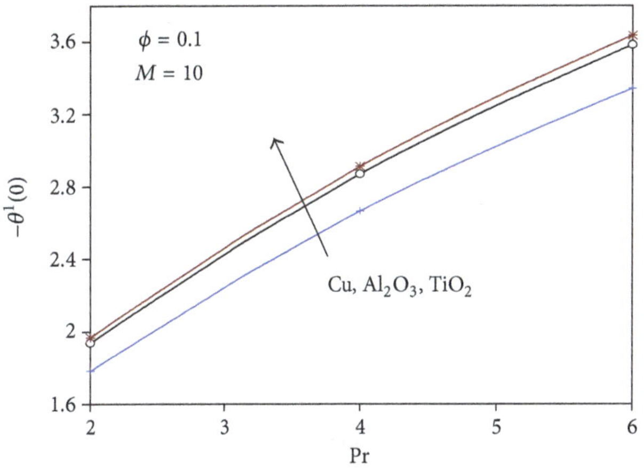

FIGURE 11: Heat transfer effect against the Prandtl number.

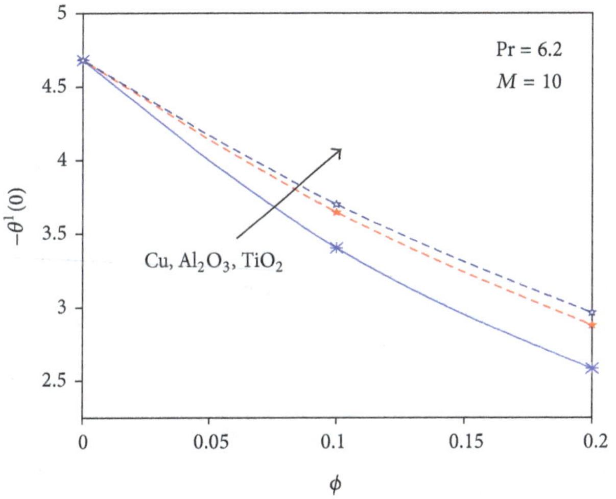

FIGURE 12: Heat transfer effect against the volume fraction.

TABLE 2: The numeric values of skin friction and the Nusselt number for various values of ϕ.

ϕ	$f''(0)$	$-\theta'(0)$
0	2	4.68429586
0.1	1.53686694	3.40361315
0.2	1.1448668	2.58644151

increase in Prandtl number and decreases with an increase in the volume fraction.

From Table 2, it is observed that skin friction decreases with the increase in the volume fraction.

4. Conclusion

In the present paper, we studied the effect of heat transfer on the Marangoni boundary layer flow past a flat plate in nanofluids in presence of transverse magnetic field. With the similarity transformation, the governing equations of motion together with boundary conditions were transformed to a set of nonlinear ordinary differential equations. The numerical solutions are then obtained for these equations by the help of MATLAB "bvp4c" programming tool. Different types of nanoparticles like Cu, Al_2O_3, and TiO_2 were taken into consideration with H_2O as base fluid. The effects of magnetic field parameter M, solid volume fraction of the nanofluid ϕ on the velocity and temperature fields for different nano particles, and the Prandtl number Pr on temperature field were plotted and analyzed. Also the effects of the Prandtl number Pr and solid volume fraction ϕ on local the Nusselt number for the different nanoparticles were discussed for a fixed value of magnetic field parameter M. It is found that the inclusion of the magnetic field parameter on the flow increased the temperature and decreased the velocity fields in all types of nanofluids. A similar profile was observed on the inclusion of solid volume fraction of the nanoparticles. It was noted that presence of the Prandtl number reduced the temperature field. Also it was observed that for a fixed Prandtl number and other parameters, the rate of heat transfer is more in TiO_2-H_2O.

References

[1] L. G. Napolitano, "Microgravity fluid dynamics," in *Proceedings of the 2nd Levitch Conference*, Washington, DC, USA, 1978.

[2] Y. Okano, M. Itoh, and A. Hirata, "Natural and marangoni convections in a two-dimensional rectangular open boat," *Journal of Chemical Engineering of Japan*, vol. 22, no. 3, pp. 275–281, 1989.

[3] D. M. Christopher and B. Wang, "Prandtl number effects for Marangoni convection over a flat surface," *International Journal of Th rmal Sciences*, vol. 40, no. 6, pp. 564–570, 2001.

[4] I. Pop, A. Postelnicu, and T. Groşan, "Thermosolutal Marangoni forced convection boundary layers," *Meccanica*, vol. 36, no. 5, pp. 555–571, 2001.

[5] E. Magyari and A. J. Chamkha, "Exact analytical solutions for thermosolutal Marangoni convection in the presence of heat and mass generation or consumption," *Heat and Mass Transfer*, vol. 43, no. 9, pp. 965–974, 2007.

[6] A. Al-Mudhaf and A. J. Chamkha, "Similarity solutions for MHD thermosolutal Marangoni convection over a flat surface in the presence of heat generation or absorption effects," *Heat and Mass Transfer*, vol. 42, no. 2, pp. 112–121, 2005.

[7] C. H. Chen, "Marangoni effects on forced convection of power-law liquids in a thin film over a stretching surface," *Physics Letters A*, vol. 370, no. 1, pp. 51–57, 2007.

[8] S. P. Jang and S. U. S. Choi, "Role of Brownian motion in the enhanced thermal conductivity of nanofluids," *Applied Physics Letters*, vol. 84, no. 21, pp. 4316–4318, 2004.

[9] N. M. Arifin, R. Nazar, and I. Pop, "Marangoni driven boundary layer flow past a flat plate in nanofluid with suction/injection," in *Proceedings of the International Conference on Mathematical Science (ICMS '10)*, pp. 94–99, Bolu, Turkey, November 2010.

[10] R. K. Tiwari and M. K. Das, "Heat transfer augmentation in a two-sided lid-driven differentially heated square cavity utilizing nanofluids," *International Journal of Heat and Mass Transfer*, vol. 50, no. 9-10, pp. 2002–2018, 2007.

[11] J. Buongiorno, "Convective transport in nanofluids," *Journal of Heat Transfer*, vol. 128, no. 3, pp. 240–250, 2006.

[12] W. Daungthongsuk and S. Wongwises, "A critical review of convective heat transfer of nanofluids," *Renewable and Sustainable Energy Reviews*, vol. 11, no. 5, pp. 797–817, 2007.

[13] V. Trisaksri and S. Wongwises, "Critical review of heat transfer characteristics of nanofluids," *Renewable and Sustainable Energy Reviews*, vol. 11, no. 3, pp. 512–523, 2007.

[14] X. Q. Wang and A. S. Mujumdar, "Heat transfer characteristics of nanofluids: a review," *International Journal of Thermal Sciences*, vol. 46, no. 1, pp. 1–19, 2007.

[15] S. Kakaç and A. Pramuanjaroenkij, "Review of convective heat transfer enhancement with nanofluids," *International Journal of Heat and Mass Transfer*, vol. 52, no. 13-14, pp. 3187–3196, 2009.

[16] R. A. Hamid, N. M. Arifin, R. M. Nazar, and I. Pop, "Radiation effects on Marangoni boundary layer flow past a flat plate in nanofluid," in *Proceedings of the International MultiConference of Engineers and Computer Scientists 2011(IMECS '11)* vol. 3, pp. 1260–1263, Hong Kong, March 2011.

[17] H. F. Oztop and E. Abu-Nada, "Numerical study of natural convection in partially heated rectangular enclosures filled with nanofluids," *International Journal of Heat and Fluid Flow*, vol. 29, no. 5, pp. 1326–1336, 2008.

4

Falkner-Skan Flow of a Maxwell Fluid with Heat Transfer and Magnetic Field

M. Qasim[1] and S. Noreen[2]

[1] *Department of Mathematics, COMSATS Institute of Information Technology, Park Road, Chak Shahzad, Islamabad 44000, Pakistan*
[2] *Department of Mathematics, COMSATS Institute of Information Technology, Attock 43600, Pakistan*

Correspondence should be addressed to M. Qasim; mq_qau@yahoo.com

Academic Editor: George S. Dulikravich

This investigation deals with the Falkner-Skan flow of a Maxwell fluid in the presence of nonuniform applied magnetic field with heat transfer. Governing problems of flow and heat transfer are solved analytically by employing the homotopy analysis method (HAM). Effects of the involved parameters, namely, the Deborah number, Hartman number, and the Prandtl number, are examined carefully. A comparative study is made with the known numerical solution in a limiting sense and an excellent agreement is noted.

1. Introduction

The Falkner-Skan problem under various aspects has attracted the attention of several researchers [1]. This problem under various aspects has been discussed extensively for viscous fluid. The interested readers may consult the studies in [2–11] for detailed information in viscous fluids. There are several materials which do not obey the Newton's law of viscosity, for example, biological products like blood and vaccines, foodstuffs like honey, ketchup, butter, and mayonnaise, certain paints, cosmetic products, pharmaceutical chemicals and so forth. These fluids are characterized as the non-Newtonian fluids. Investigation of such fluids is very useful in industrial, engineering, and biological applications. However, such fluids cannot be studied by employing a single constitutive relationship. This is due to diverse properties of non-Newtonian fluids in nature. These non-Newtonian fluid models are discussed in view of three main categories, namely, the differential, the rate, and the integral types. The simplest subclass of rate type fluids is called Maxwell. The Maxwell fluid allows for the relaxation effects which cannot be predicted in differential type fluids, namely, second, third, and fourth grades. Recently, there has been an increasing interest in the theory of rate type fluids and, in particular, a Maxwell fluid model has been accorded much attention. The Falkner-Skan wedge flow of a non-Newtonian fluid was firstly investigated by Rajagopal et al. [12]. Massoudi and Ramezan [13] discussed the effect of injection or suction on the Falkner-Skan flows of second grade fluids. The Falkner-Skan wedge flow of power-law fluids embedded in a porous medium is investigated by Kim [14]. Olagunju [15] studied this flow problem for viscoelastic fluid. In [10–15], the attention has been given to the differential type fluids. To the best of our knowledge, no one investigated the Falkner-Skan flow problem for rate type fluids.

In [10], Yao has examined the Falkner-Skan wedge flow. He established series solution for the velocity and temperature by using homotopy analysis method [16–25]. The purpose of the present contribution is to extend the flow analysis of study [10] in two directions. The first generalization is concerned with the consideration of electrically conducting fluid. Such analysis has inserted in power generators the cooling of reactors, MHD accelerators, the design of heat exchanges and electrostatic filters. Secondly, we consider the Maxwell fluid instead of viscous fluid. The rest of the paper is arranged as follows. The description of the problem is presented in Section 2. Section 3 develops the homotopy solutions for velocity and temperature. Convergence of the derived solution is examined in Section 4. Further, the variations of embedded parameters have been discussed in this section.

2. Problem Development

We study the steady two-dimensional flow of a Maxwell fluid in the presence of a magnetic field. The magnetic Reynolds number is small so that induced magnetic field is neglected. The stream velocity $U(x)$ varies according to x^n. The constant temperature of surface and free stream is T_w and T_∞, respectively. The boundary layer equations for the considered problem are

$$\frac{\partial u}{\partial x} + \frac{\partial v}{\partial y} = 0, \tag{1}$$

$$u\frac{\partial u}{\partial x} + v\frac{\partial u}{\partial y} + \lambda\left(u^2\frac{\partial^2 u}{\partial x^2} + v^2\frac{\partial^2 u}{\partial y^2} + 2uv\frac{\partial^2 u}{\partial x \partial y}\right) = \nu\frac{\partial^2 u}{\partial y^2} + U\frac{\partial U}{\partial x} + \lambda U^2\frac{\partial^2 U}{\partial x^2} - \frac{\sigma B^2}{\rho}\left(u - U + \lambda v\frac{\partial u}{\partial y}\right), \tag{2}$$

$$u\frac{\partial T}{\partial x} + v\frac{\partial T}{\partial y} = \frac{k}{\rho c_p}\frac{\partial^2 T}{\partial y^2}, \tag{3}$$

where u and v are the velocity components in the x- and y-directions, respectively, ν is kinematic viscosity, σ is the electrical conductivity, k is the thermal conductivity, ρ is the fluid density, T is the fluid temperature, λ is relaxation time, B is the magnetic field, and c_p is specific heat.

The relevant boundary conditions are prescribed as follows:

$$\begin{gathered} u = 0, \quad v = 0, \quad T = T_w \quad \text{at } y = 0, \\ u \longrightarrow U(x), \quad T \longrightarrow T_\infty \quad \text{as } y \longrightarrow \infty, \end{gathered} \tag{4}$$

with [1]

$$U(x) = ax^n, \tag{5}$$

and [7, 8]

$$B(x) = B_0 x^{(n-1)/2}. \tag{6}$$

Putting

$$\begin{gathered} \psi = \sqrt{\frac{2}{n+1}}\sqrt{\nu x U} f(\eta), \\ \eta = \sqrt{\frac{n+1}{2}}\sqrt{\frac{U}{x\nu}}y, \quad u = Uf'(\eta), \\ v = -\sqrt{\frac{n+1}{2}}\sqrt{\frac{U}{x\nu}}\left[f(\eta) + \frac{n-1}{n+1}\eta f'(\eta)\right], \\ \theta(\eta) = \frac{T_w - T}{T_w - T_\infty}, \end{gathered} \tag{7}$$

$$u = \frac{\partial \psi}{\partial y}, \quad v = -\frac{\partial \psi}{\partial x}, \tag{8}$$

(1) is satisfied identically and (2)–(4) give

$$\begin{aligned} & f''' + ff'' + \frac{2n}{n+1}\left(1 - f'^2\right) \\ & - M^2\left(f' - 1\right) + \left(M^2\beta + 1\right)ff'' \\ & + \beta\left(2n\frac{n-1}{n+1}\left(1 - f'^3\right) + (3n-1)ff'f'' \right. \\ & \left. - \frac{n+1}{2}f^2 f''' + \frac{n-1}{2}\eta f'^2 f''\right) = 0, \end{aligned} \tag{9}$$

$$\theta'' + \Pr f\theta' = 0, \tag{10}$$

$$\begin{gathered} f(0) = 0, \quad f'(0) = 0, \quad \theta(0) = 0, \\ f'(\infty) = 1, \quad \theta(\infty) = 1. \end{gathered} \tag{11}$$

Here ψ is the stream function, prime denotes the differentiation with respect to η, γ is the local Deborah number, n is the constant parameter, M is Hartmann number, and Pr is the Prandtl number. The values of β, M, and Pr are

$$\beta = \frac{\lambda U}{x}, \quad M = \frac{2\sigma B_0^2}{\rho a(n+1)}, \quad \Pr = \frac{\mu c_p}{k}. \tag{12}$$

The local Nusselt number Nu_x and heat transfer from the plate q_w are given by

$$Nu_x = \frac{x q_w}{k(T_w - T_\infty)}, \quad q_w = -k\left(\frac{\partial T}{\partial y}\right)_{y=0}. \tag{13}$$

Invoking (7) one obtains

$$\mathrm{Re}_x^{-1/2}\left(\frac{n+1}{2}\right)Nu_x = -\theta'(0) \tag{14}$$

in which the local Reynolds number $\mathrm{Re}_x = Ux/\nu$. For $M = 0 = \beta$, from (9), we have

$$f''' + ff'' + \gamma\left(1 - f'^2\right) = 0, \tag{15}$$

where $\gamma = 2n/(n+1)$.

3. Homotopy Analysis Solutions

For an interest in homotopy solutions, we express $f(\eta)$ and $\theta(\eta)$ by a set of base functions

$$\begin{gathered} \{\eta^m \exp(-2n\eta),\ m, n \geq 0\}, \\ \{\eta^m \exp(-n\eta),\ m, n \geq 0\}, \end{gathered} \tag{16}$$

as follows:

$$\begin{gathered} f(\eta) = \sum_{m=0}^{\infty}\sum_{n=0}^{\infty} a_{m,n}\eta^m \exp(-2n\eta), \\ \theta(\eta) = \sum_{m=0}^{\infty}\sum_{n=0}^{\infty} b_{m,n}\eta^m \exp(-2n\eta), \end{gathered} \tag{17}$$

where $a_{m,n}$ and $b_{m,n}$ are the coefficients. The initial approximations of $f(\eta)$ and $\theta(\eta)$ and the auxiliary linear operators $\mathscr{L}_f$ and $\mathscr{L}_\theta$ are

$$f_0(\eta) = \eta - \frac{1 - \exp(-2\eta)}{2}, \qquad \theta_0(\eta) = 1 - \exp(-\eta), \tag{18}$$

$$\mathscr{L}_f(f) = \frac{d^3 f}{d\eta^3} - 4\frac{df}{d\eta}, \qquad \mathscr{L}_\theta(\theta) = \frac{d^2\theta}{d\eta^2} + \frac{d\theta}{d\eta} \tag{19}$$

$$\mathscr{L}_f[C_1 + C_2\exp(2\eta) + C_3\exp(-2\eta)] = 0, \tag{20}$$

$$\mathscr{L}_\theta[C_4\exp(\eta) + C_5\exp(-\eta)] = 0 \tag{21}$$

in which C_i $(i = 1 - 5)$ are the arbitrary constants. If $p \in [0,1]$ is the embedding parameter and $\hbar_f$, $\hbar_\theta$ and $\hbar_\phi$, are the nonzero auxiliary parameters, then the problems at zeroth order give

$$(1-p)\mathscr{L}_f\left[\hat{f}(\eta;p) - f_0(\eta)\right] = p\hbar_f \mathscr{N}_f\left[\hat{f}(\eta;p)\right], \tag{22}$$

$$(1-p)\mathscr{L}_\theta\left[\hat{\theta}(\eta;p) - \theta_0(\eta)\right] = p\hbar_\theta \mathscr{N}_\theta\left[\hat{\theta}(\eta;p), \hat{f}(\eta;p)\right], \tag{23}$$

$$\hat{f}(0;p) = 0, \qquad \hat{f}'(0;p) = 0, \qquad \hat{f}'(\infty;p) = 1, \tag{24}$$

$$\hat{\theta}(0;p) = 0, \qquad \hat{\theta}(\infty;p) = 1, \tag{25}$$

$$\begin{aligned}
&\mathscr{N}_f\left[\hat{f}(\eta;p)\right] \\
&= \frac{\partial^3 \hat{f}(\eta;p)}{\partial\eta^3} + \hat{f}(\eta;p)\frac{\partial^2 \hat{f}(\eta;p)}{\partial\eta^2} \\
&+ \frac{2n}{n+1}\left[1 - \left(\frac{\partial \hat{f}(\eta,p)}{\partial\eta}\right)^2\right] \\
&- M^2\left(\frac{\partial \hat{f}(\eta,p)}{\partial\eta} - 1\right) \\
&+ \left(M^2\beta + 1\right)\frac{\partial \hat{f}(\eta,p)}{\partial\eta}\frac{\partial^2 \hat{f}(\eta,p)}{\partial\eta^2} \\
&+ \beta\left[2n\frac{n-1}{n+1}\left[1 - \left(\frac{\partial \hat{f}(\eta,p)}{\partial\eta}\right)^3\right]\right. \\
&+ (3n-1)\hat{f}(\eta;p)\frac{\partial \hat{f}(\eta,p)}{\partial\eta}\frac{\partial^2 \hat{f}(\eta;p)}{\partial\eta^2} \\
&- \frac{n+1}{2}\hat{f}^2(\eta;p)\frac{\partial 3\hat{f}(\eta;p)}{\partial\eta 3} \\
&\left. + \frac{n-1}{2}\left(\frac{\partial \hat{f}(\eta,p)}{\partial\eta}\right)^2\frac{\partial^2 \hat{f}(\eta,p)}{\partial\eta^2}\right],
\end{aligned} \tag{26}$$

$$\mathscr{N}_\theta\left[\hat{\theta}(\eta;p), \hat{f}(\eta;p)\right] = \frac{\partial^2\hat{\theta}(\eta,p)}{\partial\eta^2} + \Pr\hat{f}(\eta;p)\frac{\partial\hat{\theta}(\eta;p)}{\partial\eta}. \tag{27}$$

For $p = 0$ and $p = 1$, we have

$$\hat{f}(\eta;0) = f_0(\eta), \qquad \hat{f}(\eta;1) = f(\eta), \tag{28}$$

$$\hat{\theta}(\eta;0) = \theta_0(\eta), \qquad \hat{\theta}(\eta;1) = \theta(\eta). \tag{29}$$

When p increases from 0 to 1, $\hat{f}(\eta;p)$ and $\hat{\theta}(\eta;p)$ vary from $f_0(\eta)$ and $\theta_0(\eta)$ to the exact solutions $f(\eta)$ and $\theta(\eta)$. In view of Taylors theorem and (20) and (21), one arrives at

$$\hat{f}(\eta;p) = f_0(\eta) + \sum_{m=1}^{\infty} f_m(\eta)p^m, \tag{30}$$

$$\hat{\theta}(\eta;p) = \theta_0(\eta) + \sum_{m=1}^{\infty} \theta_m(\eta)p^m, \tag{31}$$

$$\begin{aligned}
f_m(\eta) &= \frac{1}{m!}\frac{\partial^m \hat{f}(\eta;p)}{\partial p^m}\bigg|_{p=0}, \\
\theta_m(\eta) &= \frac{1}{m!}\frac{\partial^m \hat{\theta}(\eta;p)}{\partial p^m}\bigg|_{p=0}.
\end{aligned} \tag{32}$$

The auxiliary parameters are so properly chosen that the series (28) and (29) converge at $p = 1$, and hence

$$f(\eta) = f_0(\eta) + \sum_{m=1}^{\infty} f_m(\eta), \tag{33}$$

$$\theta(\eta) = \theta_0(\eta) + \sum_{m=1}^{\infty} \theta_m(\eta). \tag{34}$$

The problems at mth-order deformation satisfy the following equations and boundary conditions:

$$\mathscr{L}_f\left[f_m(\eta) - \chi_m f_{m-1}(\eta)\right] = \hbar_f \mathscr{R}_m^f(\eta), \tag{35}$$

$$\mathscr{L}_\theta\left[\theta_m(\eta) - \chi_m \theta_{m-1}(\eta)\right] = \hbar_\theta \mathscr{R}_m^\theta(\eta), \tag{36}$$

$$f_m(0) = f_m'(0) = f_m'(\infty) = f_m''(\infty) = 0, \tag{37}$$

$$\theta_m(0) = \theta_m(\infty) = 0, \tag{38}$$

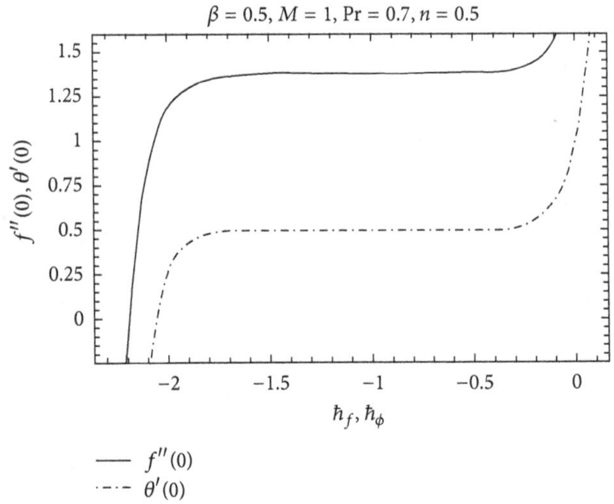

FIGURE 1: $\hbar$ curves for 15th-order approximations.

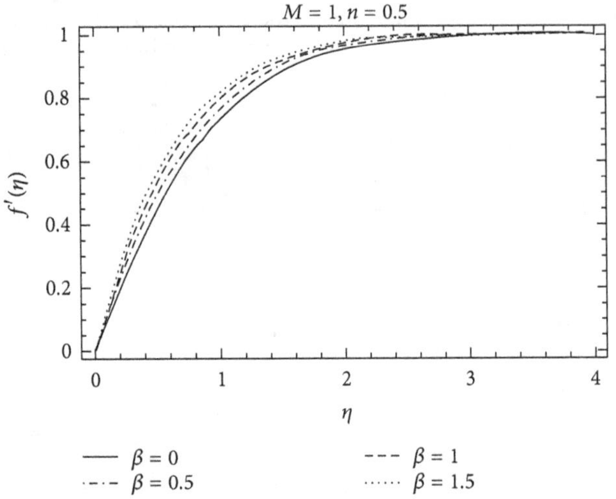

FIGURE 2: Influence of β on f'.

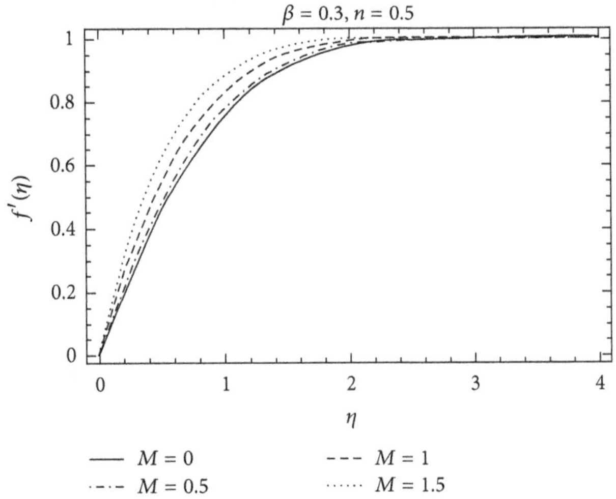

FIGURE 3: Influence of M on f'.

$$\begin{aligned}\mathcal{R}_m^f(\eta) &= f_{m-1}'''(\eta) \\ &+ \sum_{k=0}^{m-1}\Big[f_{m-1-k}f_k'' + \frac{2n}{n+1}\left(1 - f_{m-1-k}'f_k'\right) \\ &\quad + \left(M^2\beta + 1\right)f_{m-1-k}'f_k''\Big] \\ &+ \beta\sum_{k=0}^{m-1}\Big[2n\frac{n-1}{n+1}\left(1 - f_{m-1-k}'\sum_{l=0}^{k} f_{k-l}'f_l''\right) \\ &\quad + (3n-1)f_{m-1-k}\sum_{l=0}^{k} f_{k-l}'f_l'' \\ &\quad - \frac{n+1}{2}f_{m-1-k}''\sum_{l=0}^{k} f_{k-l}'f_l''' \\ &\quad + \frac{n-1}{2}f_{m-1-k}'\sum_{l=0}^{k} f_{k-l}'f_l''\Big]\end{aligned} \tag{39}$$

$$\mathcal{R}_m^\theta(\eta) = \theta_{m-1}'' + \Pr\sum_{k=0}^{m-1}\left[\theta_{m-1-k}'f_k\right], \tag{40}$$

$$\chi_m = \begin{cases}0, & m \le 1 \\ 1, & m > 1.\end{cases} \tag{41}$$

4. Convergence of the Homotopy Solutions

Obviously the auxiliary parameters $\hbar_f$ and $\hbar_\theta$ in the series solutions (31) and (32) have a definite role in adjusting and controlling the convergence. For the admissible values of $\hbar_f$ and $\hbar_\theta$, the $\hbar_f$ and $\hbar_\theta$ curves are portrayed for 15th-order of approximations. It is noticed that the ranges for the admissible values of $\hbar_f$ and $\hbar_\theta$ are $-1.5 \le \hbar_f \le -0.2$ and $-1.2 \le \hbar_\theta \le -0.6$ (Figure 1). Moreover, the series given by (31) and (32) converge in the whole region of η when $\hbar_f = -0.5$ and $\hbar_\theta = -0.8$.

5. Results and Discussion

The purpose of this section is to investigate the variations of parameters including Deborah number γ, the constant n, Hartman number M, and the Prandtl number Pr on the velocity f' and the temperature θ. The variation of the Deborah number β, Hartman number M, and the parameter n on the velocity f' can be seen in Figures 2–5. It is found from Figure 2 that the boundary layer thickness decreases with an increase in γ. Figure 3 is plotted for the effects of Hartman number M on the velocity profile f'. The effects of M on f' are qualitatively similar to those of γ. Figure 4 depicts the velocity field for different values of parameter n. Clearly f' is an increasing function of n while the boundary layer thickness decreases. Figure 5 shows the effects of Prandtl number Pr on the temperature profile θ. The thermal

Table 1: Comparison of the results $\theta'(\eta)$ when $\beta = 0 = M$.

η	$f(\eta)$ HAM	$f(\eta)$ [26]	$f'(\eta)$ HAM	$f'(\eta)$ [26]	$f''(\eta)$ HAM	$f''(\eta)$ [26]
0.0	0.000000	0.00000	0.000000	0.000000	0.469600	0.46960
0.5	0.058643	0.05864	0.234228	0.23423	0.465038	0.46503
1.0	0.232990	0.23299	0.460633	0.46063	0.434379	0.43438
2.0	0.886821	0.86680	0.816695	0.81669	0.255669	0.25567
3.0	1.795568	1.79557	0.969092	0.96905	0.067713	0.06771
4.0	2.783886	2.78388	0.997770	0.99777	0.006875	0.00687
5.0	3.783235	3.78323	0.999937	0.99994	0.000258	0.00026

Table 2: Comparison of the results when $M = 0 = \beta$.

Pr	$\gamma = 0.0$ HAM	$\gamma = 0.0$ [26]	$\gamma = 1.0$ HAM	$\gamma = 1.0$ [26]	$\gamma = 2.0$ HAM	$\gamma = 2.0$ [26]
0.001	0.024492	0.02449	0.024831	0.02483	0.024922	0.02492
0.003	0.041546	0.04154	0.042523	0.04252	0.042780	0.04278
0.100	0.198031	0.19803	0.219502	0.21950	0.226096	0.22600
0.300	0.303712	0.30371	0.351471	0.35147	0.366813	0.36681
1.000	0.469601	0.46960	0.570475	0.57047	0.605204	0.60520
2.000	0.597234	0.59723	0.743721	0.74372	0.795991	0.79599
3.000	0.685967	0.68596	0.865224	0.86522	0.930362	0.93036

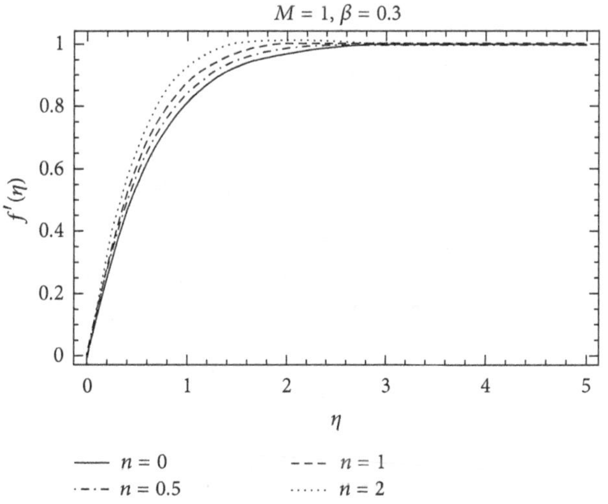

Figure 4: Influence of n on f'.

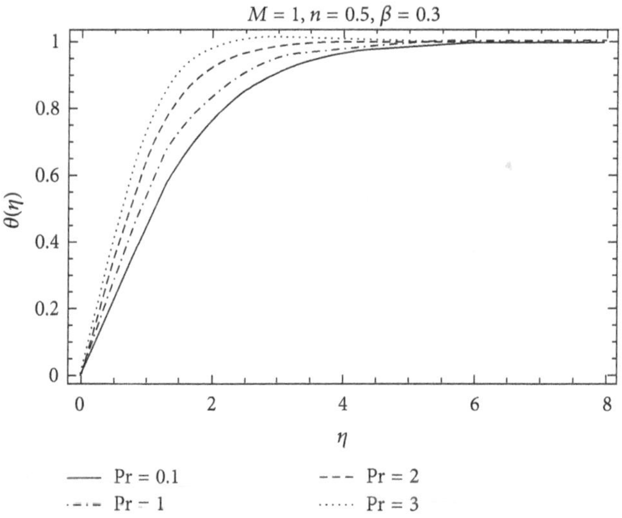

Figure 5: Influence of Pr on θ.

boundary layer thickness decreases when Pr increases. The variation of Hartman number M on the temperature profile is shown in Figure 6. It is noticed that the temperature profile increases when M is increased. The effects of parameter n are shown in Figure 7. It is observed that an increase in the value of n decreases the thermal boundary layer thickness. Figures 8 and 9 are displayed to analyze the comparison of the present analytical results with the existing numerical solutions. An excellent agreement is found between the two solutions for different values of γ. In Table 1, we have computed the numerical values of $f(\eta)$, $f'(\eta)$, and $f''(\eta)$ for the comparison of the present analytical results with the numerical solution [26]. Table 2 is also a comparison between homotopy solution and numerical solution [26] in a special case. An excellent agreement is found between the two solutions.

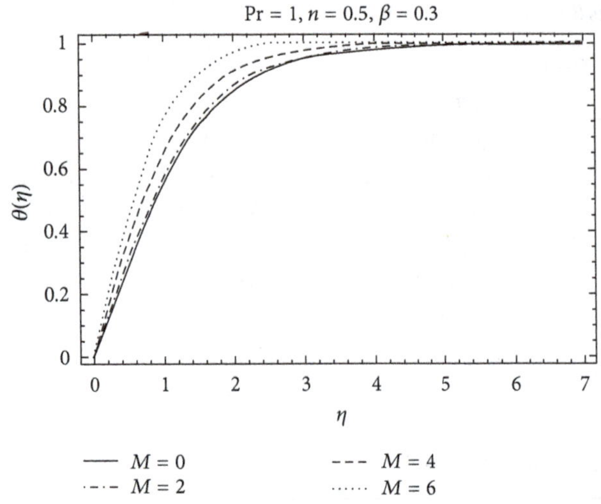

Figure 6: Influence of M on θ.

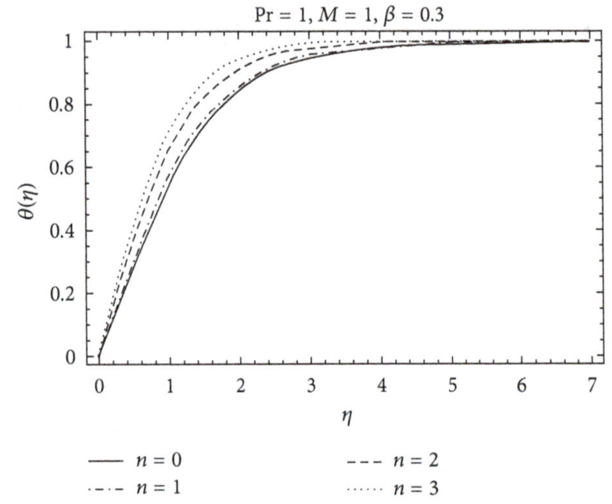

Figure 7: Influence of n on θ.

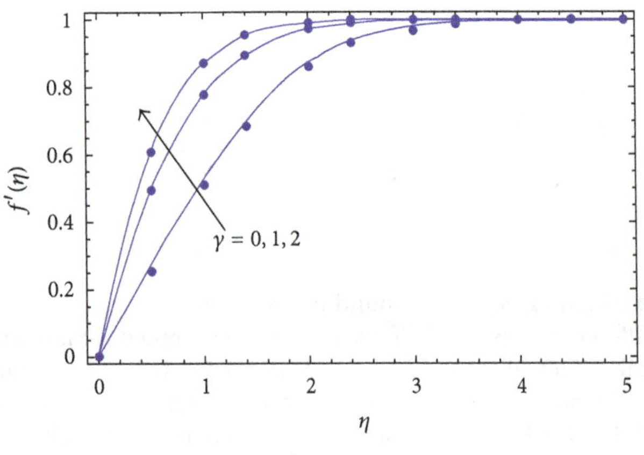

Figure 8: Influence of γ on f' when $M = \beta = 0$. Solid lines: HAM solution [10]; filled circles: numerical solution [26].

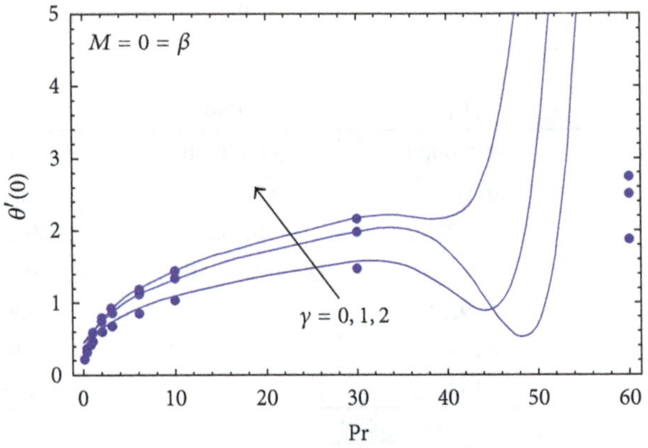

Figure 9: Influence of γ on $\theta'(0)$ when $\beta = 0$. Solid lines: HAM solution [10]; filled circles: numerical solution [26].

References

[1] H. Schlichting, *Boundary Layer Theory*, McGraw Hill, New York, NY, USA, 6th edition, 1979.

[2] V. M. Falkner and S. W. Skan, "Some approximate solutions of the boundary layer equations," *Philosophical Magazine*, vol. 12, pp. 865–896, 1931.

[3] T. Cebeci and H. B. Keller, "Shooting and parallel shooting methods for solving the Falkner-Skan boundary-layer equation," *Journal of Computational Physics*, vol. 7, no. 2, pp. 289–300, 1971.

[4] S. P. Hastings and W. Troy, "Oscillatory solutions of the Falkner-Skan equation," *Proceedings of the Royal Society of London A*, vol. 397, no. 1813, pp. 415–418, 1985.

[5] E. F. F. Botta, F. J. Hut, and A. E. P. Veldman, "The role of periodic solutions in the Falkner-Skan problem for $\lambda > 0$," *Journal of Engineering Mathematics*, vol. 20, no. 1, pp. 81–93, 1986.

[6] E. Alizadeh, M. Farhadi, K. Sedighi, H. R. Ebrahimi-Kebria, and A. Ghafourian, "Solution of the Falkner-Skan equation for wedge by Adomian Decomposition Method," *Communications in Nonlinear Science and Numerical Simulation*, vol. 14, no. 3, pp. 724–733, 2009.

[7] S. Abbasbandy and T. Hayat, "Solution of the MHD Falkner-Skan flow by Hankel-Padé method," *Physics Letters A*, vol. 373, no. 7, pp. 731–734, 2009.

[8] S. Abbasbandy and T. Hayat, "Solution of the MHD Falkner-Skan flow by homotopy analysis method," *Communications in Nonlinear Science and Numerical Simulation*, vol. 14, no. 9-10, pp. 3591–3598, 2009.

[9] E. Magyari, "Falkner-Skan flows past moving boundaries: an exactly solvable case," *Acta Mechanica*, vol. 203, no. 1-2, pp. 13–21, 2009.

[10] B. Yao, "Series solution of the temperature distribution in the Falkner-Skan wedge flow by the homotopy analysis method," *European Journal of Mechanics B*, vol. 28, no. 5, pp. 689–693, 2009.

[11] B. Yao, "Approximate analytical solution to the Falkner-Skan wedge flow with the permeable wall of uniform suction," *Communications in Nonlinear Science and Numerical Simulation*, vol. 14, no. 8, pp. 3320–3326, 2009.

[12] K. R. Rajagopal, A. S. Gupta, and T. Y. Na, "A note on the falkner-skan flows of a non-newtonian fluid," *International Journal of Non-Linear Mechanics*, vol. 18, no. 4, pp. 313–320, 1983.

[13] M. Massoudi and M. Ramezan, "Effect of injection or suction on the Falkner-Skan flows of second grade fluids," *International Journal of Non-Linear Mechanics*, vol. 24, no. 3, pp. 221–227, 1989.

[14] Y. J. Kim, "The Falkner-Skan wedge flows of power-law fluids embedded in a porous medium," *Transport in Porous Media*, vol. 44, no. 2, pp. 267–279, 2001.

[15] D. O. Olagunju, "The Falkner-Skan flow of a viscoelastic fluid," *International Journal of Non-Linear Mechanics*, vol. 41, pp. 825–829, 2006.

[16] S. J. Liao, *Beyond Perturbation: Introduction to homotopy Analysis Method*, Chapman and Hall, CRC Press, Boca Raton, Fla, USA, 2003.

[17] S. J. Liao, "Notes on the homotopy analysis method: some definitions and theorems," *Communications in Nonlinear Science and Numerical Simulation*, vol. 14, pp. 983–997, 2009.

[18] S.-J. Liao, "A general approach to get series solution of non-similarity boundary-layer flows," *Communications in Nonlinear Science and Numerical Simulation*, vol. 14, no. 5, pp. 2144–2159, 2009.

[19] S. J. Liao, "On the homotopy analysis method for nonlinear problems," *Applied Mathematics and Computation*, vol. 147, no. 2, pp. 499–513, 2004.

[20] T. Hayat, M. Qasim, and Z. Abbas, "Homotopy solution for the unsteady three-dimensional MHD flow and mass transfer in a porous space," *Communications in Nonlinear Science and Numerical Simulation*, vol. 15, no. 9, pp. 2375–2387, 2010.

[21] I. Hashim, O. Abdulaziz, and S. Momani, "The homotopy analysis method for Cauchy reaction diffusion problems," *Physics Letters A*, vol. 372, pp. 613–618, 2008.

[22] I. Hashim, O. Abdulaziz, and S. Momani, "Homotopy analysis method for fractional IVPs," *Communications in Nonlinear Science and Numerical Simulation*, vol. 14, no. 3, pp. 674–684, 2009.

[23] A. S. Bataineh, M. S. M. Noorani, and I. Hashim, "Solutions of time-dependent Emden-Fowler type equations by homotopy analysis method," *Physics Letters A*, vol. 371, no. 1-2, pp. 72–82, 2007.

[24] A. S. Bataineh, M. S. M. Noorani, and I. Hashim, "On a new reliable modification of homotopy analysis method," *Communications in Nonlinear Science and Numerical Simulation*, vol. 14, no. 2, pp. 409–423, 2009.

[25] M. Dehghan and R. Salehi, "The use of homotopy analysis method to solve the time-dependent nonlinear Eikonal partial differential equation," *Zeitschrift fur Naturforschung A*, vol. 66, no. 5, pp. 259–271, 2011.

[26] F. M. White, *Viscous Fluid Flow*, McGraw-Hill, New York, NY, USA, 2nd edition, 1991.

5

New Application of (G'/G)-Expansion Method for Generalized (2+1)-Dimensional Nonlinear Evolution Equations

Mohammad Najafi, Maliheh Najafi, and Somayeh Arbabi

Department of Mathematics, Anar Branch, Islamic Azad University, Anar, Iran

Correspondence should be addressed to Mohammad Najafi; m_najafi82@yahoo.com

Academic Editor: Shouming Zhong

We established (G'/G)-expansion method for (2+1)-dimensional nonlinear evolution equations. This method was used to construct travelling wave solutions of (2+1)-dimensional nonlinear evolution equations. (2+1)-Dimensional breaking soliton equation, (2+1)-dimensional Calogero-Bogoyavlenskii-Schiff (CBS) equation, and (2+1)-dimensional Bogoyavlenskii's Breaking soliton equation are chosen to illustrate the effectiveness of the method.

1. Introduction

In this work, we will study the generalized (2+1)-dimensional nonlinear evolution equations

$$u_{xt} + au_x u_{xy} + bu_{xx}u_y + u_{xxxy} = 0, \tag{1}$$

where a and b are parameters, for example, namely, the (2+1)-dimensional Calogero-Bogoyavlenskii-Schiff (CBS) equation for which $a = 4$ and $b = 2$, [1]:

$$u_{xt} + 4u_x u_{xy} + 2u_{xx}u_y + u_{xxxy} = 0, \tag{2}$$

and the (2+1)-dimensional breaking soliton equation for which $a = -4$ and $b = -2$, [2]:

$$u_{xt} - 4u_x u_{xy} - 2u_{xx}u_y + u_{xxxy} = 0, \tag{3}$$

and the (2+1)-dimensional Bogoyavlenskii's Breaking soliton equation for which $a = 4$ and $b = 4$, [3]:

$$u_{xt} + 4u_x u_{xy} + 4u_{xx}u_y + u_{xxxy} = 0. \tag{4}$$

In this paper, we solve (1) by the (G'/G)-expansion method and obtain some exact and new solutions for (2), (3), and (4).

2. The (G'/G)-Expansion Method

In this section we describe the (G'/G)-expansion method for finding traveling wave solutions of nonlinear evolution equations. Suppose that a nonlinear equation, say in two independent variables x, t is given by

$$P\left(u, u_t, u_x, u_{tx}, u_{xx}, \ldots\right) = 0, \tag{5}$$

where $u = u(x,t)$ and P is a polynomial of u and its derivatives, in which the highest order derivatives and nonlinear terms are involved. In the following we give the main steps of the (G'/G)-expansion method.

Firstly, suppose that

$$u(x,t) = u(\xi), \quad \xi = x + wt. \tag{6}$$

The traveling wave variable (6) permits reducing (5) to an ODE for $u = u(\xi)$

$$P\left(u, u', u'', u''', \ldots\right) = 0. \tag{7}$$

Secondly, suppose that the solution of (7) can be expressed by a polynomial in (G'/G) as follows:

$$u(\xi) = \alpha_m \left(\frac{G'}{G}\right)^m + \cdots, \tag{8}$$

where $G = G(\xi)$ satisfies the second-order LODE in the form

$$G'' + \lambda G' + \mu G = 0, \tag{9}$$

$\alpha_m, \ldots, \lambda$ and μ are constants to be determined later, $\alpha_m \neq 0$. The unwritten part in (8) is also a polynomial in (G'/G), the degree of which is generally equal to or less than $m-1$. The positive integer m can be determined by considering the homogeneous balance between the highest order derivatives and nonlinear terms appearing in (7).

Thirdly, substituting (8) into (7) and using the second-order linear ODE (9), collecting all terms with the same order of (G'/G) together, the left-hand side of (7) is converted into another polynomial in (G'/G). Equating each coefficient of this polynomial to zero yields a set of algebraic equations for $\alpha_m, \ldots, \lambda$ and μ.

Fourthly, assuming that the constants $\alpha_m, \ldots, \lambda$ and μ can be obtained by solving the algebraic equations in *Thir ly*. Since the general solutions of the second-order LODE (9) have been well known for us, then substituting $\alpha_m, \ldots, w$ and the general solutions of (9) into (8) we have traveling wave solutions of the nonlinear evolution equation (5) (for more details see [4–7]).

3. New Application of (G'/G)-Expansion Method

In this section we apply the (G'/G)-expansion method to the generalized (2+1)-dimensional nonlinear evolution equations

$$u_{xt} + au_x u_{xy} + bu_{xx}u_y + u_{xxxy} = 0. \quad (10)$$

We introduce a transformation

$$u(x, y, t) = u(\xi), \quad \xi = x + y - ct. \quad (11)$$

By substituting (11) into (10), we have

$$-cu'' + au'u'' + bu'u'' + u^{(4)} = 0, \quad (12)$$

or equivalently

$$-cu'' + (a+b)u'u'' + u^{(4)} = 0, \quad (13)$$

where prime denotes the differential with respect to ξ. Integrating (13) with respect to ξ and taking the integration constant as zero yields

$$-cu' + \frac{(a+b)}{2}(u')^2 + u^{(3)} = g, \quad (14)$$

in which (14) is obtained by integrating and neglecting the constant of integration and where prime denotes the derivative with respect to the same variable ξ and g is the integration constant that can be determined later.

Suppose that the solutions of the O.D.E (14) can be expressed by a polynomial in (G'/G) as follows:

$$u(\xi) = \sum_{i=0}^{m} a_i \left(\frac{G'}{G}\right)^i, \quad (15)$$

where a_i are constants, $G = G(\xi)$ satisfies the second-order LODE in the form:

$$G'' + \lambda G' + \mu G = 0, \quad (16)$$

where λ and μ are constants.

By balancing the order of $(u')^2$ and $u^{(3)}$ in (11), we have $2m+2 = m+3$, then $m = 1$. So we can write

$$u(\xi) = a_1\left(\frac{G'}{G}\right) + a_0, \quad a_1 \neq 0, \quad (17)$$

where a_1, a_0 are constants to be determined later. Then it follows

$$u'(\xi) = -a_1\left(\frac{G'}{G}\right)^2 - a_1\lambda\left(\frac{G'}{G}\right) - a_1\mu, \quad (18)$$

$$u''(\xi) = 2a_1\left(\frac{G'}{G}\right)^3 + 3a_1\lambda\left(\frac{G'}{G}\right)^2 + a_1(\lambda^2 - 2\mu)\left(\frac{G'}{G}\right) + a_1\lambda\mu, \quad (19)$$

$$\begin{aligned} u'''(\xi) = &- 6a_1\left(\frac{G'}{G}\right)^4 - 12a_1\lambda\left(\frac{G'}{G}\right)^3 \\ &- \left(7a_1\lambda^2 + 8a_1\mu\right)\left(\frac{G'}{G}\right)^2 \\ &- \left(a_1\lambda^3 + 8a_1\lambda\mu\right)\left(\frac{G'}{G}\right) - a_1\lambda^2\mu - 2a_1\mu^2. \end{aligned} \quad (20)$$

Substituting (18), (19), and (20) into (14) and collecting all the terms with the same power of (G'/G) together, equating each coefficient to zero yields a set of simultaneous algebraic equations as follows:

$$\begin{aligned} &\left(\frac{G'}{G}\right)^4 : a_1(a+b) - 12 = 0, \\ &\left(\frac{G'}{G}\right)^3 : 2a_1(a+b)\lambda - 24\lambda = 0, \\ &\left(\frac{G'}{G}\right)^2 : 2c + 2a_1(a+b)\mu - 14\lambda^2 + a_1(a+b)\lambda^2 - 16\mu = 0, \\ &\left(\frac{G'}{G}\right)^1 : 2a_1(a+b)\lambda\mu - 2\lambda^3 + 2c\lambda - 16\lambda\mu = 0, \\ &\left(\frac{G'}{G}\right)^0 : a_1(a+b)\mu^2 + 2c\mu - 2\lambda^2\mu - 4\mu^2 - g = 0. \end{aligned} \quad (21)$$

Solving the algebraic equations above by using the Maple, we get

$$a_0 = a_0, \quad a_1 = \frac{12}{a+b}, \quad g = 0, \quad c = \lambda^2 - 4\mu, \quad (22)$$

where c, λ, μ are arbitrary constants.

Substituting (22) into (17) and by using (11), we have

$$u(\xi) = \frac{12}{a+b}\left(\frac{G'}{G}\right) + a_0, \quad (23)$$

where $\xi = x + y - (\lambda^2 - 4\mu)t$.

Substituting the general solutions of (16) into (23), we obtain the following.

Case 1. When $\lambda^2 - 4\mu > 0$

$$u_1(\xi) = \frac{6\sqrt{\lambda^2-4\mu}}{a+b} \times \left(\frac{C_1 \sinh(1/2)\sqrt{\lambda^2-4\mu}\xi + C_2 \cosh(1/2)\sqrt{\lambda^2-4\mu}\xi}{C_1 \cosh(1/2)\sqrt{\lambda^2-4\mu}\xi + C_2 \sinh(1/2)\sqrt{\lambda^2-4\mu}\xi} - \lambda\right) + a_0, \quad (24)$$

where $\xi = x + y - (\lambda^2 - 4\mu)t$ and $C_1, C_2, c, \lambda, \mu$ are arbitrary constant.

Case 2. When $\lambda^2 - 4\mu < 0$

$$u_2(\xi) = \frac{6\sqrt{4\mu-\lambda^2}}{a+b} \times \left(\frac{-C_1 \sin(1/2)\sqrt{4\mu-\lambda^2}\xi + C_2 \cos(1/2)\sqrt{4\mu-\lambda^2}\xi}{C_1 \cos(1/2)\sqrt{4\mu-\lambda^2}\xi + C_2 \sin(1/2)\sqrt{4\mu-\lambda^2}\xi} - \lambda\right) + a_0, \quad (25)$$

where $\xi = x + y - (\lambda^2 - 4\mu)t$ and $C_1, C_2, c, \lambda, \mu$ are arbitrary constant.

Case 3. When $\lambda^2 - 4\mu = 0$

$$u_3(\xi) = \frac{12}{a+b}\left(\frac{C_2}{C_1 + C_2\xi} - \frac{\lambda}{2}\right) + a_0, \quad (26)$$

where $\xi = x + y - (\lambda^2 - 4\mu)t$ and $C_1, C_2, c, \lambda, \mu$ are arbitrary constant.

In particular, if we take $C_1 \neq 0$ and $C_2 = 0$, then $u_1(\xi)$ becomes

$$u_{11}(\xi) = \frac{6\sqrt{\lambda^2-4\mu}}{a+b}\left(\tanh\frac{1}{2}\sqrt{\lambda^2-4\mu}\xi - \lambda\right) + a_0 \quad (27)$$

and $u_2(\xi)$ becomes

$$u_{21}(\xi) = -\frac{6\sqrt{4\mu-\lambda^2}}{a+b}\left(\tan\frac{1}{2}\sqrt{4\mu-\lambda^2}\xi + \lambda\right) + a_0. \quad (28)$$

When we take $C_1 = 0$ and $C_2 \neq 0$, then $u_1(\xi)$ becomes

$$u_{12}(\xi) = \frac{6\sqrt{\lambda^2-4\mu}}{a+b}\left(\coth\frac{1}{2}\sqrt{\lambda^2-4\mu}\xi - \lambda\right) + a_0 \quad (29)$$

and $u_2(\xi)$ becomes

$$u_{22}(\xi) = \frac{6\sqrt{4\mu-\lambda^2}}{\alpha+b}\left(\cot\frac{1}{2}\sqrt{4\mu-\lambda^2}\xi - \lambda\right) + a_0. \quad (30)$$

In the following Sections 4–6, we will apply solutions $u_i(\xi)$, $i = 1,2,3$ to (2+1)-dimensional breaking soliton equation and (2+1)-dimensional Calogero-Bogoyavlenskii-Schiff (CBS) equation and (2+1)-dimensional Bogoyavlenskii's Breaking soliton equation.

4. Exact Solution of (2)

In this section, we investigate explicit formula of solutions of the following (2+1)-dimensional Calogero-Bogoyavlenskii-Schiff (CBS) equation given in [1]

$$u_{xt} + 4u_x u_{xy} + 2u_{xx}u_y + u_{xxxy} = 0. \quad (31)$$

By using Section 3, we have following exact solution.

Case 1. When $\lambda^2 - 4\mu > 0$

$$u_1(\xi) = \sqrt{\lambda^2-4\mu} \times \left(\frac{C_1 \sinh(1/2)\sqrt{\lambda^2-4\mu}\xi + C_2 \cosh(1/2)\sqrt{\lambda^2-4\mu}\xi}{C_1 \cosh(1/2)\sqrt{\lambda^2-4\mu}\xi + C_2 \sinh(1/2)\sqrt{\lambda^2-4\mu}\xi} - \lambda\right) + a_0, \quad (32)$$

where $\xi = x + y - (\lambda^2 - 4\mu)t$ and $C_1, C_2, c, \lambda, \mu$ are arbitrary constant.

Case 2. When $\lambda^2 - 4\mu < 0$

$$u_2(\xi) = \sqrt{\lambda^2-4\mu} \times \left(\frac{-C_1 \sin(1/2)\sqrt{4\mu-\lambda^2}\xi + C_2 \cos(1/2)\sqrt{4\mu-\lambda^2}\xi}{C_1 \cos(1/2)\sqrt{4\mu-\lambda^2}\xi + C_2 \sin(1/2)\sqrt{4\mu-\lambda^2}\xi} - \lambda\right) + a_0, \quad (33)$$

where $\xi = x + y - (\lambda^2 - 4\mu)t$ and $C_1, C_2, c, \lambda, \mu$ are arbitrary constant.

Case 3. When $\lambda^2 - 4\mu = 0$

$$u_3(\xi) = 2\left(\frac{C_2}{C_1 + C_2\xi} - \frac{\lambda}{2}\right) + a_0, \quad (34)$$

where $\xi = x + y - (\lambda^2 - 4\mu)t$ and $C_1, C_2, c, \lambda, \mu$ are arbitrary constant.

In particular, if we take $C_1 \neq 0$ and $C_2 = 0$, then $u_1(\xi)$ becomes

$$u_{11}(\xi) = \sqrt{\lambda^2-4\mu}\left(\tanh\frac{1}{2}\sqrt{\lambda^2-4\mu}\xi - \lambda\right) + a_0, \quad (35)$$

and $u_2(\xi)$ becomes

$$u_{21}(\xi) = -\sqrt{\lambda^2 - 4\mu}\left(\tan\frac{1}{2}\sqrt{4\mu - \lambda^2}\xi + \lambda\right) + a_0 \quad (36)$$

when we take $C_1 = 0$ and $C_2 \neq 0$, then $u_1(\xi)$ becomes

$$u_{12}(\xi) = \sqrt{\lambda^2 - 4\mu}\left(\coth\frac{1}{2}\sqrt{\lambda^2 - 4\mu}\xi - \lambda\right) + a_0 \quad (37)$$

and $u_2(\xi)$ becomes

$$u_{22}(\xi) = \sqrt{\lambda^2 - 4\mu}\left(\cot\frac{1}{2}\sqrt{4\mu - \lambda^2}\xi - \lambda\right) + a_0. \quad (38)$$

5. Exact Solution of (3)

In this section, we investigate explicit formula of solutions of the following (2+1)-dimensional Breaking soliton equation given in [2]

$$u_{xt} - 4u_x u_{xy} - 2u_{xx}u_y + u_{xxxy} = 0. \quad (39)$$

By using Section 3, we have following exact solution:

Case 1. When $\lambda^2 - 4\mu > 0$

$$u_1(\xi) = -\sqrt{\lambda^2 - 4\mu} \times \left(\frac{C_1\sinh(1/2)\sqrt{\lambda^2-4\mu}\xi + C_2\cosh(1/2)\sqrt{\lambda^2-4\mu}\xi}{C_1\cosh(1/2)\sqrt{\lambda^2-4\mu}\xi + C_2\sinh(1/2)\sqrt{\lambda^2-4\mu}\xi} - \lambda\right) + a_0, \quad (40)$$

where $\xi = x + y - (\lambda^2 - 4\mu)t$ and $C_1, C_2, c, \lambda, \mu$ are arbitrary constant.

Case 2. When $\lambda^2 - 4\mu < 0$

$$u_2(\xi) = -\sqrt{\lambda^2 - 4\mu} \times \left(\frac{-C_1\sin(1/2)\sqrt{4\mu-\lambda^2}\xi + C_2\cos(1/2)\sqrt{4\mu-\lambda^2}\xi}{C_1\cos(1/2)\sqrt{4\mu-\lambda^2}\xi + C_2\sin(1/2)\sqrt{4\mu-\lambda^2}\xi} - \lambda\right) + a_0, \quad (41)$$

where $\xi = x + y - (\lambda^2 - 4\mu)t$ and $C_1, C_2, c, \lambda, \mu$ are arbitrary constant.

Case 3. When $\lambda^2 - 4\mu = 0$

$$u_3(\xi) = -2\left(\frac{C_2}{C_1 + C_2\xi} - \frac{\lambda}{2}\right) + a_0, \quad (42)$$

where $\xi = x + y - (\lambda^2 - 4\mu)t$ and $C_1, C_2, c, \lambda, \mu$ are arbitrary constant.

In particular, if we take $C_1 \neq 0$ and $C_2 = 0$, then $u_1(\xi)$ becomes

$$u_{11}(\xi) = -\sqrt{\lambda^2 - 4\mu}\left(\tanh\frac{1}{2}\sqrt{\lambda^2 - 4\mu}\xi - \lambda\right) + a_0, \quad (43)$$

and $u_2(\xi)$ becomes

$$u_{21}(\xi) = \sqrt{\lambda^2 - 4\mu}\left(\tan\frac{1}{2}\sqrt{4\mu - \lambda^2}\xi + \lambda\right) + a_0. \quad (44)$$

When we take $C_1 = 0$ and $C_2 \neq 0$, then $u_1(\xi)$ becomes

$$u_{12}(\xi) = -\sqrt{\lambda^2 - 4\mu}\left(\coth\frac{1}{2}\sqrt{\lambda^2 - 4\mu}\xi - \lambda\right) + a_0, \quad (45)$$

and $u_2(\xi)$ becomes

$$u_{22}(\xi) = -\sqrt{\lambda^2 - 4\mu}\left(\cot\frac{1}{2}\sqrt{4\mu - \lambda^2}\xi - \lambda\right) + a_0. \quad (46)$$

6. Exact Solution of (4)

In this section, we investigate explicit formula of solutions of the following (2+1)-dimensional Bogoyavlenskii's Breaking soliton equation given in [3]:

$$u_{xt} + 4u_x u_{xy} + 4u_{xx}u_y + u_{xxxy} = 0. \quad (47)$$

By using Section 3, we have following exact solutions.

Case 1. When $\lambda^2 - 4\mu > 0$

$$u_1(\xi) = \frac{3\sqrt{\lambda^2 - 4\mu}}{4} \times \left(\frac{C_1\sinh(1/2)\sqrt{\lambda^2-4\mu}\xi + C_2\cosh(1/2)\sqrt{\lambda^2-4\mu}\xi}{C_1\cosh(1/2)\sqrt{\lambda^2-4\mu}\xi + C_2\sinh(1/2)\sqrt{\lambda^2-4\mu}\xi} - \lambda\right) + a_0, \quad (48)$$

where $\xi = x + y - (\lambda^2 - 4\mu)t$ and $C_1, C_2, c, \lambda, \mu$ are arbitrary constant.

Case 2. When $\lambda^2 - 4\mu < 0$

$$u_2(\xi) = \frac{3\sqrt{4\mu - \lambda^2}}{4} \times \left(\frac{-C_1\sin(1/2)\sqrt{4\mu-\lambda^2}\xi + C_2\cos(1/2)\sqrt{4\mu-\lambda^2}\xi}{C_1\cos(1/2)\sqrt{4\mu-\lambda^2}\xi + C_2\sin(1/2)\sqrt{4\mu-\lambda^2}\xi} - \lambda\right) + a_0, \quad (49)$$

where $\xi = x + y - (\lambda^2 - 4\mu)t$ and $C_1, C_2, c, \lambda, \mu$ are arbitrary constant.

Case 3. When $\lambda^2 - 4\mu = 0$

$$u_3(\xi) = \frac{3}{2}\left(\frac{C_2}{C_1 + C_2\xi} - \frac{\lambda}{2}\right) + a_0, \quad (50)$$

where $\xi = x + y - (\lambda^2 - 4\mu)t$ and $C_1, C_2, c, \lambda, \mu$ are arbitrary constant.

In particular, if we take $C_1 \neq 0$ and $C_2 = 0$, then $u_1(\xi)$ becomes

$$u_{11}(\xi) = \frac{3\sqrt{\lambda^2 - 4\mu}}{4}\left(\tanh\frac{1}{2}\sqrt{\lambda^2 - 4\mu}\xi - \lambda\right) + a_0, \quad (51)$$

and $u_2(\xi)$ becomes

$$u_{21}(\xi) = -\frac{3\sqrt{4\mu - \lambda^2}}{4}\left(\tan\frac{1}{2}\sqrt{4\mu - \lambda^2}\xi + \lambda\right) + a_0. \quad (52)$$

When we take $C_1 = 0$ and $C_2 \neq 0$, then $u_1(\xi)$ becomes

$$u_{12}(\xi) = \frac{3\sqrt{\lambda^2 - 4\mu}}{4}\left(\coth\frac{1}{2}\sqrt{\lambda^2 - 4\mu}\xi - \lambda\right) + a_0, \quad (53)$$

and $u_2(\xi)$ becomes

$$u_{22}(\xi) = \frac{3\sqrt{4\mu - \lambda^2}}{4}\left(\cot\frac{1}{2}\sqrt{4\mu - \lambda^2}\xi - \lambda\right) + a_0. \quad (54)$$

7. Conclusions

In this paper, by using the (G'/G)-expansion method, we obtained some explicit formulas of solutions for the generalized (2+1)-dimensional nonlinear evolution equations. We chose (2+1)-dimensional breaking soliton equation and (2+1)-dimensional Calogero-Bogoyavlenskii-Schiff (CBS) equation and (2+1)-dimensional Bogoyavlenskii's Breaking soliton equation to illustrate the effectiveness of the method. Those solutions were similar to the solutions obtained in other paper. The study reveals the power of the method.

References

[1] A. M. Wazwaz, "The (2+1)-dimensional CBS equations: multiple soliton solutions and multiple singular soliton solutions," *Zeitschrift fur Naturforschung A*, vol. 65, no. 3, pp. 173–181, 2010.

[2] A. M. Wazwaz, "Integrable (2+1)-dimensional breaking soliton equations," *Physica Scripta*, vol. 81, pp. 1–5, 2010.

[3] M. T. Darvishi and M. Najafi, "Some exact solutions of the (2+1)-dimensional breaking soliton equation using the three-wave method," *International Journal of Computational and Mathematical Sciences*, vol. 6, no. 1, pp. 13–16, 2012.

[4] M. T. Darvishi, M. Najafi, and M. Najafi, "Traveling wave solutions for the (3+1)-dimensional breaking soliton equation by (G'/G)-expansion method and modified F-expansion method," *International Journal of Computational and Mathematical Sciences*, vol. 6, no. 2, pp. 64–69, 2012.

[5] E. M. E. Zayed and S. Al-Joudi, "Applications of an Improved (G'/G)-expansion method to nonlinear PDEs in mathematical physics," *AIP Conference Proceedings*, vol. 1168, no. 1, pp. 371–376, 2009.

[6] Z. Y. Bin and L. Chao, "Application of modified (G'/G)-expansion method to traveling wave solutions for Whitham-Broer-Kaup-like equations," *Communications in Theoretical Physics*, vol. 51, no. 4, pp. 664–670, 2009.

[7] I. Aslan and T. Özis, "On the validity and reliability of the (G'/G)-expansion method by using higher-order nonlinear equations," *Applied Mathematics and Computation*, vol. 211, no. 2, pp. 531–536, 2009.

6

Analytical Treatment and Convergence of the Adomian Decomposition Method for Instability Phenomena Arising during Oil Recovery Process

Ramakanta Meher[1,2] and Srikanta K. Meher[1,2]

[1] *Department of Mathematics, Sardar Vallabhbhai National Institute of Technology, Surat 395007, India*
[2] *Department of Mathematics, Anand Agriculture University, Dahod, Gujarat, India*

Correspondence should be addressed to Ramakanta Meher; meher_ramakanta@yahoo.com

Academic Editor: George S. Dulikravich

An abstract result is proved for the convergence of Adomian decomposition method for partial differential equations that model porous medium equation. Moreover, we prove that this decomposition scheme applied to a porous medium equation arising in instability phenomena in double phase flow through porous media is convergent in a suitable Hilbert space. Furthermore, this technique is utilized to find closed-form solutions for the problem under consideration.

1. Introduction

The oil-water movement in a porous medium is an important problem of petroleum technology and water hydrology (Scheidegger [1]). The motion of two immiscible fluids in a homogenous porous medium was obtained by Buckley and Leveret without considering capillary pressure. The basic assumption underlying in the present investigation is that the oil and water form two immiscible liquid phases and water represents preferentially wetting phase.

During secondary oil recovery when a fluid (oil) contained in a porous matrix of an oil formatted region in an oil reservoir is displaced by another fluid of lesser viscosity, that is, water, instead of regular displacement of the whole front, perturberance (fingers) may occur which shoot thorough the porous medium at relatively great speed. These protuberances are called fingers. This phenomenon is called instability. Scheidegger [1] analyzed the statistical behavior of instability in a displacement process through a homogeneous porous medium with capillary pressure and pressure-dependent phase densities. This problem has great importance for oil production in petroleum technology.

Displacement of oil from a porous matrix by an external force which gives rise to a pressure gradient is known as forced instability. Instability can occur in cocurrent and counter-current flow modes.

The stability of a water flood depends on the mobility ratio between oil, water, heterogeneity of the porous medium, segregation of the fluids in the reservoir, and dissipation of fluid fronts caused by capillary pressure. Instabilities may occur in both miscible and immiscible processes and originate on the interface between oil and water. These frontal instabilities are often characterized by a number of penetrating fingers of displacing fluid.

Many of the oil fields around the world contain high viscous oil which gives a mobility ratio greater than unity ($M > 1$). Such mobility ratios are unfavorable and may cause the occurrence of frontal instabilities. If $M > 1$, the injected water may channel through the oil in an unstable manner. This viscous fingering leads to premature breakthrough of water and a poor volumetric sweep. In principle, an increase in water viscosity should give lower residual oil saturation due to an increase in the capillary number. This increase is however normally too small to significantly affect the residual oil saturation.

Many researchers have studied this phenomenon with different points of view. The problem of the flow of two immiscible phases in homogeneous porous media without capillary pressure effect has been discussed by Buckely and Leverett [2]. Oroveanu [3] has formally extended this discussion for heterogeneous porous media. Verma [4] has discussed the statistical behavior of fingering in a displacement process in heterogeneous porous medium with capillary pressure. Venkateswar [5] has discussed on the flow of immiscible liquids in a heterogeneous porous medium with capillary pressure and connate water saturation.

In this paper, we investigate the applicability of Adomian decomposition method to the nonlinear partial differential equation arising in instability phenomena in double phase flow through porous media in order to obtain the analytical solution.

The paper is organized as follows: in Section 2, we have written the statement of the problem along with some relation. Fundamental equation of instability phenomena is discussed in Section 3. In Section 4, we introduce the Adomian decomposition method applied to solve nonlinear functional equations. The novelty of this paper is in Section 5, where we develop and prove the convergence of Adomian decomposition scheme, which leads to an abstract result, and analytical solution to the equation. Section 6 is devoted to simulation results for some interesting choices of initial data. We conclude by summarizing the paper in Section 7.

2. Statement of Problem

It is well known that in secondary oil recovery process when water with constant velocity "v" is injected into a seam saturated with oil and consisting of homogenous porous medium, it is assumed that the entire oil on the initial boundary of the seam, $x = 0$ (x is measured in the direction of displacement), is displaced through a small distance due to the impact of injecting water which forms instability at the common interface where water meets the oil zone. To understand this phenomenon, we consider here a horizontal porous matrix of length L with its impermeable surface filled with oil formatted porous media.

For the definiteness of the problem, consider that there is a uniform water injection into oil saturated porous matrix of an oil formatted region having homogenous physical characteristics such that the injecting water outs through the oil formation region of the oil reservoir and gives rise to perturberance (fingers) at the interface where injected water pushes the oil from the oil formatted region. This furnishes well-developed fingers as in Figure 1. The stability of a water flood depends on the mobility ratio between oil and water, heterogeneity of the porous medium, segregation of the fluids in the reservoir, and dissipation of fluid fronts caused by capillary pressure. Instabilities may occur in both miscible and immiscible processes and originate on the interface between oil and water. These frontal instabilities are often characterized by a number of penetrating fingers of displacing fluid. Therefore, the entire oil at the initial boundary $x = 0$ (x being measured in the direction of displacement) is displaced through a distance "L" due to water injection. It is further assumed that complete saturation exists at the initial boundary, and the saturation of displaced water (fingers) in oil zone may happen up to distance $x = L$.

2.1. Relative Permeability and Phase Saturation Relation. Following Scheidegger [1], the relationship between relative permeability and phase saturation may be taken as

$$k_w = S_w, \quad k_0 = 1 - \alpha S_w \quad (\alpha = 1.11), \tag{1}$$

where k_w and k_o denote the relative permeabilities of water and oil and S_w be the saturation of water (regarded as the wetting phase).

3. Fundamental Equations

The seepage velocity of water (v_w) and oil (v_o) may be written as (by Darcy's law)

$$v_o = -\left(\frac{k_o}{\mu_o}K\right)\frac{\partial p_o}{\partial x} \qquad v_w = -\left(\frac{k_w}{\mu_w}K\right)\frac{\partial p_w}{\partial x}, \tag{2}$$

where "K" is the permeability of homogenous medium k_o and k_w are relative permeability of oil and water which are function of S_o and S_w. S_w and S_o are saturation of water and oil, respectively, and p_w and p_o denote pressure of water and oil, while μ_w, μ_o are constant kinematic viscosities of the phases in homogenous porous media.

The equations of continuity (phase densities are regarded as constant throughout the process of instability) are

$$\phi\frac{\partial S_w}{\partial t} + \frac{\partial v_w}{\partial x} = 0, \qquad \phi\frac{\partial S_o}{\partial t} + \frac{\partial v_o}{\partial x} = 0, \tag{3}$$

where "ϕ" is the porosity of the medium.

From the definition of capillary pressure (p_c) and phase saturation "S_w", it is evident that

$$p_c = p_o - p_w, \qquad S_o + S_w = 1. \tag{4}$$

Equation (2)–(4) gives

$$\phi\frac{\partial S_w}{\partial t} + \frac{1}{2}\frac{\partial}{\partial x}\left[K\frac{k_w}{\mu_w}\frac{dp_c}{dS_w}\frac{\partial S_w}{\partial x}\right] = 0. \tag{5}$$

For definiteness of the mathematical analysis, we assume standard forms of Verma [4] or the analytical relationship between the relative permeability phase saturation and a linear relationship between capillary pressure phase saturation as

$$k_w = S_w, \quad p_c = \beta\left(C_0 + S_w^{-2}\right) \quad (\beta, C_0 \text{ is constant}), \tag{6}$$

where k_w is function of S_w only (Scheidegger [1]) and the negative power of the second term is due to the fact that p_c is a decreasing function of S_w.

Substituting this value in (5), we get

$$\frac{\partial S_w}{\partial t} = \frac{K\beta}{\mu_w\phi}\frac{\partial}{\partial x}\left(\frac{1}{S_w^2}\frac{\partial S_w}{\partial x}\right). \tag{7}$$

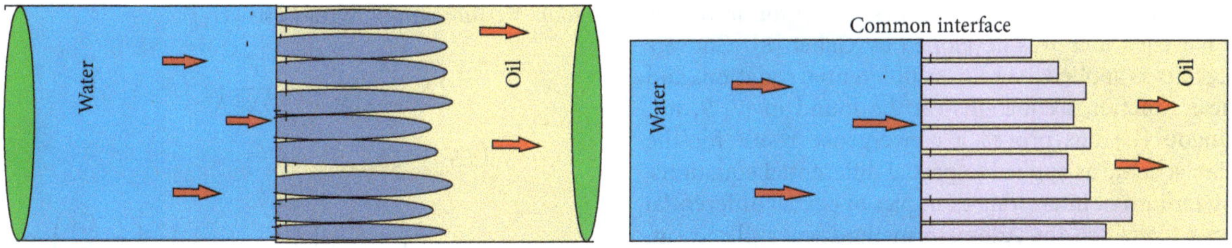

FIGURE 1: Diagram showing the formation of instability of oil zone due to the external force of injected water during secondary recovery in an oil reservoir.

Equation (7) represents a nonlinear partial differential equation describing instability phenomenon of two immiscible fluids (i.e., water and oil) flow through homogeneous porous cylindrical medium with impervious bounding surfaces on three sides of an oil reservoir.

Choosing dimensionless variables $X = x/L$, $T = (K\beta/\phi\mu_w L^2)t$.

It implies if $0 < x < L$, then $0 < X < 1$, and here we will study the behavior of injected water up to $X = 1$.

The dimensionless form of (7) is as

$$\frac{\partial S_w}{\partial T} = \frac{\partial}{\partial X}\left(\frac{1}{S_w^2}\frac{\partial S_w}{\partial X}\right) \tag{8}$$

We choose appropriate initial and Dirichlet's boundary condition due to the behavior of saturation of displaced water at the interface in instability phenomena; that is, instability of oil and water zone at the interface is high, and it becomes stable as it becomes away from the interface $X = 0$ and by Verma [4] as

$$\begin{gathered} S_w(X,0) = S_{w,0}(X) = \left(X^2+1\right)^{-1/2}, \\ S_w(0,T) = f_1(T), \qquad S_w(1,T) = f_2(T), \end{gathered} \tag{9}$$

where f_1 and f_2 are the saturation of water at common interface $X = 0$ and saturation of water at end of matrix of length $X = 1$ (i.e., $x = L$). Here, during instability phenomena saturation fingers may take place up to the end of matrix, that is, up to $x = L$. To stabilize or to find the behavior of the saturation fingers, it is necessary to discuss the behavior of saturation of displace water by solving (8) together with (9).

Equation (8) is the desired nonlinear partial differential equation with suitable initial and boundary conditions which describes the saturation of displaced water in instability phenomena arising during oil recovery process.

4. Analysis of the Adomian Method

In the early 1980s, a new numerical method was developed by Adomian [6] in order to solve nonlinear functional equations of the form

$$LS_w + RS_w + NS_w = g, \tag{10}$$

using an iterative decomposition scheme that led to elegant computation of closed-form analytical solutions or analytical approximations to solutions. In (10), L represents the linear part, N represents the nonlinear part, R represents the remainder or lower order terms, and g is the nonhomogeneous right-hand side. The solution S_w and the nonlinearity N are assumed to have the following analytic expansions, respectively:

$$S_w = \sum_{n=0}^{\infty} S_{wn}, \qquad NS_w = \sum_{n=0}^{\infty} A_n, \tag{11}$$

where the A_n's are the Adomian polynomials that depend only on $S_{w0}, S_{w1}, S_{w2}, \ldots, S_{wn}$ and are given by the following formula:

$$A_n = \frac{1}{n!}\frac{d^n}{d\lambda^n}\left[N\left(\sum_{k=0}^{\infty}\lambda^k S_{wk}\right)\right]_{\lambda=0}, \quad n \geq 0. \tag{12}$$

In order to better explain the method, we will first assume the convergence of the series in (11) and deal with the rigorous convergence issues later. The parameter k is a dummy variable introduced for ease of computation. There are several different versions of (12) that can be found in the literature that leads to easier computation of the A_n's. It should be noted that the A_n's are the terms of analytic expansion of NS_w, where $S_w = \sum_{n=0}^{\infty} S_{wn}\lambda^n$.

In [7], Adomian has shown that the expansion for NS_w in (11) is a rearrangement of the Taylor series expansion of NS_w about the initial function S_{w0} in a suitable Hilbert or Banach space. Substitution of (11) in (10) results in the following:

$$L\left(\sum_{n=0}^{\infty} S_{wn}\right) = -R\left(\sum_{n=0}^{\infty} S_{wn}\right) - \sum_{n=0}^{\infty} A_n + g. \tag{13}$$

The above equation can be rewritten in a recursive fashion, yielding iterates S_{wn}, the sum of which converges to the solution S_w satisfying (13) if it exists

$$\begin{gathered} \sum_{n=0}^{\infty} S_{wn} = -L^{-1}R\left(\sum_{n=0}^{\infty} S_{wn}\right) - L^{-1}\sum_{n=0}^{\infty} A_n + L^{-1}(g), \\ S_{w0} = L^{-1}(g); \qquad S_{w,n+1} = -L^{-1}R(S_{wn}) - L^{-1}(A_n). \end{gathered} \tag{14}$$

Typically, the symbol L^{-1} represents a formal inverse of the linear operator L. In the case of partial differential equations, L is the highest order partial derivative operator for which the formal inverse can be computed using integrations.

A general theory of decomposition schemes for nonlinear functional equations was developed by Gabet [8]. Convergence results as applied to ordinary differential equations and nonlinear functional equations can be found in [7, 9, 10]. Mavoungou [11] has proved a convergence result for the Adomian scheme as applied to partial differential equations. A compendium of interesting examples of partial differential equations for which the Adomian method was utilized can be found in [12]. In general, the iterates in the Adomian decomposition scheme (14) converge very rapidly to the unique solution of the functional equation (10) provided that the scheme satisfies the property of strong convergence as discussed in [8].

5. Convergence Analysis of the Adomian Decomposition Method

We recall the following theorem from [11] which guarantees the convergence of Adomian's method for the general operator equation given by $LS_w + RS_w + NS_w = g$.

Consider the Hilbert space $H = L^2((\alpha,\beta)\times[0,T])$ defined by the set of applications:

$$S_w : (\alpha,\beta)\times[0,T] \longrightarrow R$$
$$\text{with} \int_{(\alpha,\beta)\times[0,T]} S_W^2(\eta,\xi)\,d\eta\,d\xi < +\infty. \tag{15}$$

Let us denote

$$LS_w = \frac{\partial S_w}{\partial T}, \qquad NS_w = S_w^{-2}\frac{\partial S_w}{\partial X},$$
$$TS_w = RS_w + NS_w = \frac{\partial}{\partial X}\left(S_w^{-2}\frac{\partial S_w}{\partial X}\right). \tag{16}$$

Theorem 1. *Let $TS_w = -RS_w - NS_w$ be a hemicontinuous operator in a Hilbert space H and satisfy the following hypothesis:*

(H_1)

$$\left(TS'_w - TS^{*\prime}_w, S'_w - S^{*\prime}_w\right) \geq k\left\|S'_w - S^{*\prime}_w\right\|^2,$$
$$k > 0,\ \forall S'_w, S^{*'}_w \in H, \tag{17}$$

(H_2) *whatever may be $M > 0$, there exist constant $C(M) > 0$ such that for $S'_w, S^*_w \in H$ with $\|S'_w\| \leq M$, $\|S^{*\prime}_w\| \leq M$, we have*

$$\left(TS'_w - TS^{*\prime}_w, w\right) \leq C(M)\left\|S'_w - S^{*\prime}_w\right\|\|w\|$$
$$\text{for every } w \in H. \tag{18}$$

Th n, for every $g \in H'$, the nonlinear functional equation $LS_w + RS_w + NS_w = g$ admits a unique solution $S_w \in H$. Furthermore, if the solution S_w can be represented in a series form given by $S_w = \sum_{n=0}^{\infty} S_{wn}\lambda^n$, then the Adomian decomposition scheme corresponding to the functional equation under consideration converges strongly to a $S_w \in H$, which is the unique solution to the functional equation.

Proof. Verification of hypothesis (H_1)

$$TS'_w - TS^{*\prime}_w = -\frac{\partial^2}{\partial X^2}\left(S_w^{-1'} - S_w^{-1'}\right)$$
$$\left(TS'_w - TS^{*\prime}_w, S'_w - S^{*\prime}_w\right) \tag{19}$$
$$= \left(-\frac{\partial^2}{\partial X^2}\left(S_w^{-1'} - S_w^{-1*'}\right), S'_w - S^{*'}_w\right).$$

Since $\partial^2/\partial X^2$ is a differential operator in H, then there exist constant "δ" such that,

according to Schwartz inequality, we get

$$\left(\frac{\partial^2}{\partial X^2}\left(S_w^{-1\prime} - S_w^{-1*\prime}\right), S'_w - S^{*\prime}_w\right)$$
$$\leq \delta\left\|S_w^{-1'} - S_w^{-1*'}\right\|\left\|S'_w - S^{*'}_w\right\|. \tag{20}$$

Now, we use mean value theorem, then we have

$$\left(\frac{\partial^2}{\partial X^2}\left(S_w^{-1\prime} - S_w^{-1*\prime}\right), S'_w - S^{*\prime}_w\right)$$
$$\leq \delta\left\|S_w^{-1'} - S_w^{-1*'}\right\|\left\|S'_w - S^{*'}_w\right\| \leq \frac{\delta}{M^2}\left\|S'_w - S^{*'}_w\right\|^2, \tag{21}$$

for $\|S'_w\| \leq M$ and $\|S^{*\prime}_w\| \leq M$.

Therefore,

$$\left(-\frac{\partial^2}{\partial X^2}\left(S_w^{-1\prime} - S_w^{-1*\prime}\right), S'_w - S^{*\prime}_w\right) \geq \frac{\delta}{M^2}\left\|S'_w - S^{*'}_w\right\|^2. \tag{22}$$

Substituting (22) in (19),

$$\left(TS'_w - TS^{*\prime}_w, S'_w - S^{*\prime}_w\right) \geq k\left\|S'_w - S^{*\prime}_w\right\|^2, \tag{23}$$

where $k = \delta/M^2$. Hence we find the hypothesis (H_1).

For hypothesis (H_2),

$$\left(TS'_w - TS^{*\prime}_w, V\right) = \left(-\frac{\partial^2}{\partial X^2}\left(S_w^{-1\prime} - S_w^{-1*\prime}\right), V\right)$$
$$\leq \frac{1}{M^2}\left\|S_w - S^{*\prime}_w\right\|\|V\| \tag{24}$$
$$= C(M)\left\|S_w - S^{*'}_w\right\|\|V\|,$$

where $C(M) = 1/M^2$, and, therefore, (H_2) holds. The proof is complete.

Remark 2. We note that the constant $C(M)$ is function of M, and the linearity of T allows us to prove (H_2). Furthermore, since every linear continuous operator is hemicontinuous, the operator T is hemicontinuous.

6. Simulations Results

Using the analysis of Adomian Decomposition Method, (8) can be written in operator form $L_T S_w$ as

$$L_T S_w(X,T) = L_X\left(NS_w(X,T)\right). \tag{25}$$

Operating the inverse operators on both sides of (25), it gives

$$S_w(X,T) = S_{w0}(X) + L_T^{-1}\left[L_X\left(NS_w(X,T)\right)\right], \quad (26)$$

where $NS_w(X,T) = S_w(\partial S_w/\partial X)$ and $S_{w0}(X)$ can be solved subject to the corresponding initial condition (9).

It is well known from (11) that the solution of (8) can be written in series form as follows:

$$S_w(X,T) = \sum_{n=0}^{\infty} S_{wn}(X,T), \quad (27)$$

where $S_{w0}, S_{w1}, S_{w2}, \ldots$ are the saturations of different fingers at any distance X and at any time $T > 0$ and the nonlinear term can be represented as $NS_w(X,T) = \sum_{n=0}^{\infty} A_n$, where A_n's are the Adomian's special polynomials to be determined and defined by (12).

Following the analysis of Adomian decomposition method as discussed in [7, 8], for the determination of the components $S_{w,n}(X,T)$ of $S_w(X,T)$, we set the recursive relation as

$$\sum_{n=0}^{\infty} S_{w,n}(X,T) = \left(X^2+1\right)^{-1/2} + L_T^{-1}\left[L_X\left(\sum_{n=0}^{\infty} A_n\right)\right], \quad (28)$$

where $S_{w0} = (X^2+1)^{-1/2} = S_w(X,0)$ from (9) and $S_{w,k+1} = L_T^{-1}((A_k)_x)$.

6.1. Adomian Polynomials Are as Follows

$$\begin{aligned} A_0 &= S_{w,0}^{-2}(S_{w,0})_X, \\ A_1 &= S_{w,0}^{-2}(S_{w,1})_X - 2S_{w,0}^{-3}S_{w,1}(S_{w,0})_X, \\ A_2 &= S_{w,0}^{-2}(S_{w,2})_X - 2S_{w,0}^{-3}S_{w,1}(S_{w,1})_X \\ &\quad -2S_{w,0}^{-3}S_{w,2}(S_{w,0})_X + 3S_{w,0}^{-4}S_{w,1}^{2}(S_{w,0})_X \\ &\ \ \vdots \end{aligned} \quad (29)$$

On Substituting (29) into (28), it gives the components of saturation of water as

$$\begin{aligned} S_{w,0}(X,T) &= \left(X^2+1\right)^{-1/2}, \\ S_{w,1}(X,T) &= L_T^{-1}\left((A_0)_X\right) = -\left(X^2+1\right)^{-3/2}T, \\ S_{w,2}(X,T) &= L_T^{-1}\left((A_1)_X\right) = \left(X^2+1\right)^{-5/2}\left(1-2X^2\right)\frac{T^2}{2!}, \\ S_{w,3}(X,T) &= L_T^{-1}\left((A_2)_X\right) \\ &= -\left(X^2+1\right)^{-7/2}\left(1-10X^2+4X^4\right)\frac{T^3}{3!} \\ &\ \ \vdots \end{aligned} \quad (30)$$

and so on.

In view of (30), the approximate solution in a series form is given by

$$\begin{aligned} \sigma_n(X,T) &= \left(X^2+1\right)^{-1/2} \\ &\quad - \left(X^2+1\right)^{-3/2}T + \left(X^2+1\right)^{-5/2}\left(1-2X^2\right)\frac{T^2}{2!} \\ &\quad - \left(X^2+1\right)^{-7/2}\left(1-10X^2+4X^4\right)\frac{T^3}{3!} + \cdots, \end{aligned} \quad (31)$$

which converges to $S_w(X,T)$, and the closed-form solution of (8) as

$$S_w(X,T) = \left(X^2+e^{2T}\right)^{-1/2}. \quad (32)$$

It is interesting to note that $S_w(X,T)$ in (32) has the following asymptotic behavior:

$$\lim_{X\to\infty} S_w(X,T) = 0, \qquad \lim_{T\to\infty} S_w(X,T) = 0, \quad (33)$$

and the flux satisfying

$$\lim_{X\to\infty} S_w^{-2}(S_w)_X = -1 \quad (34)$$

implies that the saturation of water decreases with distance and time increases. Moreover, this solution vanishes as X approaches infinity.

An explicit general exact solution of the form

$$S_w(X,T) = C^{1/2}\left(C^2(X+B)^2 + Ae^{2CT}\right)^{-1/2} \quad (35)$$

by using a generalized conditional symmetry method, where A, B, and $C > 0$ are arbitrary constants. Moreover, an implicit solution was obtained for $C < 0$.

7. Conclusion

In this paper, we have proved the convergence of the Adomian decomposition scheme for the case of instability phenomena. The solution (31) of (8) gives the saturation of water $S_w(X,T)$ as a function of distance and time in displaced oil zone

TABLE 1: Saturation versus time keeping distance fixed: Adomian decomposition method.

X/T	0.1	0.2	0.3	0.4	0.5	0.6	0.7	0.8	0.9	1
0.1	0.901153	0.815959	0.738585	0.668161	0.603819	0.544687	0.489897	0.43858	0.389865	0.342883
0.2	0.890376	0.807984	0.732875	0.664522	0.602394	0.545964	0.494703	0.448083	0.405574	0.366648
0.3	0.873241	0.795169	0.723512	0.658171	0.59905	0.54605	0.499073	0.458021	0.422796	0.393301
0.4	0.850831	0.778173	0.7108	0.649006	0.593088	0.543343	0.500065	0.463552	0.434098	0.412001
0.5	0.8244	0.757806	0.69522	0.637214	0.584359	0.537229	0.496395	0.462431	0.435908	0.417399
0.6	0.795209	0.734933	0.677375	0.623243	0.573248	0.528099	0.488505	0.455177	0.428824	0.410155
0.7	0.764408	0.710386	0.657894	0.60766	0.560411	0.516877	0.477784	0.443861	0.415836	0.394437
0.8	0.732959	0.684903	0.637365	0.591013	0.546511	0.504525	0.465723	0.430769	0.400329	0.375069
0.9	0.701618	0.659097	0.616291	0.573761	0.532069	0.491775	0.45344	0.417626	0.384893	0.355803
1	0.670941	0.63345	0.595075	0.556257	0.51744	0.479065	0.441573	0.405408	0.37101	0.338822

TABLE 2: Generalized conditional symmetry method.

X/T	0.1	0.2	0.3	0.4	0.5	0.6	0.7	0.8	0.9	1
0.1	0.901156	0.816	0.738794	0.668819	0.605418	0.547987	0.495974	0.448876	0.406234	0.367631
0.2	0.890375	0.80797	0.732818	0.664376	0.602117	0.545535	0.494154	0.447526	0.405232	0.366888
0.3	0.873237	0.795098	0.723174	0.657164	0.596732	0.541521	0.491165	0.445301	0.403579	0.365659
0.4	0.850824	0.778069	0.710289	0.64745	0.589432	0.536047	0.487069	0.442243	0.401298	0.36396
0.5	0.824393	0.757701	0.694693	0.635572	0.580427	0.529247	0.481951	0.438401	0.398421	0.36181
0.6	0.795204	0.734852	0.676957	0.621905	0.569962	0.521278	0.47591	0.433839	0.394987	0.359232
0.7	0.764406	0.710342	0.65765	0.606837	0.558295	0.512308	0.469055	0.428627	0.391041	0.356257
0.8	0.732959	0.684895	0.637302	0.59074	0.545684	0.502511	0.461501	0.42284	0.386633	0.352913
0.9	0.70162	0.659119	0.616378	0.57396	0.532376	0.492061	0.453367	0.416557	0.381812	0.349235
1	0.670944	0.633492	0.595268	0.556799	0.518596	0.481119	0.444765	0.409856	0.376631	0.345258

causing instability at the interface up to distance $X = 1$ due to an external force of injected water through injecting well during oil recovery process. Equation (31) is an approximate solution containing four terms of an infinite series. The solution contains negative power terms which shows that the saturation of displaced water, that is instability of fingers decreases as the distance X from the interface increases, and it is obvious for the instability phenomena. The instability of oil and water zone is the maximum at the interface $X = 0$ and becomes stable as it away from the interface $X = 0$, that is, likely to be stable at the boundary point $X = 1$. The absolute errors for saturation of water S_w have been calculated. As seen from Tables 1 and 2, Adomian decomposition method gives a very good approximation to the partial exact solution by using only three terms of the decomposition series.

References

[1] A. E. Scheidegger, *Th Physics of Flow Through Porous Media*, University of Toronto Press, 1960.

[2] S. E. Buckley and M. C. Leverett, "Mechanism of fluid displacement in sands," *Transactions of the American Institute of Mining*, vol. 146, pp. 107–110, 1942.

[3] T. Oroveanu, "Scurgerea fluiidelor prin medii poroase neomogene," *Editura Academiei Republicii Populare Romine*, vol. 92, no. 328, 1963.

[4] A. P. Verma, "Statistical behavior of fingering in a displacement process in heterogeneous porous medium with capillary pressure," *Canadian Journal of Physics*, vol. 47, no. 319, 1969.

[5] G. Venkateswar, "On the flow of immiscible liquids in a heterogeneous porous medium with capillary pressure and connate water saturation," *International Journal of Engineering Science*, vol. 13, pp. 161–171, 1975.

[6] G. Adomian, *Solving Frontier Problems in Physics: the Decomposition Method*, Kluwer Academic, Dordrecht, The Netherlands, 1994.

[7] K. Abbaoui and Y. Cherruault, "Convergence of Adomian's method applied to differential equations," *Computers and Mathematics with Applications*, vol. 28, no. 5, pp. 103–109, 1994.

[8] L. Gabet, "The theoretical foundation of the Adomian method," *Computers and Mathematics with Applications*, vol. 27, no. 12, pp. 41–52, 1994.

[9] Y. Cherruault, "Convergence of Adomian's method," *Kybemetes*, vol. 18, no. 2, pp. 31–38, 1989.

[10] Y. Cherruault and G. Adomian, "Decomposition methods: a new proof of convergence," *Mathematical and Computer Modelling*, vol. 18, no. 12, pp. 103–106, 1993.

[11] T. Mavoungou, "Convergence of Adomian's method and applications to non-linear partial differential equations," *Kybernetes*, vol. 21, no. 6, pp. 13–25, 1994.

[12] A. M. Wazwaz, *Partial Diff rential Equations: Methods and Applications*, Balkema, Tokyo, Japan, 2002.

7

Interval Arithmetic for Nonlinear Problem Solving

Benito A. Stradi-Granados

Department of Materials Science and Engineering, Institute of Technology of Costa Rica, Cartago 07050, Costa Rica

Correspondence should be addressed to Benito A. Stradi-Granados; bstradi@itcr.ac.cr

Academic Editor: A. Nazli Gundes

Implementation of interval arithmetic in complex problems has been hampered by the tedious programming exercise needed to develop a particular implementation. In order to improve productivity, the use of interval mathematics is demonstrated using the computing platform INTLAB that allows for the development of interval-arithmetic-based programs more efficiently than with previous interval-arithmetic libraries. An interval-Newton Generalized-Bisection (IN/GB) method is developed in this platform and applied to determine the solutions of selected nonlinear problems. Cases 1 and 2 demonstrate the effectiveness of the implementation applied to traditional polynomial problems. Case 3 demonstrates the robustness of the implementation in the case of multiple specific volume solutions. Case 4 exemplifies the robustness and effectiveness of the implementation in the determination of multiple critical points for a mixture of methane and hydrogen sulfide. The examples demonstrate the effectiveness of the method by finding all existing roots with mathematical certainty.

1. Introduction

There are a large number of problems that require the computation of stationary points and roots of equations. These are found in the fields of optimization [1], economics and finance [2, 3], thermodynamics [4, 5], applied mathematics [6], and similar contributions over the last thirty years. The application of interval arithmetic involves from algebraic equations with known solutions to more complicated systems representing physical phenomena. Among the earlier problems, second- and third-order polynomial problems serve to illustrate the effectiveness of the implementation. Similarly, more complicated multidimensional problems serve to illustrate stability and robustness of the implementation. In particular, the determination of critical points is of interest. Critical point computations are a well-known highly nonlinear problem that has been studied for a long time. The literature is endowed with some worthy analyses of the problem. Michelsen and Heidemann [7] discuss the calculation of critical points from cubic two-constant equations of state. They numerically determine the critical points of mixtures solving the highly nonlinear critical point equations. Hoteit et al. [8] claim an efficient algorithm based on bisection, secant, and inverse quadratic programming methods where their method is apparently faster but incapable of handling the presence of multiple critical points without restarting the program after each critical point determination. Nichita and Gomez [9] discuss the use of the tunneling global optimization method to determine all the of global minima where they concentrate on the determination of the critical points of mixtures. They use temperature and molar volume as primary variables. Their method lacks the strength of being able to stop when all critical points have been found, and this follows from the fact that at the beginning of the process the number of critical points to be determined needs to be specified. Michelsen [10] develops some earlier ideas that deal with the computation of phase envelopes with use of finite differences. Michelsen [11] discusses the isothermal flash separation and critical point computations with the use of tangent plane analysis. Michelsen and Kistenmacher [12] describe the disadvantage of using composition-dependent binary interaction coefficients in computations. Michelsen and Heidemann [13] present the calculation of tricritical points using tangent-plane distance analysis with some extra algebra in the mathematical development. The results are generated using the Peng-Robinson and Soave-Redlich-Kwong (SRK) equations of state. Michelsen [14] illustrates the determination of the critical points and phase boundaries using the real-arithmetic Newton-Raphson method with temperature and pressure as variables; in his analysis rather close initial estimates to the solution are needed for convergence.

There are also other more general contributions such as that of Cai et al. [15] who developed expressions for the spinodal criterion, critical criterion, and various stability tests for systems containing one discrete component and one polydisperse polymer. Lindvig et al. [16] propose the EFV (Entropic-FV) model for predicting the miscibility behavior of paints. Cismondi and Michelsen [17] use a Newton procedure for the calculation of critical lines, critical endpoints, and three-phase lines for binary mixtures. Carstensen and Petković [18] propose the use of a hybrid method combining normal (Nourein's method) and interval arithmetic to arrive to the solution of polynomial equations efficiently. Maranas and Floudas [19] propose an approach based on convex lower bounding and domain partitioning to achieve convergence to all solutions within a domain. Stradi et al. [4] discuss the computation of critical points using the INTBIS library [20].

This paper presents the solution of nonlinear problems using interval arithmetic in INTLAB (INTerval LABoratory) [21] by using the interval-Newton Generalized-Bisection method (IN/GB). The advantage of this approach is that INTLAB is developed over MATLAB [22], a well-known and powerful computational package widely used in mathematics and engineering applications [23–25]. Although, with the cost of interpretation overhead [26], INTLAB facilitates interval computations by using the MATLAB user interface and debugging tools, this makes production times smaller and results generation faster [27]. Section 2 of the paper provides a description of the problems to be solved in the paper, Section 3 proposes the methodology to solve the problems using the IN/GB method, Section 4 presents the results and discussion comparing the IN/GB method with the simple bisection method, and Section 5 highlights the conclusions derived from this study.

2. Problem Description

This section starts with two examples to demonstrate the use and performance of the IN/GB method. The first deals with two second-order equations where the solution is to be found over a relatively large domain. The second illustrates a particularly important case of a repeated root that causes the first derivative to be zero, and consequently it would traditionally create problems during the process of solution for the real-arithmetic Newton-Raphson method [23]. The third and fourth examples are related to the behavior of fluids in the two phase and critical regions.

Consider the general case where all of the roots to the nonlinear problem $f(x) = 0$ have to be found within a given domain in the absence of an initial guess.

Case 1. The first problem is a set of two second-order equations as follows:

$$\begin{aligned}(xy)^2 + 4x - 4 &= 0,\\ y^2 + 4xy - 4 &= 0.\end{aligned} \tag{1}$$

The domain for the search is the same for both variables: $x \in [-100, 100]$, $y \in [-100, 100]$. Three solutions are found for these two equations with mathematical certainty in a single run of the IN/GB method implemented over the INTLAB program.

Case 2. The second problem is a third-order polynomial with three real roots, where one of the roots is repeated:

$$x^3 - 5x^2 + 7x - 3 = 0. \tag{2}$$

This is an important example because the derivative of the function is zero at the repeated root. The traditional real-arithmetic Newton-Raphson method [23] would fail if we needed to apply the method near the root where the derivative is close to zero. The domain of the search is $x \in [-100, 100]$.

The interval Newton is expected to have problems because of its reliance on the interval derivative of the function to find a solution, but the combination with the bisection method should solve the problem.

Case 3. The third problem deals with the determination of volume solutions for the Peng-Robinson equation of state [28] for carbon dioxide at a given temperature and pressure, which is given by the following expression that related pressure, temperature, and volume:

$$P - \frac{RT}{(v-b)} + \frac{a}{(v*(v+b)+b*(v-b))} = 0. \tag{3}$$

This can also be written in the form

$$\begin{aligned}&P(v-b)(v*(v+b)+b*(v-b))\\ &\quad - RT(v*(v+b)+b*(v-b)) + a*(v-b) = 0,\\ a &= \frac{0.45724R^2T_C^2}{P_C\left(1+\left(0.3764+1.54226w-0.2699w^2\right)\left(1-T_r^{1/2}\right)\right)^2},\\ b &= \frac{0.07780RT_C}{P_C}.\end{aligned} \tag{4}$$

The necessary conditions for this problem are the following:

$$\begin{aligned}T_C &= 304.21\ \mathrm{K},\\ P_C &= 73.83\ \mathrm{bar},\\ w &= 0.224,\\ T &= 301\ \mathrm{K},\\ P &= 68.18 \ * 10^5\ \mathrm{Pa},\\ R &= 83.14\,(\mathrm{Pa\,cm^3/gmol\,K}),\end{aligned}$$

and the interval for the volume search is given by $v \in [1.1b, 16b]$.

This is a classic chemical engineering problem where, for a given equation of state, there may be more than one molar volume that satisfies the equation of state at a given temperature and pressure. In particular, if the conditions are such that two phases coexist, like in this case, then there will be three solutions. The traditional interpretation is that the solutions represent the liquid molar volume (smallest volume solution) and the vapor molar volume (largest volume solution). The middle volume solution is an unstable solution, and thus not observable.

Case 4. This is a highly nonlinear problem subject to analysis by different research groups for a number of years [4, 7, 14, 17]. In this case, there is a mixture of methane and hydrogen sulfide where the problem is to determine all of the critical points for a given composition and domain ranges for temperature and molar volume using the Redlich-Kwong equation of state [28]. The problem was solved in a previous development with the use of the INTBIS library and some extensive programming.

The domains of the variables under study are indicated as follows:

$$T \in [180, 600]K,$$
$$v \in [1.1, 4]b,$$
$$\Delta x_1 \in [0, 1],$$
$$\Delta x_2 \in [-1, 1].$$

The criticality conditions are written as follows:

$$\sum_{j=1}^{C} A_{ij}\Delta n_j = \frac{RT}{n}\left(\frac{\Delta n_i}{y_i} + F_1\left(\beta_i\overline{N} + \overline{\beta}\right) + \beta_i F_1^2\overline{\beta}\right) + \frac{a}{bn}\left(\beta_i\overline{\beta}F_3 - \frac{F_5}{a}\sum_{j=1}^{C} a_{ij}\Delta n_j + F_6\left(\beta_i\overline{\beta} - \alpha_i\overline{\beta} - \overline{\alpha}\beta_i\right)\right), \quad i = 1,\ldots,C, \tag{5}$$

$$\sum_{i=1}^{C}\sum_{j=1}^{C}\sum_{k=1}^{C} A_{ijk}\Delta n_i\Delta n_j\Delta n_k = \frac{RT}{n^2}\left(-\sum_{i=1}^{C}\frac{\Delta n_i^3}{y_i^2} + 3\overline{N}\left(\overline{\beta}F_1\right)^2 + 2\left(\overline{\beta}F_1\right)^3\right) + \frac{a}{n^2 b}\left(3\overline{\beta}^2\left(2\overline{\alpha} - \overline{\beta}\right)\left(F_3 + F_6\right) - 2\overline{\beta}^3 F_4 - 3\overline{\beta}\overline{a}F_6\right). \tag{6}$$

The mass balance for a mixture of constant composition is given by

$$\sum_{i=1}^{C}\Delta n_i^2 - 1 = 0. \tag{7}$$

Meanings are given in the appendix.

3. Methodology

We employed the interval mathematics platform INTLAB to implement the interval-Newton Generalized-Bisection (IN/GB) method for the solution of nonlinear problems. The advantage of this procedure is that INTLAB is user friendly and runs over MATLAB; this makes INTLAB a very promising platform by putting the tools of MATLAB at the disposal of the INTLAB programmer. In the past, one of the major problems with the use of interval mathematics was the use of libraries that required extensive programming and lacked utilities for fast debugging. As a result, large time investments needed to be made in programming and debugging prior to the implementation and generation of first results, a situation that obviously derives in making interval arithmetic less attractive to a larger scientific audience. However, with INTLAB, the situation is different: program sizes are much smaller, debugging is more efficient, and results are generated more rapidly.

3.1. The Interval-Newton Generalized-Bisection Method. An interval-Newton Generalized-Bisection method (IN/GB) is presented to demonstrate the use of the INTLAB platform for nonlinear problems of different levels of complexity.

The interval IN/GB method is a powerful computational approach that allows for the computation of all of the roots of a nonlinear problem without the need for initial guesses for the variables [1, 29, 30].

The IN/GB method requires only that we specify the domain of the variables of interest. The program tests a sequence of enclosures determined through the IN/GB method. If there is a root, then the IN/GB method determines a very thin slice in which the root lies. On the other hand, if there is no root, then the algorithm also determines with mathematical certainty the absence of solutions.

The most important component part of the IN/GB method is the Newton method in interval arithmetic. This is written as follows:

$$J(X)(N - x) = -f(x). \tag{8}$$

This equation is of the form $AX = b$, and consequently suitable for solution with simultaneous linear equations solvers.

In this equation, capital letters are intervals and small case letters are real numbers. $J(X)$ is the interval extension of the Jacobian matrix, which is determined by evaluating the Jacobian matrix, J, using the interval domain X, rather than the real variable x, where x is generally the midpoint of X. This system of equations is solved using the Gauss-Seidel method [1], or a similar algorithm to determine N, where N is the image X generated by the application of the IN/GB method. If N and X intersect, then the process is repeated, noting that in successive applications it is the intersection of N and X ($N \cap X$) that serves as initial interval. The process continues until a sufficiently thin interval around the solution is derived. Other implementations, when sufficiently closed to the solution, switch to the real-arithmetic Newton-Raphson method to accelerate convergence [20].

TABLE 1: Interval-Newton Generalized-Bisection method results and actions.

Operation	Result	Comment	Action
$N \cap X$	I	I smaller in size	Continue with interval Newton
$N \cap X$	X, other	No reduction in the original domain size	Bisect

TABLE 2: Results for Case 1.

Domain	Equations	Roots	Subset	Method
[−100, 100]	$(x_1x_2)^2 + 4x_1 - 4 = 0$	$x_1 = -0.68443924$	647	N
	$x_2^2 + 4x_1x_2 - 4 = 0$	$x_2 = 3.79247516$	809	B
		$x_1 = 0.465822965$	654	N
		$x_2 = -3.13799232$	814	B
		$x_1 = 0.84600538$	677	N
		$x_2 = 0.92770308$	819	B
			Total	
			685	N
			820	B

N: IN/GB method, B: Bisection method.

If N is a subset of X, then there is a single root in the domain of interest [29]. Similarly, if N does not intersect X, then there is no root in the domain, and a new subdomain is tested for the presence of roots.

It is important to indicate at this point that there are several problems not generally mentioned in the literature that may occur when using the IN/GB method. If the interval X is too large, then the image N will contain and be larger than the original interval X, generating the problem where the image is larger than the original interval and no convergence is achieved with the interval-Newton portion of the method alone. The other problem is where the image N is equal to X; in this case the search will not advance, and no solution will be achieved within the permissible execution time. In both cases, the action taken is to proceed with the partition of the box, as shown in Table 1, and restart the search for each subdomain generated. The partition is done with the Bisection method [23] over each dimension of the domain followed by a range test applied to all subdomains generated.

Application of the Bisection method generates two intervals from each interval that is bisected. Consequently, in order to explore all of the possibilities, the combinations of these intervals are needed to determine the existence and values of all possible solutions.

The existence of a root is ascertained by evaluating the original function over the interval, or subdomain, of interest, using interval arithmetic [30]. If the resulting interval contains zero, $0 \subset f(X)$, then there may exist at least one root in that domain. This process of testing for the existence of a root in the domain of interest is called *range test*. If the evaluation of the function over the interval of interest, X, results in an interval image that does not contain zero, then there is a mathematical guarantee that there is no root within the interval X.

For example, if three intervals are bisected, then six subintervals are generated, and eight combinations of them would need to be tested first in the *range test* to ascertain whether there may be a root in the subdomain, and second in the IN/GB method search for the solution.

Generally, multiple applications of the Bisection method, part of the IN/GB method, are needed when large variable domains are used. This situation is particularly true when zeros are contained in diagonal elements of the interval Jacobian expression, J. The presence of these zeros generates interval images, N, in the IN/GB method with infinite spans for some or all variables. Consequently, bisection is needed to divide and eliminate those subdomains where infinite spans are generated. These computational problems are managed efficiently in INTLAB. In INTLAB, the programming environment allows for processing warnings identified by the flags infinite (inf) and not-a-number (NaN) without an error that would stop the program execution. The flag inf would identify quantities that exceed real number representation such as e^{1000} and also results generated by the division of a number by zero: 1.0/0.0. The flag NaN would identify quantities that are mathematically undefined such as 0.0/0.0 or intervals that do not intersect. Consequently, if these flags occur, then bisection is applied because they are indicative of problems with the interval-Newton portion of the algorithm. This capacity to handle numerically undefined quantities saves extensive programming time and debugging effort invested on the solution of a particular problem.

The search for a solution can only end with one of two outcomes, either a root is found within an interval or no root is found with mathematical certainty. The main steps of the algorithm are represented in Figure 8.

4. Results and Discussion

4.1. Case 1. Table 2 presents the results obtained using the IN/GB and the Bisection methods applied to Case 1. The IN/GB method determined all of the roots and searched the

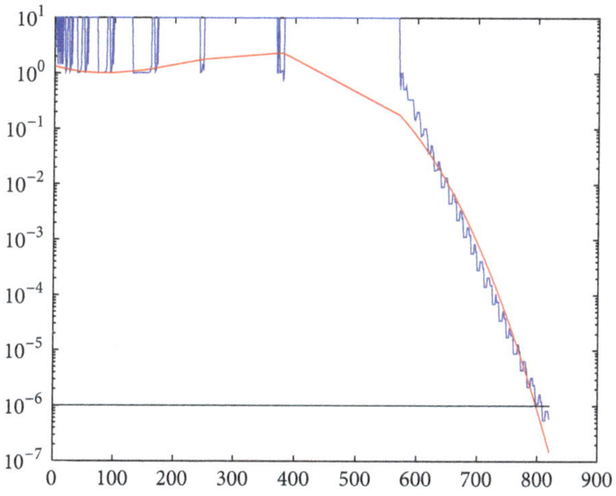

FIGURE 1: Evolution of the relative error (y-axis) with the subinterval number for the Bisection method for Case 1.

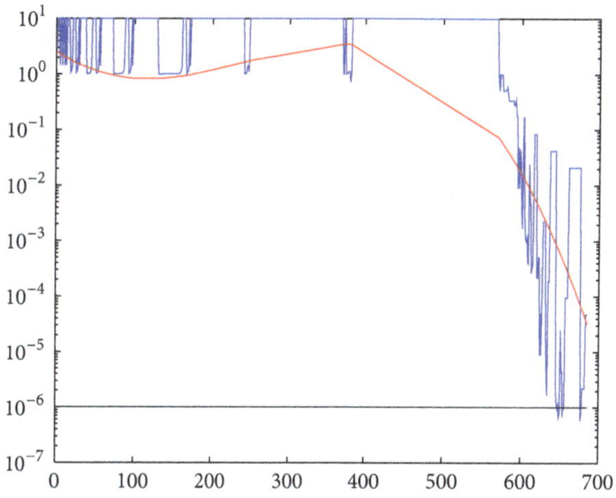

FIGURE 2: Evolution of the relative error (y-axis) with the subinterval number for the IN/GB method for Case 1.

entire domain with 685 subdivisions. The Bisection method with interval arithmetic was applied, and convergence to all roots was achieved with 820 subdivisions of the original interval. It is important to mention that, before applying either the IN/GB method or the Bisection method, a range test is applied to the subdomain under consideration to determine whether a root may be found within the subdomain. Three solutions are found for these two equations. The algorithm developed found three solutions without the need for initial guesses over a fairly large domain. This case is a baseline case where the solutions can be calculated also with the real-arithmetic Newton method, but then at least three different initial guesses would need to be provided.

Figure 1 shows the evolution of the error with the subinterval number for the Bisection method; Figure 2 shows the evolution of the error with the subinterval number for the IN/GB method. The error limit is represented with a horizontal line at $1 * 10^{-6}$, and the trends are approximated with a third-order polynomial in a semilog plot for the Bisection and the IN/GB methods. The interval-Newton part of the IN/GB method may generate some very large errors particularly when the derivative, or the Jacobian matrix, contains a zero in the diagonal-element intervals. Consequently, in order to keep the scale to a legible size, large errors described as inf and nondefined quantities described as NaN appear with an error value of 10 on the graph's y-axis. For both the bisection method and the IN/GB method, the solutions are found toward the end of the search of subdomains.

4.2. Case 2. Table 3 presents the results for Case 2, where the IN/GB method determines all of the roots to this third-order equation using 17723 subdivisions to complete the search over the entire domain. The bisection method requires 17981 subdivisions to search the entire domain and find all of the roots. It is important to notice that the first root is found much faster by the IN/GB method by a factor of more than a hundred times over the bisection method. However, the convergence speed is smaller at the second

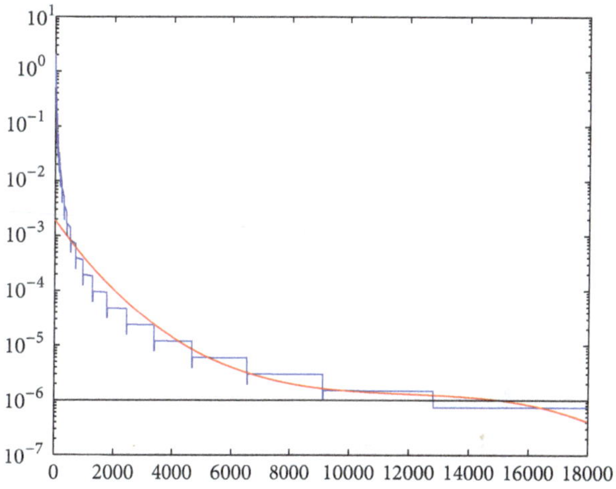

FIGURE 3: Evolution of the relative error (y-axis) with the subinterval number for the Bisection method for Case 2.

root where both algorithms need closely the same number of subdivisions. For this simple example, the roots are evident and can be determined by inspection. However, if using the real-arithmetic Newton-Raphson method, there would be no immediate way to determine that one of the roots is repeated, and this situation would require additional time to test additional initial guesses or execute a polynomial deflation procedure with the roots already found.

Figure 3 shows the evolution of the error with the subinterval number for the Bisection method. Figure 4 shows the evolution of the error with the subinterval number for the IN/GB method.

In the case of the Bisection method, it is particularly different the convergence profile, compared to Case 1; both cases have been approximated with a third-order polynomial determined with a least-squares fit. The IN/GB method presents some very large errors at the beginning of the search, and a log-log plot is used to better depict this behavior. Large error values are constrained to a value of 10 for graphical

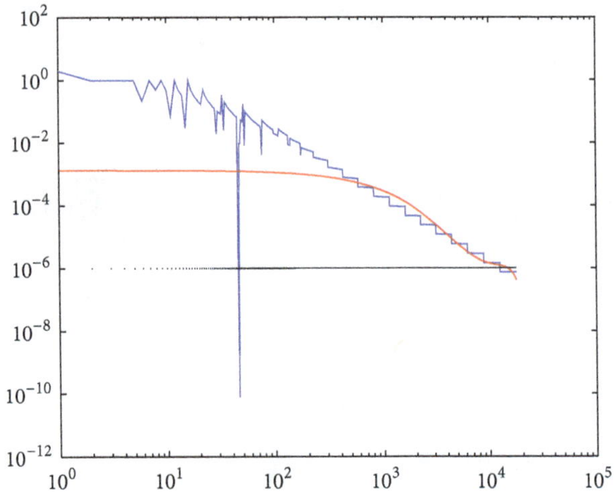

FIGURE 4: Evolution of the relative error (y-axis) with the subinterval number for the IN/GB method for Case 2.

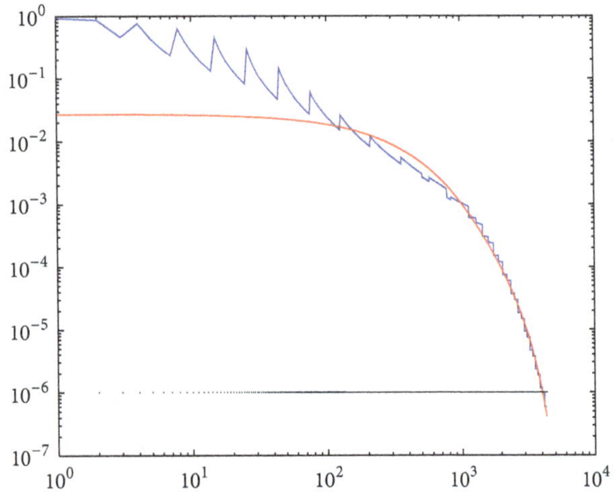

FIGURE 5: Evolution of the relative error (y-axis) with the subinterval number for the Bisection and IN/GB methods for Case 3.

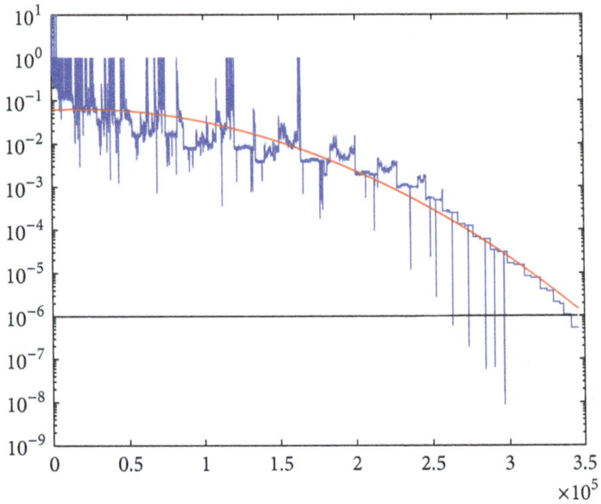

FIGURE 6: Evolution of the relative error (y-axis) with the subinterval number for the IN/GB method for Case 4.

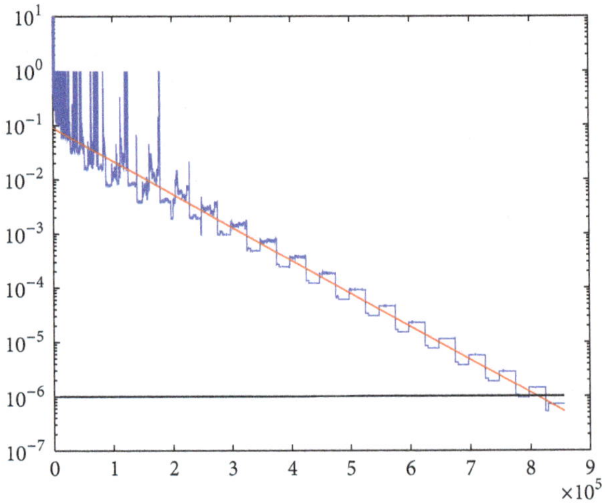

FIGURE 7: Evolution of the relative error (y-axis) with the subinterval number for the Bisection method for Case 4.

clarity, and they are indicative of inf and NaN results; in these cases, the algorithm switches from the interval-Newton to the Bisection method. In this case, the roots are found by IN/GB method both at the beginning and at the end of the process of testing all generated subintervals as indicated by the values below $1 * 10^{-6}$.

4.3. Case 3. Table 4 presents the results for Case 3. This is a particularly interesting example because it exposes some of the difficulties in interval computations. In this example, the interval-Newton method does not result in an interval, N, with the desired characteristics: (1) that N be of smaller size than the original interval, X, and (2) that N and X intersect. During the search process with the interval-Newton, this portion of the IN/GB method generates NaN or inf messages, and the program switches to the Bisection method. This point is very important because it clearly demonstrates that if for some reason the interval-Newton portion of the algorithm is incapable of dealing with the problem then the other half of the algorithm (i.e., the Bisection method) takes over and provides a solution, making the IN/GB a very efficient method.

Once an interval is bisected, a range test on each subdomain created is performed, and if a root is detected (i.e., $0 \in f(X)$), then that interval is stacked for further processing. This process will continue until either each subdomain is eliminated for not containing zero (i.e., $0 \notin f(X)$) or a root is found within the subdomain with a prescribed error tolerance. In this example, the IN/GB method finds the solutions and scans the entire domain by dividing the domain in 4332 subsets. Consequently, the graph in Figure 5 for the Bisection and the IN/GB methods is the same.

4.4. Case 4. Table 5 presents the results for Case 4. This is a more elaborate example because for a multicomponent mixture there is no procedure to determine *a priori* whether

TABLE 3: Results for Case 2.

Domain	Equation	Roots	Subset	Method
$[-100, 100]$	$x^3 - 5x^2 + 7x - 3 = 0$	$x = 3.00000000$	46	N
		$x = 2.99997777$	9121	B
		$x = 0.99807270$	12543	N
		$x = 0.99807270$	12801	B
		(double root)		
			Total	
			17723	N
			17981	B

TABLE 4: Results for Case 3.

Domain	Equations	Roots	Subset	Method
$[29.31, 426.43]$	$\text{alpha} = (v * (v + b) + b * (v - b))$	$v = 100.0632$	4031	N
	$\text{beta} = (v - b)$	$v = 100.0632$	4031	B
	$P * \text{alpha} * \text{beta} - \text{RT} * \text{alpha} + a * \text{beta} = 0$	$v = 160.6770$	4181	N
	$T = 301K$	$v = 160.6770$	4181	B
	$P = 68.18\,\text{bar}$	$v = 79.6437$	4238	N
	$b = 26.6520264\,\text{cm}^3/\text{gmol}$	$v = 79.6437$	4238	B
			Total	
			4332	N
			4332	B

TABLE 5: Results for Case 4.

Domain	Equations	Roots	Subset	Method
$[0.30, 1]\ T$	Redlich-Kwong-Soave equation of state	$T = 0.48143566$	168781	N
$[0.275, 1]\ v$	Mixture of methane and hydrogen sulfide	$v = 0.56426799$	829620	B
$[0, 1]\ \Delta n_1$	Find all of the critical points in the domain	$\Delta n_1 = 0.08460040$		
$[-1, 1]\ \Delta n_2$	T scaled by $600K$	$\Delta n_2 = -0.99641496$		
$x_1 = 0.49\ CH_4$	v scaled by $4b$			
$x_2 = 0.51\ H_2S$				
		$T = 0.34691664$	237185	N
		$v = 0.29124484$	824978	B
		$\Delta n_1 = 0.6814699$		
		$\Delta n_2 = -0.73189541$		
		$T = 0.38612396$	285239	N
		$v = 0.36953418$	779696	B
		$\Delta n_1 = 0.52293204$		
		$\Delta n_2 = -0.85237438$		
			Total	
			345733	N
			856970	B

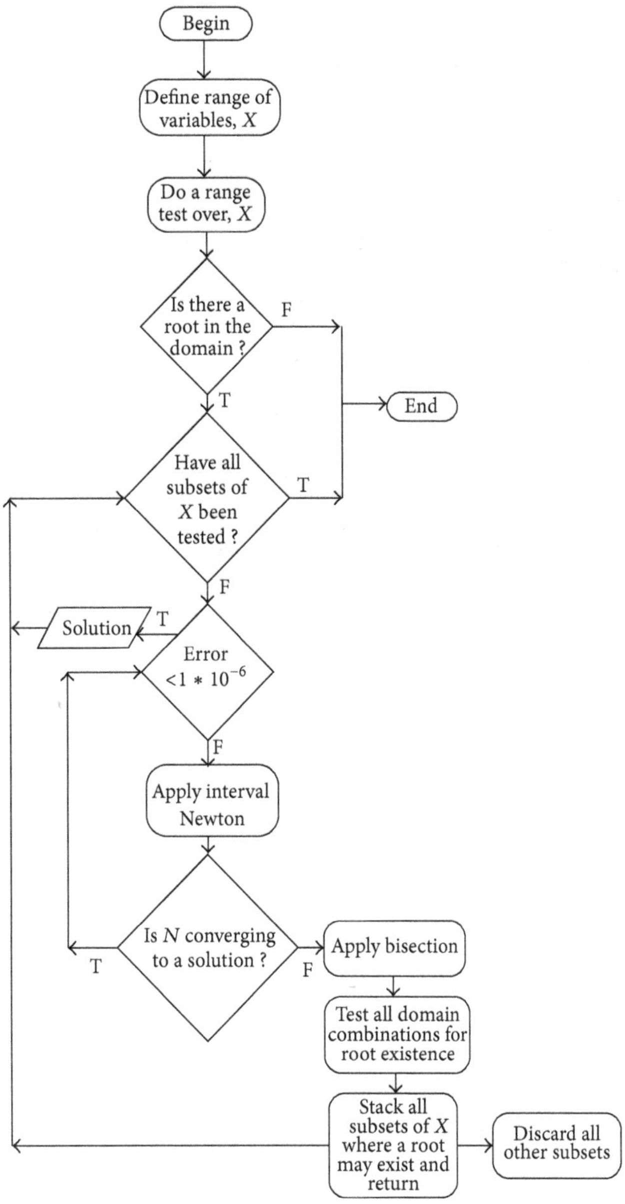

Figure 8

there will be none, one, or more than one critical points. The IN/GB method is much faster (Figure 6) than the Bisection method (Figure 7), and the order for finding the solutions differs between the methods. The IN/GB finds the solutions and scans the entire domain by dividing the domain in 345733 subsets, while the Bisection method finds the solutions and scans the entire domain by dividing the domain in 865970 subsets.

It is illustrative to plot the error in the IN/GB method versus the number of subdivisions of the initial domain. There are some very large errors, particularly at the beginning, which are cut to a value of 10 units, in order to preserve a reasonable scaling in the figures. As it was mentioned previously, for those cases with very large errors (inf) or not-a-number warnings (NaN), the algorithm proceeded to bisecting those domains in order to continue with the search. The relative error used to determine solutions was of $1 * 10^{-6}$, which is indicated in the graphs with a flat line.

The Bisection method converges linearly in a semilog plot toward the solutions which are found at the end of the process, as it is expected. However, the IN/GB finds solutions prior to the end of the process and with fewer subdivisions of the original interval.

5. Conclusions

This paper has introduced the use of INTLAB in computations for the determination of several solutions without the need for initial guesses and with a single run of the algorithm. The IN/GB method was implemented in INTLAB with four examples. The first two examples were algebraic equations that serve as a reference to verify the effectiveness of the algorithm and describe some important findings of the IN/GB method. The third and fourth cases are related to the computation of multiple solutions that represent properties of substances and fluids. Convergence to a solution is faster with the IN/GB method in general, but there are exceptions to this rule particularly when a singular Jacobian matrix may occur.

Appendix

$$
\begin{aligned}
\sum_{j=1}^{C} A_{ij}\Delta n_j
&= \frac{RT}{n}\left(\frac{\Delta n_i}{y_i} + F_1\left(\beta_i\overline{N} + \overline{\beta}\right) + \beta_i F_1^2 \overline{\beta}\right) \\
&\quad + \frac{a}{bn}\left(\beta_i\overline{\beta}F_3 - \frac{F_5}{a}\sum_{j=1}^{C} a_{ij}\Delta n_j + F_6\left(\beta_i\overline{\beta} - \alpha_i\overline{\beta} - \overline{\alpha}\beta_i\right)\right), \\
&\qquad i = 1,\dots,C,
\end{aligned}
$$

$$
\begin{aligned}
\sum_{i=1}^{C}\sum_{j=1}^{C}\sum_{k=1}^{C} A_{ijk}\Delta n_i\Delta n_j\Delta n_k
&= \frac{RT}{n^2}\left(-\sum_{i=1}^{C}\left(\frac{\Delta n_i^3}{y_i^2}\right) + 3\overline{N}\left(\overline{\beta}F_1\right)^2 + 2\left(\overline{\beta}F_1\right)^3\right) \\
&\quad + \frac{a}{n^2 b}\left(3\overline{\beta}^2\left(2\overline{\alpha} - \overline{\beta}\right)\left(F_3 + F_6\right) - 2\overline{\beta}^3 F_4 - 3\overline{\beta}\overline{\alpha}F_6\right),
\end{aligned}
\tag{A.1}
$$

where the following are considered.

C is the total number of components.

A_{ij} is the second-order derivative of the Helmholtz free energy with respect to the number of moles of species i and j.

Δn_j is the change in total number of moles of species j. This is used in the context of a Taylor series expansion of the Helmholtz free energy in terms of composition.

R is the ideal gas constant.

T is the absolute temperature.

n is the total number of moles.

n_i is the number of moles of species i, analogous meaning for other subindices.

$\overline{N}$ is the parameter used in the equations for the computation of critical points; it is defined as

$$\overline{N} = \sum_{i=1}^{C} \Delta n_i. \tag{A.2}$$

y_i is the mole fraction of component i, analogous meaning for other subindices.

a is the average energy parameter of the equation of state for the mixture of components. This is computed using standard van der Waals mixing rules:

$$a = \sum_{i=1}^{C}\sum_{j=1}^{C} \frac{n_i n_j}{n^2} a_{ij}, \quad a_{ij} = \left(a_i a_j\right)^{0.5}\left(1 - k_{ij}\right). \tag{A.3}$$

a_{ij} is the energy of interaction parameter between species i and j.

a_i is the energy parameter of species i. The meaning is the same for other subindices. It is computed using the following correlation:

$$a_i = \left(\frac{\left(RT_{c_i}\right)^2 \eta}{P_{c_i}}\right)\left[1 + c_i\left(1 - \left(\frac{T}{T_{c_i}}\right)^{0.5}\right)\right],$$

$\eta = 0.42748$

for the Redlich-Kwong-Soave equation of state,

$\eta = 0.45724$ (A.4)

for the Peng-Robinson equation of state,

$c_i = 0.48 + 1.574w_i - 0.176w_i^2$

for the Redlich-Kwong-Soave equation of state,

$c_i = 0.37464 + 1.54226w_i - 0.26992w_i^2$

for the Peng-Robinson equation of state.

w_i is the accentric factor of species i.

T_{c_i} is the critical temperature of species i.

P_{c_i} is the critical pressure of species i.

w_i is the accentric factor of species i.

k_{ij} is the binary interaction parameter between species i and j.

α_k is the parameter used in the equations for the computation of critical points; it is defined as

$$\alpha_k = \frac{\sum_{i=1}^{C} y_i a_{ik}}{a}. \tag{A.5}$$

$\overline{\alpha}$ is the parameter used in the equations for the computation of critical points; it is defined as

$$\overline{\alpha} = \sum_{i=1}^{C} \Delta n_i \alpha_i. \tag{A.6}$$

$\overline{a}$ is the parameter used in the equations for the computation of critical points; it is defined as

$$\overline{a} = \frac{1}{a}\sum_{i=1}^{C}\sum_{j=1}^{C} \Delta n_i \Delta n_j a_{ij}. \tag{A.7}$$

b is the average van der Waals mole volume of the equation of state. This is computed using standard van der Waals mixing rules:

$$b = \sum_{i=1}^{nc}\left(\frac{n_i}{n}\right) b_i. \tag{A.8}$$

b_i is the van der Waals mole volume of species i. This is computed as follows:

$$b_i = \frac{0.08664 RT_{c_i}}{P_{c_i}} \tag{A.9}$$

for the Redlich-Kwong-Soave equation of state, and

$$b_i = \frac{0.07780 RT_{c_i}}{P_{c_i}} \tag{A.10}$$

for the Peng-Robinson equation of state.

β_i is the parameter used in the equations for the computation of critical points; it is defined as

$$\beta_i = \frac{b_i}{b}. \tag{A.11}$$

$\overline{\beta}$ is the parameter used in the equations for the computation of critical points; it is defined as

$$\overline{\beta} = \sum_{i=1}^{C} \Delta n_i \beta_i. \tag{A.12}$$

F_{1-6} are the auxiliary functions used in the computation of critical points; they are defined as

$$F_1 = \frac{1}{K-1},$$

$$F_2 = \frac{2}{D_1 - D_2}\left(\frac{D_1}{K+D_1} - \frac{D_2}{K+D_2}\right),$$

$$F_3 = \frac{1}{(D_1 - D_2)}\left(\left(\frac{D_1}{K+D_1}\right)^2 - \left(\frac{D_2}{K+D_2}\right)^2\right),$$

$$F_4 = \frac{1}{(D_1 - D_2)}\left(\left(\frac{D_1}{K+D_1}\right)^3 - \left(\frac{D_2}{K+D_2}\right)^3\right), \quad \text{(A.13)}$$

$$F_5 = \frac{2}{(D_1 - D_2)}\ln\left(\frac{K+D_1}{K+D_2}\right),$$

$$F_6 = \left(\frac{2}{(D_1 - D_2)}\right) \times \left(\left(\frac{D_1}{K+D_1} - \frac{D_2}{K+D_2}\right) - \ln\left(\frac{K+D_1}{K+D_2}\right)\right).$$

K is the dimensionless volume; it is defined as

$$K = \frac{V}{nb}. \quad \text{(A.14)}$$

$D_{1\text{-}2}$ are the parameters of the equation of state. Their values are calculated with the following formulas:

$$D_1 = \frac{u_o + \sqrt{u_o^2 - 4w_o}}{2},$$
$$D_2 = \frac{u_o - \sqrt{u_o^2 - 4w_o}}{2}, \quad \text{(A.15)}$$

where $u_o = 1$, $w_o = 0$ for the Redlich-Kwong-Soave equation of state and $u_o = 2$, $w_o = -1$ for the Peng-Robinson equation of state.

Acknowledgment

The author wishes to acknowledge the helpful comments and review provided by Professor Emmanuel Haven at the University of Leicester, UK.

References

[1] R. B. Kearfott, *Rigorous Global Search: Continuous Problem*, Kluwer Academic Publishers, Dordrecht, The Netherlands, 1996.

[2] B. Stradi and E. Haven, "Optimal investment strategy via interval arithmetic," *International Journal of Theoretical and Applied Finance*, vol. 8, no. 2, pp. 185–206, 2005.

[3] B. A. Stradi-Granados and E. Haven, "The use of interval arithmetic in solving a non-linear rational expectation based multiperiod output-inflation process model: the case of the IN/GB method," *European Journal of Operational Research*, vol. 203, no. 1, pp. 222–229, 2010.

[4] B. A. Stradi, J. F. Brennecke, P. Kohn, and M. A. Stadtherr, "Reliable computation of mixture critical points," *AIChE Journal*, vol. 47, no. 1, pp. 212–221, 2001.

[5] H. Gecegormez and Y. Demirel, "Phase stability analysis using interval Newton method with NRTL model," *Fluid Phase Equilibria*, vol. 237, no. 1-2, pp. 48–58, 2005.

[6] Z. Galias, "Proving the existence of long periodic orbits in 1D maps using interval Newton method and backward shooting," *Topology and its Applications*, vol. 124, no. 1, pp. 25–37, 2002.

[7] M. L. Michelsen and R. A. Heidemann, "Calculation of critical points from cubic two-constant equations of state," *AIChE Journal*, vol. 27, no. 2, pp. 521–523, 1981.

[8] H. Hoteit, E. Santiso, and A. Firoozabadi, "An efficient and robust algorithm for the calculation of gas-liquid critical point of multicomponent petroleum fluids," *Fluid Phase Equilibria*, vol. 241, no. 1-2, pp. 186–195, 2006.

[9] D. V. Nichita and S. Gomez, "Efficient and reliable mixture critical points calculation by global optimization," *Fluid Phase Equilibria*, vol. 291, no. 2, pp. 125–140, 2010.

[10] M. L. Michelsen, "Calculation of phase envelopes and critical points for multicomponent mixtures," *Fluid Phase Equilibria*, vol. 4, no. 1-2, pp. 1–10, 1980.

[11] M. L. Michelsen, "Phase equilibrium calculations. What is easy and what is difficult?" *Computers and Chemical Engineering*, vol. 17, no. 5-6, pp. 431–439, 1993.

[12] M. L. Michelsen and H. Kistenmacher, "On composition-dependent interaction coefficeints," *Fluid Phase Equilibria*, vol. 58, no. 1-2, pp. 229–230, 1990.

[13] M. L. Michelsen and R. A. Heidemann, "Calculation of tri-critical points," *Fluid Phase Equilibria*, vol. 39, no. 1, pp. 53–74, 1988.

[14] M. L. Michelsen, "Calculation of critical points and phase boundaries in the critical region," *Fluid Phase Equilibria*, vol. 16, no. 1, pp. 57–76, 1984.

[15] J. Cai, H. Liu, Y. Hu, and J. M. Prausnitz, "Critical properties of polydisperse fluid mixtures from an equation of state," *Fluid Phase Equilibria*, vol. 168, no. 1, pp. 91–106, 2000.

[16] T. Lindvig, L. L. Hestkjar, A. F. Hansen, M. L. Michelsen, and G. M. Kontogeorgis, "Phase equilibria for complex polymer solutions," *Fluid Phase Equilibria*, vol. 194–197, pp. 663–673, 2002.

[17] M. Cismondi and M. L. Michelsen, "Global phase equilibrium calculations: critical lines, critical end points and liquid-liquid-vapour equilibrium in binary mixtures," *Journal of Supercritical Fluids*, vol. 39, no. 3, pp. 287–295, 2007.

[18] C. Carstensen and M. S. Petković, "On iteration methods without derivatives for the simultaneous determination of polynomial zeros," *Journal of Computational and Applied Mathematics*, vol. 45, no. 3, pp. 251–266, 1993.

[19] C. D. Maranas and C. A. Floudas, "Finding all solutions of nonlinearly constrained systems of equations," *Journal of Global Optimization*, vol. 7, no. 2, pp. 143–182, 1995.

[20] R. B. Kearfott and M. Novoa III, "Algorithm 681 INTBIS, a portable interval Newton/bisection package," *ACM Transactions on Mathematical Software*, vol. 16, no. 2, pp. 152–157, 1990.

[21] S. M. Rump, "INTLAB: INTerval LABoratory," in *Developments in Reliable Computing*, T. Csendes, Ed., pp. 77–104, Kluwer Academic Publishers, Dodrecht, The Netherlands, 1999.

[22] "Matlab R2011b (64 bit)," Mathworks, Natick, Mass, USA, 2011.

[23] S. C. Chapra and R. P. Canale, *Numerical Methods for Engineers*, McGraw-Hill, New York, NY, USA, 6th edition, 2010.

[24] S. Nakamura, *Numerical Analysis and Graphical Visualization*, Prentice-Hall, Upper Saddle River, NJ, USA, 2nd edition, 2001.

[25] S. Attaway, *MAtlab: A Practical Introduction to Programming and Problem Solving*, Butterworth-Heinemann, New York, NY, USA, 2nd edition, 2011.

[26] K. Ozaki, T. Ogita, S. M. Rump, and S. Oishi, "Fast algorithms for floating-point interval matrix multiplication," *Journal of Computational and Applied Mathematics*, vol. 236, no. 7, pp. 1795–1814, 2012.

[27] S. M. Rump and T. Ogita, "Super-fast validated solution of linear systems," *Journal of Computational and Applied Mathematics*, vol. 199, no. 2, pp. 199–206, 2007.

[28] R. C. Reid, J. M. Prausnitz, and B. E. Poling, *Th Properties of Gases and Liquids*, McGraw-Hill, New York, NY, USA, 4th edition, 1987.

[29] A. Neumaier, *Interval Methods for Systems of Equations*, Cambridge University Press, Cambridge, UK, 1990.

[30] R. Moore, *Interval Analysis*, Prentice-Hall, Upper Saddle River, NJ, USA, 1966.

Existence and Global Attractivity of Positive Periodic Solutions for The Neutral Multidelay Logarithmic Population Model with Impulse

Zhenguo Luo,[1,2] Jianhua Huang,[1] Liping Luo,[2] and Binxiang Dai[3]

[1] *Department of Mathematics, National University of Defense Technology, Changsha 410073, China*
[2] *Department of Mathematics, Hengyang Normal University, Hengyang, Hunan 421008, China*
[3] *School of Mathematical Sciences and Statistics, Central South University, Changsha, Hunan 410075, China*

Correspondence should be addressed to Zhenguo Luo; robertl86@163.com

Academic Editor: Shouming Zhong

Suffiicient and realistic conditions are established in this paper for the existence and global attractivity of a positive periodic solution to the neutral multidelay logarithmic population model with impulse by using the theory of abstract continuous theorem of k-set contractive operator and some inequality techniques. The results improve and generalize the known ones in Li 1999, Lu and Ge 2004, Y. Luo and Z. G. Luo 2010, and Wang et al. 2009. As an application, we also give an example to illustrate the feasibility of our main results.

1. Introduction

In this paper, we investigate the existence and uniqueness of the positive periodic solution of the following neutral population system with multiple delays and impulse:

$$\begin{aligned} \frac{dN(t)}{dt} = N(t)\Bigg[& r(t) - a(t)\ln N(t) \\ & - \sum_{i=1}^{n} b_i(t)\ln N(t-\sigma_i(t)) \\ & - \sum_{j=1}^{m} c_j(t)\int_{-\infty}^{t} K_j(t-s)\ln N(s)\,ds \\ & - \sum_{l=1}^{p} d_l(t)\frac{d\ln N(t-\tau_l(t))}{dt}\Bigg], \quad t \neq t_k, \\ N(t_k^+) = e^{(1+\theta_k)} & N(t_k), \quad k = 1,2,\ldots, \end{aligned} \tag{1}$$

with the following initial conditions:

$$\begin{gathered} N(\xi) = \varphi(\xi), \quad N'(\xi) = \varphi'(\xi), \quad \xi \in [-\tau, 0], \quad \varphi(0) > 0, \\ \varphi \in C([-\tau,0), R^+) \bigcap C^1([-\tau,0)), R^+, \end{gathered} \tag{2}$$

where $r(t)$, $a(t)$, $b_i(t)$, $c_j(t)$, $d_l(t)$, $\sigma_i(t)$, $\tau_l(t)$ are positive continuous ω-periodic functions with $\sigma_i(t) \geq 0$, $\tau_l(t) \geq 0$, $t \in [0,\omega]$, $\tau = \max\{\sigma_i(t), \tau_l(t)\}$, $\sigma_i'(t) < 1$, $\tau_l'(t) < 1$, for all $i = \{1,2,\ldots,n\}$, for all $l = \{1,2,\ldots,p\}$. Furthermore, $d_l(t) \in C^1(R,R)$, $\tau_l(t) \in C^2(R,R)$, $\int_0^\infty K_j(s)ds = 1$, $\int_0^{+\infty} sK_j(s)ds < +\infty$, for all $j = \{1,2,\ldots,m\}$, for all $l = \{1,2,\ldots,p\}$. For the ecological justification of (1) and similar types refer to [1–7]. In recent years, Gopalsamy [1] and Kirlinger [2] had proposed the following single species logarithmic model:

$$N'(t) = N(t)[a - b\ln N(t) - c\ln N(t-\tau)]. \tag{3}$$

In [3], Li considered the following nonautonomous single species logarithmic model:

$$N'(t) = N(t)[a(t) - b(t)\ln N(t) - c(t)\ln N(t-\tau(t))]. \tag{4}$$

He used the continuation theorem of the coincidence degree theory to establish sufficient conditions for the existence and attractivity of positive periodic solutions of the system (4).

For more works on the periodic solution of the neutral type logistic model or the Lotka-Volterra model, see [8–12] for details. Only little scholars considered the neutral logarithmic model (see [4–7]). Li [4] had studied the following single species neutral logarithmic model:

$$N'(t) = N(t)\left[r(t) - a(t)\ln N(t-\sigma) - b(t)\left(\ln N(t-\eta)\right)'\right]. \tag{5}$$

Lu and Ge [5] and Y. Luo and Z. G. Luo [6] employed an abstract continuous theorem of k-set contractive operator to investigate the following equation:

$$N'(t) = N(t)\left[r(t) - \sum_{j=1}^{n} a_j(t)\ln N\left(t-\sigma_j(t)\right) - \sum_{i=1}^{m} b_i(t)\left(\ln N\left(t-\tau_i(t)\right)\right)'\right]. \tag{6}$$

They established some criteria to guarantee the existence of positive periodic solutions of the system (6), respectively.

In [7], Wang et al. had investigated the existence and uniqueness of the positive periodic solution of the following neutral multispecies logarithmic population model:

$$\begin{aligned} N'(t) = N(t)\Bigg[&r(t) - a(t)\ln N(t) - \sum_{j=1}^{n} b_j(t)\ln N\left(t-\tau_j(t)\right) \\ &- \sum_{j=1}^{n} c_j(t)\int_{-\infty}^{t} k_j(t-s)\ln N((s))\,ds - \sum_{j=1}^{n} d_j(t)\left(\ln N\left(t-\eta_j(t)\right)\right)'\Bigg]. \end{aligned} \tag{7}$$

By using an abstract continuous theorem of a k-set contractive operator, the criteria are established for the existence and global attractivity of positive periodic solutions for model (7).

On the other hand, there are some other perturbations in the real world such as fires and floods, which are not suitable to be considered continually. These perturbations bring sudden changes to the system. Systems with such sudden perturbations involving impulsive differential equations have attracted the interest of many researchers in the past twenty years [13–19], since they provide a natural description of several real processes subject to certain perturbations whose duration is negligible in comparison with the duration of the process. Such processes are often investigated in various fields of science and technology such as physics, population dynamics, ecology, biological systems, and optimal control. For details, see [20–22].

However, to this day, no scholars had done works on the existence, uniqueness, and global stability of the positive periodic solution of (1). One could easily see that the systems (5)–(7) are all special cases of the system (1). Therefore, we propose and study the system (1) in this paper.

Throughout this paper, we make the following notations. Let $\omega > 0$ be a constant,

$C_\omega = \{x \mid x \in C(R,R), x(t+\omega) = x(t)\}$, with the norm defined by $|x|_0 = \max_{t\in[0,\omega]}|x(t)|$,

$C_\omega^1 = \{x \mid x \in C^1(R,R), x(t+\omega) = x(t)\}$, with the norm defined by $\|x\| = \max_{t\in[0,\omega]}\{|x|_0, |x'|_0\}$.

Then, those spaces are both Banach spaces. We also denote that

$$\bar{h} = \frac{1}{\omega}\int_0^\omega h(t)\,dt, \quad \text{for any } h \in PC_\omega. \tag{8}$$

For the sake of generality and convenience, we always make the following fundamental assumptions:

(H_1) $r(t)$, $a(t)$, $b_i(t)$, $c_j(t)$, $d_l(t)$, $\sigma_i(t)$, $\tau_l(t)$ $(i = 1,2,\ldots,n;\ j = 1,2,\ldots,m;\ l = 1,2,\ldots,p)$ are all positive periodic continuous functions with period $\omega > 0$;

(H_2) $0 < t_1 < t_2 < \cdots < t_k < \cdots$ are fixed impulsive points with $\lim_{k\to\infty} t_k = +\infty$;

(H_3) $\{\theta_k\}$ is a real sequence such that $\theta_k + 1 > 0$, and $\prod_{0<t_k<t}(1+\theta_k)$ is an ω-periodic function.

In the following section, some definitions and some useful lemmas are listed. In the third section, by using an abstract continuous theorem of k-set contractive operator and some inequality techniques, we acquired some sufficient conditions which ensure the existence and uniqueness of the positive periodic solution of the systems (1) and (2). Finally, we give an example to show our results.

2. Preliminaries

In order to obtain the existence and uniqueness of a periodic solution for the systems (1) and (2), we first give some definitions and lemmas.

Definition 1 (see [21]). A function $N : R \to (0,+\infty)$ is said to be a positive solution of (1) and (2), if the following conditions are satisfied:

(a) $N(t)$ is absolutely continuous on each (t_k, t_{k+1});

(b) for each $k \in Z_+$, $N(t_k^+)$ and $N(t_k^-)$ exist and $N(t_k^-) = N(t_k)$;

(c) $N(t)$ satisfies the first equation of (1) and (2) for almost everywhere (for short a.e.) in $[0,\infty] \setminus \{t_k\}$ and satisfies $N(t_k^+) = (1+\theta_k)N(t_k)$ for $t = t_k$, $k \in Z_+ = \{1,2,\ldots\}$.

Definition 2. The system (1) is said to be globally attractive, if there exists a positive solution $x(t)$ of (1) such that $\lim_{t\to+\infty}|x(t)-y(t)| = 0$ for any other positive solution $y(t)$ of (1).

We can easily get the following Lemma 3.

Lemma 3. *Th region* $R = \{N(t) : N(0) > 0, t \geq 0\}$ *is invariant with respect to* (1).

Proof. In view of biological population, we obtain $N(0) > 0$. By the system (1), we have

$$\begin{aligned} N(t) &= N(0) \\ &\times \exp\left\{\int_0^t \left[r(\eta) - a(\eta)\ln N(\eta) - \sum_{i=1}^{n} b_i(\eta)\ln N(\eta - \sigma_i(\eta)) - \sum_{j=1}^{m} c_j(\eta)\int_{-\infty}^{t} K_j(\eta - s)\ln N(s)\,ds - \sum_{l=1}^{p} d_l(\eta)\frac{d\ln N(\eta - \tau_l(\eta))}{d\eta}\right] d\eta\right\} \\ &> 0, \quad t \in [0, t_1], \\ N(t) &= N(t_k) \\ &\times \exp\left\{\int_{t_k}^t \left[r(\eta) - a(\eta)\ln N(\eta) - \sum_{i=1}^{n} b_i(\eta)\ln N(\eta - \sigma_i(\eta)) - \sum_{j=1}^{m} c_j(\eta)\int_{-\infty}^{t} K_j(\eta - s)\ln N(s)\,ds - \sum_{l=1}^{p} d_l(\eta)\frac{d\ln N(\eta - \tau_l(\eta))}{d\eta}\right] d\eta\right\} \\ &> 0, \quad t \in (t_k, t_{k+1}], \\ N(t_k^+) &= e^{(1+\theta_k)}N(t_k) > 0, \quad k \in N. \end{aligned} \tag{9}$$

Then, the solution of (1) and (2) is positive.

Under the above hypotheses (H_1)–(H_3), we consider the following neutral nonimpulsive system:

$$\frac{dy(t)}{dt} = y(t)\left[r(t) - A(t)\ln y(t) - \sum_{i=1}^{n} B_i(t)\ln y(t - \sigma_i(t)) - \sum_{j=1}^{m} C_j(t)\int_{-\infty}^{t} K_j(t - s)\ln y(s)\,ds - \sum_{l=1}^{p} D_l(t)\frac{d\ln y(t - \tau_l(t))}{dt}\right], \tag{10}$$

with the following initial conditions:

$$\begin{gathered} y(\xi) = \varphi(\xi), \quad y'(\xi) = \varphi'(\xi), \quad \xi \in [-\tau, 0], \quad \varphi(0) > 0, \\ \varphi \in C([-\tau, 0), R^+)\bigcap C^1([-\tau, 0)), R^+, \end{gathered} \tag{11}$$

where

$$\begin{aligned} A(t) &= a(t)\prod_{0<t_k<t}(1 + \theta_k), \\ B_i(t) &= b_i(t)\prod_{0<t_k<t-\sigma_i(t)}(1 + \theta_k), \\ C_j(t) &= c_j(t)\prod_{0<t_k<t}(1 + \theta_k), \\ D_l(t) &= d_l(t)\prod_{0<t_k<t-\tau_l(t)}(1 + \theta_k). \end{aligned} \tag{12}$$

By a solution $y(t)$ of (10) and (11), it means an absolutely continuous function $y(t)$ defined on $[-\tau, 0]$ that satisfies (10) a.e., for $t \geq 0$, and $y(\xi) = \varphi(\xi)$, $y'(\xi) = \varphi'(\xi)$ on $[-\tau, 0]$.

The following lemmas will be used in the proofs of our results, and the proof of the lemma is similar to that of Theorem 1 in [13].

Lemma 4. *Suppose that* (H_1)–(H_4) *hold. Then,*

(1) *if* $y(t)$ *is a solution of* (10) *and* (11) *on* $[-\tau, +\infty)$, *then* $N(t) = \prod_{0<t_k<t} e^{(1+\theta_k)} y(t)$ *is a solution of* (1) *and* (2) *on* $[-\tau, +\infty)$;

(2) *if* $N(t)$ *is a solution of* (1) *and* (2) *on* $[-\tau, +\infty)$, *then* $y(t) = \prod_{0<t_k<t}(e^{(1+\theta_k)})^{-1}N(t)$ *is a solution of* (10) *and* (11) *on* $[-\tau, +\infty)$.

Proof. (1) It is easy to see that $N(t) = \prod_{0<t_k<t} e^{(1+\theta_k)} y(t)$ is absolutely continuous on every interval $(t_k, t_{k+1}]$, $t \neq t_k$, $k = 1, 2, \ldots$,

$$\begin{aligned} &N'(t) - N(t)\left[r(t) - a(t)\ln N(t) - \sum_{i=1}^{n} b_i(t)\ln N(t - \sigma_i(t)) - \sum_{j=1}^{m} c_j(t)\int_{-\infty}^{t} K_j(t - s)\ln N(s)\,ds - \sum_{l=1}^{p} d_l(t)\frac{dN(t - \tau_l(t))}{dt}\right] \\ &= \prod_{0<t_k<t} e^{(1+\theta_k)} y'(t) - \prod_{0<t_k<t} e^{(1+\theta_k)} y(t) \end{aligned}$$

$$\times\left[r(t)-a(t)\prod_{0<t_k<t}(1+\theta_k)\ln y(t)\right.$$
$$-\sum_{i=1}^{n}b_i(t)\prod_{0<t_k<t-\sigma_i(t)}(1+\theta_k)$$
$$\times\ln y(t-\sigma_i(t))$$
$$-\sum_{j=1}^{m}c_j(t)\int_{-\infty}^{t}K_j(t-s)$$
$$\times\prod_{0<t_k<t}(1+\theta_k)\ln y(s)\,ds$$
$$-\sum_{l=1}^{p}d_l(t)\prod_{0<t_k<t-\tau_l(t)}(1+\theta_k)$$
$$\left.\times\frac{dy(t-\tau_l(t))}{dt}\right]$$
$$=\prod_{0<t_k<t}e^{(1+\theta_k)}$$
$$\times\left\{y'(t)-y(t)\right.$$
$$\times\left[r(t)-A(t)\ln y(t)\right.$$
$$-\sum_{i=1}^{n}B_i(t)\ln y(t-\sigma_i(t))$$
$$-\sum_{j=1}^{m}C_j(t)\int_{-\infty}^{t}K_j(t-s)\ln y(s)\,ds$$
$$\left.\left.-\sum_{j=1}^{n}D_l(t)\frac{dy(t-\tau_l(t))}{dt}\right]\right\}=0. \tag{13}$$

On the other hand, for any $t=t_k$, $k=1,2,\ldots$,

$$N(t_k^+)=\lim_{t\to t_k^+}\prod_{0<t_j<t}e^{(1+\theta_k)}y(t)=\prod_{0<t_j\le t_k}e^{(1+\theta_k)}y(t_k), \tag{14}$$
$$N(t_k)=\prod_{0<t_j<t_k}e^{(1+\theta_k)}y(t_k),$$

thus,

$$N(t_k^+)=e^{(1+\theta_k)}N(t_k),\quad k=1,2,\ldots. \tag{15}$$

It follows from (13)–(15) that $N_i(t)$ is a solution of (1) and (2).

(2) Since $N(t)=\prod_{0<t_k<t}e^{(1+\theta_k)}y(t)$ is absolutely continuous on every interval $(t_k,t_{k+1}]$, $t\neq t_k$, $k=1,2,\ldots$, and in view of (15), it follows that for any $k=1,2,\ldots$,

$$\begin{aligned}y(t_k^+)&=\prod_{0<t_j\le t_k}\left(e^{(1+\theta_k)}\right)^{-1}N(t_k^+)\\&=\prod_{0<t_j<t_k}\left(e^{(1+\theta_k)}\right)^{-1}N(t_k)=y(t_k),\\y(t_k^-)&=\prod_{0<t_j<t_k}\left(e^{(1+\theta_k)}\right)^{-1}N(t_k^-)\\&=\prod_{0<t_j\le t_k^-}\left(e^{(1+\theta_k)}\right)^{-1}N(t_k^-)=y(t_k),\end{aligned} \tag{16}$$

which implies that $y(t)$ is continuous on $[-\tau,+\infty)$. It is easy to prove that $y(t)$ is absolutely continuous on $[-\tau,+\infty)$. Similar to the proof of (1), one can check that $y(t)=\prod_{0<t_k<t}(e^{(1+\theta_k)})^{-1}N(t)$ are solutions of (10) and (11) on $[-\tau,+\infty)$. The proof of Lemma 3 is completed.

Here, we take the transformation $y(t)=e^{x(t)}$; then, (10) can be rewritten in the following form:

$$\begin{aligned}\frac{dx(t)}{dt}&=r(t)-A(t)x(t)\\&-\sum_{i=1}^{n}B_i(t)x(t-\sigma_i(t))\\&-\sum_{j=1}^{m}C_j(t)\int_{-\infty}^{t}K_j(t-s)x(s)\,ds\\&-\sum_{l=1}^{p}D_{0l}(t)x'(t-\tau_l(t)),\end{aligned} \tag{17}$$

where $D_{0l}(t)=D_l(t)(1-\tau_l'(t))$, $l=1,2,\ldots,p$. Therefore, we only discuss the existence and uniqueness of a periodic solution for (17).

Definition 5 (see [23]). Let U be a bounded subset in X. Define that

$$\begin{aligned}\alpha_X(U)&=\inf\left\{\delta>0:\text{there is a finite number of subsets}\right.\\&\left.U_i\subset U\text{ such that }U=\bigcup_i U_i\text{ and }\operatorname{diam}(U_i)\le\delta\right\},\end{aligned} \tag{18}$$

where $\operatorname{diam}(U_i)$ denotes the diameter of the set U_i, obviously, $0\le\alpha_X(U)<\infty$. So, $\alpha_X(U)$ is called the (Kuratowski) measure of noncompactness of X.

Definition 6 (see [23]). Let X, Y be two Banach spaces and $D \subset X$; a continuous and bounded map $T : D \to Y$ is called k-set contractive if for any bounded set $U \subset D$ one has

$$\alpha_Y (T(U)) \leq k\alpha_X (U), \quad (19)$$

T is called strict-set-contractive if it is k-set contractive for some $0 \leq k < 1$.

For a Fredholm operator $L : X \to Y$ with index zero, according to [9, 24], we define

$$l(L) = \sup \{r\alpha_X (U) \leq \alpha_Y (L(U)) \text{ for any bounded set } U \subset X\}. \quad (20)$$

Lemma 7 (see [9, 24]). *Let $L : X \to Y$ be a Fredholm operator with zero index and $a \in Y$ be a fixed point. Suppose that $N : \Omega \to Y$ is called a k-set contractive with $k < l(L)$, where $\Omega \subset X$ is bounded, open, and symmetric about $0 \in \Omega$. Further, one also assumes that*

(1) *$Lx \neq \lambda Nx + \lambda a$, for $x \in \partial\Omega$, $\lambda \in (0, 1)$;*

(2) *$[QN(x) + Qa, x]\,[QN(-x) + Qa, x] < 0$, for $x \in \operatorname{Ker} L \bigcap \partial\Omega$;*

where $[\cdot, \cdot]$ is a bilinear form on $Y \times X$ and Q is the projection of Y onto Coker(L), where Coker(L) is the cokernel of the operator L. Then, there is an $x \in \overline{\Omega}$ such that $Lx - Nx = a$.

In order to use Lemma 7 to study (17), we set $X = C^1_\omega$, $Y = C_\omega$,

$$L : C^1_\omega \longrightarrow C_\omega, \qquad Lx = \frac{dx}{dt}, \quad (21)$$

and $N : C^1_\omega \to C_\omega$ defined by

$$\begin{aligned} Nx = & -A(t)(t)x(t) - \sum_{i=1}^{n} B_i(t)x(t - \sigma_i(t)) \\ & - \sum_{j=1}^{m} C_j(t) \int_{-\infty}^{t} K_j(t-s)x(s)\,ds \\ & - \sum_{l=1}^{p} D_{0l}(t)x'(t - \tau_l(t)). \end{aligned} \quad (22)$$

It is easy to see from [25] that L is a Fredholm operator with index zero. Thus, (17) has a positive ω-periodic solution if and only if $Lx = Nx + r$ has a solution $x \in C^1_\omega$, where $r = r(t)$.

Lemma 8 (see [25]). *The differential operator L is a Fredholm operator with index zero and satisfies $l(L) \geq 1$.*

Lemma 9. *Let r_0, r_1 be two positive constants and $\Omega = \{x_i \in C^1_\omega, |x_i|_o < r_0, |x_i'|_o < r_1\}$. If $k = \sum_{l=1}^{p} |D_{0l}|_0$, then $N : \Omega \to C^1_\omega$ is a k-set contractive map.*

As Lemma 9 can be proved in the same way as in the proof of Lemma 2.4 in [7], we omit it here.

Lemma 10 (see [7, 9]). *Suppose that $\tau \in C^1_\omega$ and $\tau'(t) < 1$, $t \in [0, \omega]$, then the function $t - \tau(t)$ has a unique inverse $\mu(t)$ satisfying $\mu \in C(R, R)$ with $\mu(a + \omega) = \mu(a) + \omega$ for all $a \in R$, and if $g \in PC_\omega$, $\tau'(t) < 1$, $t \in [0, \omega]$, then $g(\mu(t)) \in PC_\omega$.*

Lemma 11 (see [26, 27]). *Suppose that $x(t)$ is a differently continuous ω-periodic function on R with ($\omega > 0$), then to any $t_* \in R$, $\max_{t_* \leq t \leq t_* + \omega} |x(t)| \leq |x(t_*)| + (1/2) \int_0^\omega |x'(t)| dt$.*

3. Main Theorem

Since $\sigma_i'(t) < 1$, $\tau_l'(t) < 1$, $t \in [0, \omega]$, we see that $\sigma_i(t)$, $\tau_l(t)$ all have its inverse function. Throughout the following part, we set $\zeta_i(t)$, $\xi_l(t)$ that represent the inverse function of $t - \sigma_i(t)$, $t - \tau_l(t)$, respectively. We also denote

$$\begin{aligned} \Gamma(t) := & A(t) + \sum_{i=1}^{n} \frac{B_i(\zeta_i(t))}{1 - \sigma_i'(\zeta_i(t))} \\ & + \sum_{j=1}^{m} C_j(t) - \sum_{l=1}^{p} \frac{D_l'(\xi_l(t))}{1 - \tau_l'(\xi_l(t))}. \end{aligned} \quad (23)$$

Remark 12. From Lemma 9, we get that $\zeta_i(\omega) = \zeta_i(0) + \omega$, $\xi_l(\omega) = \xi_l(0) + \omega$, $i = 1, 2, \ldots, n$; $l = 1, 2, \ldots, p$; then,

$$\begin{aligned} \int_0^\omega \frac{B_i(\zeta_i(s))}{1 - \sigma_i'(\zeta_i(s))} ds & = \int_{\zeta_i(0)}^{\zeta_i(\omega)} \frac{B_i(t)\left(1 - \sigma_i'(t)\right)}{1 - \sigma_i'(t)} dt \\ & = \int_{\zeta_i(0)}^{\zeta_i(0)+\omega} B_i(t)\,dt = \overline{B_i}\omega, \quad i = 1, 2, \ldots, n. \end{aligned} \quad (24)$$

Similarly,

$$\begin{aligned} \int_0^\omega \frac{D_l'(\xi_l(s))}{1 - \tau_l'(\xi_l(s))} ds & = \int_{\xi_l(0)}^{\xi_l(\omega)} \frac{D_l'(t)\left(1 - \tau_l'(t)\right)}{1 - \tau_l'(t)} dt \\ & = \int_{\xi_l(0)}^{\xi_l(0)+\omega} D_l'(t)\,dt = 0, \quad l = 1, 2, \ldots, p. \end{aligned} \quad (25)$$

Thus,

$$\overline{\Gamma}\omega = \int_0^\omega \Gamma(t)\,dt = \left(\overline{A} + \sum_{i=1}^{n} \overline{B_i} + \sum_{j=1}^{m} \overline{C_j}\right)\omega. \quad (26)$$

Theorem 13. *In addition to (H_1)–(H_3), suppose that the following conditions hold:*

(H_4) *there exists a constant $\eta > 0$ such that $|\Gamma(t)| \geq \eta$, for all $t \in [0, \omega]$, where $\Gamma(t)$ is defined by (23);*

(H_5) *if $(1/2)(|A|_0\omega + \sum_{i=1}^{n} |B_i|_0\omega + \sum_{j=1}^{m} |C_j|_0\omega^{1/2}) + \sum_{l=1}^{p} |1 - \tau_l'|_0^{1/2} |D_l|_0 < 1$, and $\sum_{l=1}^{p} |1 - \tau_l'|_0 |D_l|_0 < 1$.*

Th n, (1) has at least one positive ω-periodic solution.

Proof. Let $u(t)$ be an arbitrary ω-periodic solution of the operator equation as follows:

$$Lu = \lambda Nu + \lambda r, \quad \lambda \in (0,1), \tag{27}$$

where L, N defined by (21) and (22), respectively. Then, $u(t)$ satisfies the following operator equation:

$$\begin{aligned}\frac{du(t)}{dt} = \lambda \Bigg[& r(t) - A(t)u(t) \\ & - \sum_{i=1}^{n} B_i(t)u(t-\sigma_i(t)) \\ & - \sum_{j=1}^{m} C_j(t)\int_{-\infty}^{t} K_j(t-s)u(s)\,ds \\ & - \sum_{l=1}^{p} D_{0l}(t)u'(t-\tau_l(t))\Bigg].\end{aligned} \tag{28}$$

Integrating both sides of (28) over $[0,\omega]$, we have

$$\begin{aligned}\bar{r}\omega = \int_0^{\omega} \Bigg[& A(t)u(t) \\ & + \sum_{i=1}^{n} B_i(t)u(t-\sigma_i(t)) \\ & + \sum_{j=1}^{m} C_j(t)\int_{-\infty}^{t} K_j(t-s)u(s)\,ds \\ & + \sum_{l=1}^{p} D_{0l}(t)u'(t-\tau_l(t))\Bigg]dt \\ = \int_0^{\omega} & A(t)u(t)\,dt \\ & + \int_0^{\omega}\sum_{i=1}^{n} B_i(t)u(t-\sigma_i(t))\,dt \\ & + \int_0^{\omega}\sum_{j=1}^{m} C_j(t)\int_{-\infty}^{t} K_j(t-s)u(s)\,ds\,dt \\ & - \int_0^{\omega}\sum_{l=1}^{p} D_l'(t)u(t-\tau_l(t))\,dt.\end{aligned} \tag{29}$$

Let $t-\sigma_i(t) = s$, then $t = \zeta_i(s)$ and

$$\begin{aligned}&\int_0^{\omega} B_i(t)u(t-\sigma_i(t))\,dt \\ &\quad = \int_{-\sigma_i(0)}^{\omega-\sigma_i(\omega)} \frac{B_i(\zeta_i(s))}{1-\sigma_i'(\zeta_i(s))}u(s)\,ds.\end{aligned} \tag{30}$$

By Lemma 10, we have

$$\begin{aligned}&\int_0^{\omega} B_i(t)u(t-\sigma_i(t))\,dt \\ &\quad = \int_0^{\omega} \frac{B_i(\zeta_i(s))}{1-\sigma_i'(\zeta_i(s))}u(s)\,ds.\end{aligned} \tag{31}$$

Similarly, we have

$$\begin{aligned}&\int_0^{\omega} D_l'(t)u_i(t-\tau_l(t))\,dt \\ &\quad = \int_0^{\omega} \frac{D_l'(\xi_l(s))}{1-\tau_l'(\xi_l(s))}u(s)\,ds.\end{aligned} \tag{32}$$

Substituting (31) and (32) into (29), we have

$$\int_0^{\omega} \Gamma(s)u(s)\,ds = \bar{r}\omega. \tag{33}$$

Considering assumption (H_4), we know that $|\Gamma(t)| \geq \eta > 0$, and then, it follows from the integro mean value theorem that there exists a $\alpha \in [0,\omega]$ satisfying

$$|u(\alpha)| = \frac{\overline{r_i}}{|\Gamma(\alpha)|} \leq \frac{\overline{r_i}}{\eta}. \tag{34}$$

By Lemma 11, we can get that

$$|u|_0 \leq \frac{\overline{r_i}}{\eta} + \frac{1}{2}\int_0^{\omega}|u'(t)|\,dt. \tag{35}$$

Multiplying both sides of (28) by $u'(t)$ and integrating them over $[0,\omega]$, we have

$$\begin{aligned}&\int_0^{\omega}|u'(t)|^2 dt \\ &= \lambda \Bigg| \int_0^{\omega} r(t)u'(t)\,dt - \int_0^{\omega} A(t)u(t)u'(t)\,dt \\ &\quad - \int_0^{\omega}\sum_{i=1}^{n} B_i(t)u(t-\sigma_i(t))u'(t)\,dt \\ &\quad - \int_0^{\omega}\sum_{j=1}^{m} C_j(t)\int_{-\infty}^{t} K_j(t-s)u(s)u'(t)\,ds\,dt \\ &\quad - \int_0^{\omega}\sum_{l=1}^{p} D_{0l}(t)u'(t-\tau_l(t))u'(t)\,dt \Bigg| \\ &\leq \left[|r|_0 + \left(|A|_0 + \sum_{i=1}^{n}|B_i|_0\right)|u|_0\right]\int_0^{\omega}|u'(t)|\,dt \\ &\quad + \int_0^{\omega}\sum_{j=1}^{m}\left|C_j(t)\int_{-\infty}^{t} K_j(t-s)u(s)\,ds\,u'(t)\right|dt \\ &\quad + \int_0^{\omega}\sum_{l=1}^{p}\left|D_{0l}(t)u'(t-\tau_l(t))u'(t)\right|dt.\end{aligned} \tag{36}$$

By using the Cauchy-Schwarz inequality, we get that

$$\begin{aligned}
&\int_0^\omega |u'(t)|^2 dt \\
&\le \left[|r|_0 + \left(|A|_0 + \sum_{i=1}^{n} |B_i|_0 \right) |u|_0 \right] \\
&\quad \times \left(\int_0^\omega |u'(t)|^2 dt \right)^{1/2} \omega^{1/2} \\
&\quad + \left(\sum_{j=1}^{m} \left| C_j(t) \int_{-\infty}^{t} K_j(t-s) u(s) ds \right|^2 dt \right)^{1/2} \\
&\quad \times \left(\int_0^\omega |u'(t)|^2 dt \right)^{1/2} \\
&\quad + \sum_{l=1}^{p} \left(\int_0^\omega |D_{0l}(t) u'(t-\tau_l(t))|^2 dt \right)^{1/2} \\
&\quad \times \left(\int_0^\omega |u'(t)|^2 dt \right)^{1/2} \\
&= \left[|r|_0 \omega^{1/2} + \left(|A|_0 \omega^{1/2} + \sum_{i=1}^{n} |B_i|_0 \omega^{1/2} + \sum_{j=1}^{m} |C_j|_0 \right) |u|_0 \right] \\
&\quad \times \left(\int_0^\omega |u'(t)|^2 dt \right)^{1/2} \\
&\quad + \sum_{l=1}^{p} \left(\int_0^\omega |D_{0l}(t) u'(t-\tau_l(t))|^2 dt \right)^{1/2} \\
&\quad \times \left(\int_0^\omega |u'(t)|^2 dt \right)^{1/2} .
\end{aligned} \tag{37}$$

Meanwhile, we see that

$$\begin{aligned}
&\left(\int_0^\omega |D_{0l}(t) u'(t-\tau_l(t))|^2 dt \right)^{1/2} \\
&\quad = \left(\int_0^\omega \frac{1}{1-\tau_l'(\xi_l(t))} |D_{0l}(\xi_l(t)) u'(t)|^2 dt \right)^{1/2} \\
&\quad = \left(\int_0^\omega (1-\tau_l'(\xi_l(t))) |D_l(\xi_l(t)) u'(t)|^2 dt \right)^{1/2} \\
&\quad \le |1-\tau_l'|_0^{1/2} |D_l|_0 \left(\int_0^\omega |u'(t)|^2 dt \right).
\end{aligned} \tag{38}$$

Substituting (38) and (35) into (37), we can find that

$$\begin{aligned}
&\int_0^\omega |u'(t)|^2 dt \\
&\le \Bigg[|r|_0 \omega^{1/2} \\
&\quad + \left(|A|_0 \omega^{1/2} + \sum_{i=1}^{n} |B_i|_0 \omega^{1/2} + \sum_{j=1}^{m} |C_j|_0 \right) \\
&\quad \times \left(\frac{\overline{r_i}}{\eta} + \frac{1}{2} \int_0^\omega |u'(t)| dt \right) \Bigg] \left(\int_0^\omega |u'(t)|^2 dt \right)^{1/2} \\
&\quad + \sum_{l=1}^{p} |1-\tau_l'|_0^{1/2} |D_l|_0 \int_0^\omega |u'(t)|^2 dt \\
&\le \left[|r|_0 \omega^{1/2} + \left(|A|_0 \omega^{1/2} + \sum_{i=1}^{n} |B_i|_0 \omega^{1/2} + \sum_{j=1}^{m} |C_j|_0 \right) \frac{\overline{r_i}}{\eta} \right] \\
&\quad \times \left(\int_0^\omega |u'(t)| dt \right)^{1/2} \\
&\quad + \Bigg[\frac{1}{2} \left(|A|_0 \omega^{1/2} + \sum_{i=1}^{n} |B_i|_0 \omega^{1/2} + \sum_{j=1}^{m} |C_j|_0 \right) \omega^{1/2} \\
&\qquad + \sum_{l=1}^{p} |1-\tau_l'|_0^{1/2} |D_l|_0 \Bigg] \\
&\quad \times \int_0^\omega |u'(t)|^2 dt,
\end{aligned} \tag{39}$$

which gives that

$$\begin{aligned}
&\left(\int_0^\omega |u'(t)|^2 dt \right)^{1/2} \\
&\quad \le \left[|r|_0 \omega^{1/2} + \left(|A|_0 \omega^{1/2} + \sum_{i=1}^{n} |B_i|_0 \omega^{1/2} + \sum_{j=1}^{m} |C_j|_0 \right) \frac{\overline{r_i}}{\eta} \right] \\
&\quad + \Bigg[\frac{1}{2} \left(|A|_0 \omega + \sum_{i=1}^{n} |B_i|_0 \omega + \sum_{j=1}^{m} |C_j|_0 \omega^{1/2} \right) \\
&\qquad + \sum_{l=1}^{p} |1-\tau_l'|_0^{1/2} |D_l|_0 \Bigg] \left(\int_0^\omega |u'(t)| dt \right)^{1/2} .
\end{aligned} \tag{40}$$

From (H_5), we get that $(1/2)(|A|_0\omega + \sum_{i=1}^{n} |B_i|_0 \omega + \sum_{j=1}^{m} |C_j|_0 \omega^{1/2}) + \sum_{l=1}^{p} |1-\tau_l'|_0^{1/2} |D_l|_0 < 1$. Then, there exists a constant $N > 0$ such that

$$\left(\int_0^\omega |u_i'(t)|^2 dt \right)^{1/2} \le N. \tag{41}$$

From (35) and the *Hölder* inequality, we obtain that

$$|u|_0 \le \frac{\overline{r}}{\eta} + \frac{1}{2}\int_0^{\omega} |u'(t)|\, dt \le \frac{\overline{r}}{\eta} + \frac{\omega^{1/2}}{2}\left(\int_0^{\omega} |u'(t)|^2 dt\right)^{1/2} \le \frac{\overline{r}}{\eta} + \frac{\omega^{1/2}}{2} N := N_1. \tag{42}$$

Again from (28), we get that

$$|u'|_0 \le |r|_0 + \left(|A|_0 + \sum_{i=1}^{n}|B_i|_0 + \sum_{j=1}^{m}|C_j|_0\right)|u|_0 + \sum_{l=1}^{p}|D_{0l}|_0 |u'|_0. \tag{43}$$

From condition $\sum_{l=1}^{p}|D_{0l}|_0 \le \sum_{l=1}^{p}|1-\tau_l'|_0|D_l|_0 < 1$, it is easy to see that

$$|u'|_0 \le \frac{|r|_0 + \left(|A|_0 + \sum_{i=1}^{n}|B_i|_0 + \sum_{j=1}^{m}|C_j|_0\right)N_1}{1-\sum_{l=1}^{p}|D_{0l}|_0} := N_2. \tag{44}$$

Now, we take $N_3 > \max\{N_1, N_2, |\overline{r}/(\overline{A} + \sum_{i=1}^{n}\overline{B_i} + \sum_{j=1}^{m}\overline{C_j})|\}$ and $\Omega = \{u \mid u \in C_{\omega}^1, \|u\| < N_3\}$. Then, $k = \sum_{l=1}^{p}|D_{0l}|_0 < 1 \le l(L)$. So by (42) and (44), we can find that all the conditions of Lemma 4 except (2) hold. In what follows, we will prove that condition (2) of Lemma 4 is also satisfied. In order to do this, we defined a bounded bilinear form on $C_{\omega}, C_{\omega}^{!}$ by $[\cdot,\cdot]$ as the following $[y,x] = \int_0^{\omega} y(t)x(t)dt$. Also, we defined $Q : y \to \mathrm{Coker}(L)$ as $y :\to (1/\omega)\int_0^{\omega} y(t)dt$. It is obvious that

$$\left\{u \mid u \in \mathrm{Ker}\, L \bigcap \partial\Omega\right\} = \{u \mid u \equiv N_3, \text{or } -N_3\}. \tag{45}$$

Without loss of generality, suppose that $u \equiv N_3$; then,

$$\begin{aligned}
&[QN(u) + Qr, u]\,[QN(-u) + Qr, u] \\
&= N_3^2\left[\int_0^{\omega} r(t)\, dt - N_3\int_0^{\omega}\left(A(t) + \sum_{i=1}^{n}B_i(t) + \sum_{j=1}^{m}C_j(t)\right)dt\right] \\
&\quad\times\left[\int_0^{\omega} r(t)\, dt + N_3\int_0^{\omega}\left(A(t) + \sum_{i=1}^{n}B_i(t) + \sum_{j=1}^{m}C_j(t)\right)dt\right] \\
&= \omega^2 N_3^2\left[\overline{r} - N_3\left(\overline{A} + \sum_{i=1}^{n}\overline{B_i} + \sum_{j=1}^{m}\overline{C_j}\right)\right] \\
&\quad\times\left[\overline{r} + N_3\left(\overline{A} + \sum_{i=1}^{n}\overline{B_i} + \sum_{j=1}^{m}\overline{C_j}\right)\right],
\end{aligned} \tag{46}$$

since $N_3 > |\overline{r}/(\overline{A} + \sum_{i=1}^{n}\overline{B_i} + \sum_{j=1}^{m}\overline{C_j})|$, then

$$\begin{aligned}
N_3 &> \frac{\overline{r}}{\overline{A} + \sum_{i=1}^{n}\overline{B_i} + \sum_{j=1}^{m}\overline{C_j}}, \\
-N_3 &< -\frac{\overline{r}}{\overline{A} + \sum_{i=1}^{n}\overline{B_i} + \sum_{j=1}^{m}\overline{C_j}}.
\end{aligned} \tag{47}$$

By (46), we get that

$$[QN(u) + Qr, u]\,[QN(-u) + Qr, u] < 0. \tag{48}$$

Therefore, by using Lemma 4, we obtain that (1) has at least one positive ω-periodic solution; the proof of Theorem 13 is completed.

Since $|1-\tau_l'|_0 < 1$, for all $l = 1, 2, \ldots, p$, it follows that $\sum_{l=1}^{p}|1-\tau_l'|_0|D_l|_0 \le \sum_{l=1}^{p}|1-\tau_l'|_0^{1/2}|D_l|_0 < 1$. So from Theorem 13, we have the following result.

Corollary 14. *Suppose that (H_1)–(H_3) and the following conditions hold:*

(H_4') *there exists a constant $\eta > 0$ such that $|\Gamma(t)| \ge \eta$, for all $t \in [0,\omega]$, where $\Gamma(t)$ is defined by (23);*

(H_5') *if $(1/2)(|A|_0\omega + \sum_{i=1}^{n}|B|_0\omega + \sum_{j=1}^{m}|C_j|_0\omega^{1/2}) + \sum_{l=1}^{p}|1-\tau_l'|_0^{1/2}|D_l|_0 < 1$.*

Th n (1) has at least one positive ω-periodic solution.

On the other hand, if $|1-\tau_l'|_0 > 1$, for all $l = 1, 2, \ldots, p$, it follows that $\sum_{l=1}^{p}|1-\tau_l'|_0|D_l|_0 \ge \sum_{l=1}^{p}|1-\tau_l'|_0^{1/2}|D_l|_0$. So from Theorem 13, we also have the following result.

Corollary 15. *Suppose that (H_1)–(H_3) and the following conditions hold:*

(H_4^*) *there exists a constant $\eta > 0$ such that $|\Gamma(t)| \ge \eta$, for all $t \in [0,\omega]$, where $\Gamma(t)$ is defined by (23);*

(H_5^*) *if $(1/2)(|A|_0\omega + \sum_{i=1}^{n}|B|_0\omega + \sum_{j=1}^{m}|C_j|_0\omega^{1/2}) + \sum_{l=1}^{p}|1-\tau_l'|_0|D_l|_0 < 1$.*

Th n, (1) has at least one positive ω-periodic solution.

Our next theorem is concerned with the global attractivity of periodic solution of the system (1).

Theorem 16. *Suppose that* (H_1)–(H_5) *and the following conditions hold:*

(H_6) *there is a positive constant* μ *such that*

$$\sum_{l=1}^{p}|D_l|_0+\int_0^t\left[\sum_{l=1}^{p}|H_l|_0+\sum_{i=1}^{n}|B_i|_0+\sum_{j=1}^{m}|C_j|_0\right]\times\exp\left(-\int_u^t A(\xi)\,d\xi\right)du\le\mu<1; \tag{49}$$

(H_7) $\exp(-\int_0^t A(\xi)d\xi)\rightarrow 0$, *as* $t\rightarrow+\infty$.

Th n, the positive ω*-periodic solution of* (1) *is globally attractive, where* $H_l(t)=[D_l'(t)+D_l(t)A(t)][1-\tau_l'(t)]+D_l(t)\tau_l''(t)/[1-\tau_l'(t)]^2$.

Proof. Suppose that $y(t)=e^{v(t)}$ is a positive ω-periodic solution of (10), $y^*(t)=e^{v^*(t)}$ is another positive solution of (10). Similar to (17), we have

$$\begin{aligned}\frac{dv}{dt}&=r(t)-A(t)v(t)-\sum_{i=1}^{n}B_i(t)v(t-\sigma_i(t))\\&\quad-\sum_{j=1}^{m}C_j(t)\int_{-\infty}^{t}K_j(t-s)v(s)\,ds\\&\quad-\sum_{l=1}^{p}D_{0l}(t)v'(t-\tau_l(t)),\end{aligned} \tag{50}$$

$$\begin{aligned}\frac{dv^*}{dt}&=r(t)-A(t)v^*(t)\\&\quad-\sum_{i=1}^{n}B_i(t)v^*(t-\sigma_i(t))\\&\quad-\sum_{j=1}^{m}C_j(t)\int_{-\infty}^{t}K_j(t-s)v^*(s)\,ds\\&\quad-\sum_{l=1}^{p}D_{0l}(t)v'^*(t-\tau_l(t)).\end{aligned} \tag{51}$$

Let $v(t)-v^*(t)=w(t)$; then,

$$\begin{aligned}\frac{dw}{dt}&=-A(t)w(t)-\sum_{i=1}^{n}B_i(t)w(t-\sigma_i(t))\\&\quad-\sum_{j=1}^{m}C_j(t)\int_{-\infty}^{t}K_j(t-s)w(s)\,ds\\&\quad-\sum_{l=1}^{p}D_{0l}(t)w'(t-\tau_l(t)).\end{aligned} \tag{52}$$

Multiply both sides of (51) with $\exp(\int_0^t A(\xi)d\xi)$ and then integrate from 0 to t to obtain that

$$\begin{aligned}&\int_0^t\left[w(u)\exp\left(\int_0^u A(\xi)\,d\xi\right)\right]'du\\&=-\int_0^t\left[\sum_{i=1}^{n}B_i(u)w(t-\sigma_i(u))\right.\\&\quad+\sum_{j=1}^{m}C_j(u)\int_{-\infty}^{t}K_j(u-s)w(s)\,ds\\&\quad\left.+\sum_{l=1}^{p}D_{0l}(u)w'(u-\tau_l(u))\right]\\&\quad\times\exp\left(\int_0^u A(\xi)\,d\xi\right)du,\end{aligned} \tag{53}$$

Then,

$$\begin{aligned}w(t)&=w(0)\exp\left(-\int_0^t A(\xi)\,d\xi\right)\\&\quad-\int_0^t\left[\sum_{i=1}^{n}B_i(u)w(t-\sigma_i(u))\right.\\&\quad+\sum_{j=1}^{m}C_j(u)\int_{-\infty}^{t}K_j(u-s)w(s)\,ds\\&\quad\left.+\sum_{l=1}^{p}D_{0l}(u)w'(u-\tau_l(u))\right]\\&\quad\times\exp\left(-\int_u^t A(\xi)\,d\xi\right)du.\end{aligned} \tag{54}$$

Meanwhile, we see that

$$\begin{aligned}&\int_0^t\sum_{l=1}^{p}D_{0l}(u)w'(u-\tau_l(u))\exp\left(-\int_u^t A(\xi)\,d\xi\right)du\\&=\int_0^t\sum_{l=1}^{p}\frac{D_{0l}(u)w'(u-\tau_l(u))\left(1-\tau_l'(u)\right)}{1-\tau_l'(u)}\\&\quad\times\exp\left(-\int_u^t A(\xi)\,d\xi\right)du\\&=\int_0^t\sum_{l=1}^{p}\left[\frac{D_{0l}(u)\exp\left(-\int_u^t A(\xi)\,d\xi\right)}{1-\tau_l'(u)}\right]\\&\quad\times\left[w'(u-\tau_l(u))\left(1-\tau_l'(u)\right)\right]du\end{aligned}$$

$$
\begin{aligned}
&= \sum_{l=1}^{p}\left[\frac{D_{0l}(t)}{1-\tau_l'(t)} w(t-\tau_l(t)) - \frac{D_{0l}(0)}{1-\tau_l'(0)} w(-\tau_l(0)) \exp\left(-\int_0^t A(\xi)\,d\xi\right)\right] \\
&\quad - \sum_{l=1}^{p}\int_0^t H_l(u)\exp\left(-\int_u^t A(\xi)\,d\xi\right) w(u-\tau_l(u))\,du \\
&= \sum_{l=1}^{p}\left[D_l(t) w(t-\tau_l(t)) - D_l(0) w(-\tau_l(0)) \exp\left(-\int_0^t A(\xi)\,d\xi\right)\right] \\
&\quad - \sum_{l=1}^{p}\int_0^t H_l(u)\exp\left(-\int_u^t A(\xi)\,d\xi\right) w(u-\tau_l(u))\,du,
\end{aligned} \tag{55}
$$

where

$$
H_l(u) = \frac{\left[D_l'(u) + D_l(t) A(t)\right]\left[1-\tau_l'(t)\right] + D_l(t)\tau_l''(t)}{\left[1-\tau_l'(t)\right]^2}. \tag{56}
$$

Substituting (55) into (54), we get that

$$
\begin{aligned}
w(t) &= \left[w(0) + \sum_{l=1}^{p} D_l(0) w(-\tau_l(0))\right] \times \exp\left(-\int_0^t A(\xi)\,d\xi\right) \\
&\quad + \int_0^t \left[\sum_{l=1}^{p} H_l(u) w(u-\tau_l(u)) - \sum_{i=1}^{n} B_i(u) w(u-\sigma_i(u)) - \sum_{j=1}^{m} C_j(u)\int_{-\infty}^{u} K_j(u-s) w(s)\,ds\right] \\
&\quad \times \exp\left(-\int_u^t A(\xi)\,d\xi\right) du - \sum_{l=1}^{p} D_l(t) w(t-\tau_l(t)).
\end{aligned} \tag{57}
$$

Therefore, we have

$$
\begin{aligned}
|w|_0 &\le \left|w(0) + \sum_{l=1}^{p} D_l(0) w(-\tau_l(0))\right| \times \exp\left(-\int_0^t A(\xi)\,d\xi\right) \\
&\quad + \int_0^t \left[\sum_{l=1}^{p}|H_l|_0 + \sum_{i=1}^{n}|B_i|_0 + \sum_{j=1}^{m}|C_j|_0\right] \\
&\quad \times |w|_0 \exp\left(-\int_u^t A(\xi)\,d\xi\right) du + \sum_{l=1}^{p}|D_l|_0 |w|_0.
\end{aligned} \tag{58}
$$

From (H_6), we have

$$
\begin{aligned}
|w|_0 &\le \left|w(0) + \sum_{l=1}^{p} D_l(0) w(-\tau_l(0))\right| \exp\left(-\int_0^t A(\xi)\,d\xi\right) \\
&\quad \times \left(1 - \sum_{l=1}^{p}|D_l|_0 - \int_0^t \left[\sum_{l=1}^{p}|H_l|_0 + \sum_{i=1}^{n}|B_i|_0 + \sum_{j=1}^{m}|C_j|_0\right] \times \exp\left(-\int_u^t A(\xi)\,d\xi\right) du\right)^{-1} \\
&\le \frac{\left|w(0) + \sum_{l=1}^{p} D_l(0) w(-\tau_l(0))\right| \exp\left(-\int_0^t A(\xi)\,d\xi\right)}{1-\mu}.
\end{aligned} \tag{59}
$$

From (H_7), we have

$$
|w|_0 = |v - v^*|_0 = 0, \quad \text{as } t \longrightarrow +\infty, \tag{60}
$$

thus, $v(t) \rightarrow v^*(t)$, as $t \rightarrow +\infty$; that is, the positive ω-periodic solution of (10) is globally attractive; by Definition 2, the positive ω-periodic solution of (1) is globally attractive. The proof is completed.

Consider the following equation:

$$
\begin{aligned}
\frac{dN(t)}{dt} &= N(t)\left[r(t) - a(t)\ln N(t) - \sum_{i=1}^{n} b_i(t)\ln N(t-\sigma_i(t)) \right. \\
&\quad - \sum_{j=1}^{m} c_j(t)\int_{-\infty}^{t} K_j(t-s)\ln N(s)\,ds \\
&\quad \left. - \sum_{l=1}^{p} d_l(t)\frac{d\ln N(t-\tau_l(t))}{dt}\right],
\end{aligned} \tag{61}
$$

which is a special case of the system (1) without impulse. Similarly, we can get the following conclusions.

Corollary 17. *Suppose that the following conditions hold:*

(H_1^*) *$r(t)$, $a(t)$, $b_i(t)$, $c_j(t)$, $d_l(t)$, $\sigma_i(t)$, $\tau_l(t)$ ($i = 1,2,\ldots,n$; $j = 1,2,\ldots,m$; $l = 1,2,\ldots,p$) are all positive ω-periodic continuous functions with $\sigma_i(t) \geq 0$, $\tau_l(t) \geq 0$, $t \in [0,\omega]$, $\sigma_i'(t) < 1$, $\tau_l'(t) < 1$, for all $i = \{1,2,\ldots,n\}$, for all $l = \{1,2,\ldots,p\}$; furthermore, $d_l(t) \in C^1(R,R)$, $\tau_l(t) \in C^2(R,R)$, $\int_0^{\infty} K_j(s)ds = 1$, $\int_0^{+\infty} sK_j(s)ds < +\infty$, for all $j = \{1,2,\ldots,m\}$, for all $l = \{1,2,\ldots,p\}$;*

(H_2^*) *there exists a constant $\eta > 0$ such that $|\Gamma^*(t)| \geq \eta$, for all $t \in [0,\omega]$, where $\Gamma^*(t)$ is defi ed by the following:*

$$\Gamma^*(t) := a(t) + \sum_{i=1}^{n} \frac{a_i(\zeta_i(t))}{1-\sigma_i'(\zeta_i(t))} + \sum_{j=1}^{m} c_j(t) - \sum_{l=1}^{p} \frac{d_l'(\xi_l(t))}{1-\tau_l'(\xi_l(t))}; \tag{62}$$

(H_3^*) *if $(1/2)(|a|_0\omega + \sum_{i=1}^{n} |b_i|_0\omega + \sum_{j=1}^{m} |c_j|_0\omega^{1/2}) + \sum_{l=1}^{p} |1-\tau_l'|_0^{1/2}|d_l|_0 < 1$, and $\sum_{l=1}^{p} |1-\tau_l'|_0|d_l|_0 < 1$.*

Th n, (61) has at least one positive ω-periodic solution.

Corollary 18. *Suppose that (H_1^*) and the following conditions hold:*

(H_4^*) *there is a positive constant μ such that*

$$\sum_{l=1}^{p} |d_l|_0 + \int_0^t \left[\sum_{l=1}^{p} |h_l|_0 + \sum_{i=1}^{n} |b_i|_0 + \sum_{j=1}^{m} |c_j|_0\right] \times \exp\left(-\int_u^t a(\xi)\,d\xi\right) du \leq \mu < 1; \tag{63}$$

(H_5^*) $\exp(-\int_0^t A(\xi)d\xi) \to 0$, *as $t \to +\infty$.*

Th n, the positive ω-periodic solution of (61) is globally attractive, where $h_l(t) := [d_l'(t) + d_l(t)a(t)][1-\tau_l'(t)] + d_l(t)\tau_l''(t)/[1-\tau_l'(t)]^2$.

Remark 19. One could easily see that the systems (5)–(7) are all special cases of the system (61); we can get the similar results, and we omit them here. Hence, our results improve and generalize the corresponding results in [4–7].

4. An Example

Consider the following impulsive model:

$$\frac{dN(t)}{dt} = N(t)\left[r(t) - a(t)\ln N(t) - b(t)\ln N(t-\sigma(t)) - c(t)\int_{-\infty}^{t} K(t-s)\ln N(s)\,ds - d(t)\frac{d\ln N(t-\tau(t))}{dt}\right], \quad t \neq t_k,$$

$$\Delta N(t_k) = N(t_k^+) - N(t_k) = \theta_k N(t_k), \quad k = 1,2,\ldots, \tag{64}$$

where $r(t)$, $a(t)$, $b(t)$, $c(t)$, $d(t)$, $\sigma(t)$, $\tau(t)$ are all positive ω-periodic continuous functions with $\sigma(t) \geq 0$, $\tau(t) \geq 0$, $t \in [0,\omega]$, $\sigma'(t) < 1$, $\tau'(t) < 1$; furthermore, $d(t) \in C^1(R,R)$, $\tau(t) \in C^2(R,R)$, $\int_0^{\infty} K(s)ds = 1$, $\int_0^{+\infty} sK(s)ds < +\infty$.

Corollary 20. *Suppose that (H_1)–(H_3) and the following conditions hold:*

(H_4) *there exists a constant $\eta > 0$ such that $|\Gamma(t)| \geq \eta$, for all $t \in [0,\omega]$, where $\Gamma(t)$ is defi ed by the following:*

$$\Gamma(t) := A(t) + \frac{B(\zeta(t))}{1-\sigma'(\zeta(t))} + C(t) - \frac{D'(\xi(t))}{1-\tau'(\xi(t))}; \tag{65}$$

(H_5) *if $(1/2)[(|A|_0 + |B|_0)\omega + |C|_0\omega^{1/2}) + |1-\tau'|_0^{1/2}|D|_0 < 1$, and $|1-\tau'|_0|D|_0 < 1$.*

Th n, (64) has at least one positive ω-periodic solution, where

$$A(t) = a(t)\prod_{0<t_k<t}(1+\theta_k),$$
$$B(t) = b(t)\prod_{0<t_k<t-\sigma(t)}(1+\theta_k),$$
$$C(t) = c(t)\prod_{0<t_k<t}(1+\theta_k),$$
$$D(t) = d(t)\prod_{0<t_k<t-\tau(t)}(1+\theta_k). \tag{66}$$

Corollary 21. *Suppose that (H_1)–(H_5) and the following conditions hold:*

(H_6) *there is a positive constant μ such that*

$$|D|_0 + \int_0^t [|H|_0 + |B|_0 + |C|_0] \times \exp\left(-\int_u^t A(\xi)\,d\xi\right) du \leq \mu < 1; \tag{67}$$

(H_7) $\exp(-\int_0^t A(\xi)d\xi) \to 0$, *as $t \to +\infty$.*

Th n, the positive ω-periodic solution of (64) is globally attractive, where $H(t) := [D'(t) + D(t)A(t)][1-\tau'(t)] + D(t)\tau''(t)/[1-\tau'(t)]^2$.

Remark 22. The results in the work show that by means of appropriate impulsive perturbations, we can control the dynamics of these equations.

Acknowledgments

This work was supported by the Construct Program of the Key Discipline in Hunan Province. Research was supported by the National Natural Science Foundation of China (10971229, 11161015), the China Postdoctoral Science Foundation (2012M512162), and Hunan Provincial Natural Science-Hengyang United Foundation of China (11JJ9002).

References

[1] K. Gopalsamy, *Stability and Oscillation in Delay Differential Equations of Population Dynamics*, vol. 74 of *Mathematics and Its Applications*, Kluwer Academic, Dordrecht, The Netherlands, 1992.

[2] G. Kirlinger, "Permanence in Lotka-Volterra equations: linked prey-predator systems," *Mathematical Biosciences*, vol. 82, no. 2, pp. 165–191, 1986.

[3] Y. K. Li, "Attractivity of a positive periodic solution for all other positive solution in a delay population model," *Applied Mathematics*, vol. 12, no. 3, pp. 279–282, 1997 (Chinese).

[4] Y. K. Li, "On a periodic neutral delay logarithmic population model," *Journal of Systems Science and Complexity*, vol. 19, no. 1, pp. 34–38, 1999 (Chinese).

[5] S. P. Lu and W. G. Ge, "Existence of positive periodic solutions for neutral logarithmic population model with multiple delays," *Journal of Computational and Applied Mathematics*, vol. 166, no. 2, pp. 371–383, 2004.

[6] Y. Luo and Z. G. Luo, "Existence of positive periodic solutions for neutral multi-delay logarithmic population model," *Applied Mathematics and Computation*, vol. 216, no. 4, pp. 1310–1315, 2010.

[7] Q. Wang, Y. Wang, and B. X. Dai, "Existence and uniqueness of positive periodic solutions for a neutral Logarithmic population model," *Applied Mathematics and Computation*, vol. 213, no. 1, pp. 137–147, 2009.

[8] F. Hui and L. Jibin, "On the existence of periodic solutions of a neutral delay model of single-species population growth," *Journal of Mathematical Analysis and Applications*, vol. 259, no. 1, pp. 8–17, 2001.

[9] S. P. Lu and W. G. Ge, "Existence of positive periodic solutions for neutral population model with multiple delays," *Applied Mathematics and Computation*, vol. 153, no. 3, pp. 885–902, 2004.

[10] Z. H. Yang and J. D. Cao, "Positive periodic solutions of neutral Lotka-Volterra system with periodic delays," *Applied Mathematics and Computation*, vol. 149, no. 3, pp. 661–687, 2004.

[11] H. F. Huo, "Existence of positive periodic solutions of a neutral delay Lotka-Volterra system with impulses," *Computers and Mathematics with Applications*, vol. 48, no. 12, pp. 1833–1846, 2004.

[12] F. D. Chen, "On a nonlinear nonautonomous predator-prey model with diffusion and distributed delay," *Journal of Computational and Applied Mathematics*, vol. 180, no. 1, pp. 33–49, 2005.

[13] J. R. Yan and A. M. Zhao, "Oscillation and stability of linear impulsive delay differential equations," *Journal of Mathematical Analysis and Applications*, vol. 227, no. 1, pp. 187–194, 1998.

[14] J. S. Yu, "Explicit conditions for stability of nonlinear scalar delay differential equations with impulses," *Nonlinear Analysis: The ry, Methods and Applications*, vol. 46, no. 1, pp. 53–67, 2001.

[15] Y. H. Xia, "Positive periodic solutions for a neutral impulsive delayed Lotka-Volterra competition system with the effect of toxic substance," *Nonlinear Analysis: Real World Applications*, vol. 8, no. 1, pp. 204–221, 2007.

[16] Q. Wang and B. X. Dai, "Existence of positive periodic solutions for a neutral population model with delays and impulse," *Nonlinear Analysis: The ry, Methods and Applications*, vol. 69, no. 11, pp. 3919–3930, 2008.

[17] Y. Zhang and J. Sun, "Stability of impulsive functional differential equations," *Nonlinear Analysis: Theo y, Methods and Applications*, vol. 68, no. 12, pp. 3665–3678, 2008.

[18] G. P. Pang, F. Y. Wang, and L. S. Chen, "Extinction and permanence in delayed stage-structure predator-prey system with impulsive effects," *Chaos, Solitons and Fractals*, vol. 39, no. 5, pp. 2216–2224, 2009.

[19] J. O. Alzabut, "Almost periodic solutions for an impulsive delay Nicholson's blowflies model," *Journal of Computational and Applied Mathematics*, vol. 234, no. 1, pp. 233–239, 2010.

[20] M. Benchohra, J. Henderson, and S. K. Ntouyas, *Impulsive Differential Equations and Inclusions*, vol. 2, Hindawi Publishing Corporation, New York, NY, USA, 2006.

[21] D. D. Bainov and P. Simeonov, *Impulsive Differential Equations: Periodic Solutions and Applications*, Longman, Essex, UK, 1993.

[22] A. M. Samoikleno and N. A. Perestyuk, *Impulsive Differential Equations*, World Scientific, Singapore, 1995.

[23] D. J. Guo, *Nonlinear Functional Analysis*, Shandong Science and Technology Press, 2001.

[24] W. V. Petryshyn and Z. S. Yu, "Existence theorems for higher order nonlinear periodic boundary value problems," *Nonlinear Analysis*, vol. 6, no. 9, pp. 943–969, 1982.

[25] Z. D. Liu and Y. P. Mao, "Existence theorem for periodic solutions of higher order nonlinear differential equations," *Journal of Mathematical Analysis and Applications*, vol. 216, no. 2, pp. 481–490, 1997.

[26] Y. G. Zhou and X. H. Tang, "On existence of periodic solutions of Rayleigh equation of retarded type," *Journal of Computational and Applied Mathematics*, vol. 203, no. 1, pp. 1–5, 2007.

[27] Q. Wang, B. X. Dai, and Y. M. Chen, "Multiple periodic solutions of an impulsive predator-prey model with Holling-type IV functional response," *Mathematical and Computer Modelling*, vol. 49, no. 9-10, pp. 1829–1836, 2009.

9

Non-Darcy Mixed Convection in a Doubly Stratified Porous Medium with Soret-Dufour Effects

D. Srinivasacharya and O. Surender

Department of Mathematics, National Institute of Technology, Warangal Andhra Pradesh 506 004, India

Correspondence should be addressed to D. Srinivasacharya; dsrinivasacharya@yahoo.com

Academic Editor: Viktor Popov

This paper presents the nonsimilarity solutions for mixed convection heat and mass transfer along a semi-infinite vertical plate embedded in a doubly stratified fluid saturated porous medium in the presence of Soret and Dufour effects. The flow in the porous medium is described by employing the Darcy-Forchheimer based model. The nonlinear governing equations and their associated boundary conditions are initially cast into dimensionless forms and then solved numerically. The influence of pertinent parameters on dimensionless velocity, temperature, concentration, heat, and mass transfer in terms of the local Nusselt and Sherwood numbers is discussed and presented graphically.

1. Introduction

The study of mixed convective transport in a doubly stratified (thermal and/or solutal stratification) fluid saturated porous medium has been a topic of continuing interest in the past decades owing to its importance in many industrial and engineering applications. These applications include heat rejection into the environment such as lakes, rivers, and seas; thermal energy storage systems such as solar ponds; and heat transfer from thermal sources such as the condensers of power plants. Numerous studies on mixed convection heat and mass transfer have been reported in the past several decades using both Darcian and non-Darcian models for the porous medium. Comprehensive reviews of convective heat transfer in porous medium can be found in the books by Nield and Bejan [1], Pop and Ingham [2], and Bejan [3]. Non-Darcian models are the extensions of the classical Darcy formulation to incorporate inertial drag effects, vorticity diffusion, and combinations of these effects. Different models such as Brinkman-extended Darcy, Forchheimer-extended Darcy, and the generalized flow models were proposed in the literature to analyze the non-Darcian flow in porous media. The Darcy-Forchheimer model is an extension of classical Darcian formulation obtained by adding a velocity squared term in the momentum equation to account for the inertial effects. Several authors have reported the study of mixed convection heat and mass transfer in porous medium for which the Forchheimer-extended Darcy model is employed.

Stratification of fluid is a deposition/formation of layers and occurs due to temperature variations, concentration differences, or the presence of different fluids. It is important to examine the temperature stratification and concentration differences of hydrogen and oxygen in lakes and ponds as they may directly affect the growth rate of all cultured species. Also, the analysis of thermal stratification is important for solar engineering because higher energy efficiency can be achieved with better stratification. It has been shown by scientists that thermal stratification in energy storage may considerably increase system performance. Although the effect of stratification of the medium on the heat removal process in a porous medium is important, very little work has been reported in the literature. Mukhopadhyay and Ishak [4] presented an analysis for the axisymmetric laminar boundary layer mixed convection flow of a viscous and incompressible fluid towards a stretching cylinder immersed in a thermally stratified medium. The influence of thermal dispersion and stratification on the flow and temperature fields for mixed convection from a vertical plate embedded in a porous medium has been investigated by Hassanien et al. [5]. Steady, laminar, hydromagnetic simultaneous heat and mass transfer by mixed convection flow over a vertical plate embedded in a uniform porous medium with a stratified free stream and taking into account the presence of thermal dispersion has been investigated for the case of power-law variations of both the wall temperature and concentration by Chamkha and Khaled [6]. Ishak et al. [7] investigated the mixed convection boundary layer flow through a stable stratified porous medium bounded by a vertical surface. The effects of variable viscosities and thermal stratification on the MHD mixed convective heat and mass transfer of a viscous, incompressible, and electrically conducting fluid past a porous wedge in the presence of a chemical reaction have been investigated by Muhaimin et al. [8]. Using Galerkin finite element method, numerical investigation of mixed convection flow in a concentration-stratified fluid-saturated vertical square porous enclosure has been investigated by Rathish Kumar and Krishna Murthy [9].

In all the aforementioned papers, the significance of Dufour and Soret was neglected on the basis that they are of a smaller order of magnitude than the effects described by Fourier's and Fick's laws. However, Eckert and Drake [10] reported several cases when the Dufour effect cannot be neglected. Diffusion-thermal or Dufour effect corresponds to the energy flux caused by a concentration gradient. On the other hand, thermal diffusion or Soret effect corresponds to species differentiation occurring in an initial homogeneous mixture submitted to a thermal gradient. Due to the importance of Dufour and Soret effects for the fluids with very light molecular weight as well as medium molecular weight, many investigators have studied and reported results for these flows. Seddeek [11] analyzed the thermal-diffusion and the diffusion-thermal effects on the mixed free-forced convective and mass transfer steady laminar boundary-layer flow over an accelerating surface with a heat source in the presence of suction and blowing. The influence of a magnetic field on heat and mass transfer by mixed convection from vertical surfaces in the presence of Hall, radiation, Soret, and Dufour effects has been investigated by Shateyi et al. [12]. The Soret and Dufour effects on the steady, laminar mixed convection heat and mass transfer along a semi-infinite vertical plate embedded in a non-Darcy porous medium saturated with micropolar fluid have been studied by Srinivasacharya and Ramreddy [13]. Cheng [14] studied the Soret and Dufour effects on the boundary layer flow due to mixed convection heat and mass transfer over a downward-pointing vertical wedge in a porous medium saturated with Newtonian fluids with constant wall temperature and concentration.

In this paper, we made an attempt to obtain the nonsimilar solutions for mixed convection on a vertical plate with constant and uniform wall temperature and concentration in a stable doubly stratified non-Darcian fluid in which the ambient temperature and concentration vary linearly. Soret and Dufour effects are considered. The Keller-box method given in Cebeci and Bradshaw [15] is employed to solve the nonlinear system of this particular problem. The influence of stratification parameters, Lewis number, Forchheimer number, Buoyancy parameter, mixed convection parameter, Soret, and Dufour parameters on physical quantities are examined and displayed graphically.

2. Mathematical Formulation

Consider non-Darcian mixed convective heat and mass transfer along a semi-infinite vertical plate in a stable, doubly stratified viscous fluid saturated porous medium with uniform velocity U_∞ far away from the plate with Soret and Dufour effects. In the formulation of the present problem, the following assumptions are made.

(i) The flow is steady, laminar, incompressible, two dimensional.

(ii) The porous medium is homogeneous and isotropic (i.e., uniform with a constant porosity and permeability).

(iii) The fluid has constant properties except the density in the buoyancy term of the balance of momentum equation.

(iv) The fluid flow is moderate, so the pressure drop is proportional to the linear combination of fluid velocity and the square of velocity.

(v) The Boussinesq and boundary-layer approximations are applicable.

The x coordinate is taken along the plate, in the ascending direction and the y coordinate is measured normal to the plate, while the origin of the reference system is considered at the leading edge of the vertical plate. The physical model and the coordinate system are shown in Figure 1. The plate is maintained at uniform wall temperature and concentration T_w and C_w, respectively. The ambient medium is assumed to be vertically linearly stratified with respect to both temperature and concentration in the form $T_\infty(x) = T_{\infty,0} + Ax$, $C_\infty(x) = C_{\infty,0} + Bx$, where A and B are constants varied to alter the intensity of stratification in the medium and $T_{\infty,0}$ and $C_{\infty,0}$ are ambient temperature and concentration, respectively. With the above assumptions and using the Darcy-Forchheimer model, the governing equations for flow are given by

$$\frac{\partial u}{\partial x} + \frac{\partial v}{\partial y} = 0, \tag{1}$$

$$\frac{\partial u}{\partial y} + \frac{2c\sqrt{K}}{\nu} u \frac{\partial u}{\partial y} = \frac{Kg\beta_T}{\nu}\frac{\partial T}{\partial y} + \frac{Kg\beta_C}{\nu}\frac{\partial C}{\partial y}, \tag{2}$$

$$u\frac{\partial T}{\partial x} + v\frac{\partial T}{\partial y} = \alpha\frac{\partial^2 T}{\partial y^2} + \frac{DK_T}{C_s C_p}\frac{\partial^2 C}{\partial y^2}, \tag{3}$$

$$u\frac{\partial C}{\partial x} + v\frac{\partial C}{\partial y} = D\frac{\partial^2 C}{\partial y^2} + \frac{DK_T}{T_m}\frac{\partial^2 T}{\partial y^2}, \tag{4}$$

where u and v are the average velocity components in x and y directions, respectively, T is the temperature, C is the concentration, β_T and β_C are the thermal and solutal expansion coefficients, respectively, ν is the kinematic viscosity of the fluid, K is the permeability, g is the acceleration due to gravity, α is the thermal diffusivity of the porous medium and D is the solutal diffusivity of the porous medium, K_T is thermal diffusion ratio, C_s is concentration susceptibility, C_p

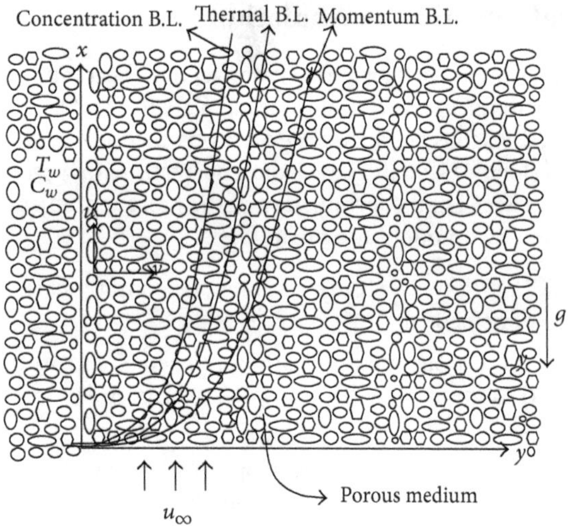

FIGURE 1: Physical model and coordinate system.

is specific heat capacity, and T_m is mean fluid temperature. The last terms in (3) and (4) are due to Dufour and Soret effects, respectively.

The boundary conditions are

$$v = 0, \quad T = T_w, \quad C = C_w \quad \text{at } y = 0, \tag{5a}$$

$$u = u_\infty, \quad T = T_\infty(x), \quad C = C_\infty(x) \quad \text{as } y \longrightarrow \infty, \tag{5b}$$

where the subscripts w, $(\infty, 0)$, and ∞ indicate the conditions at the wall, at some reference point in the medium, and at the outer edge of the boundary layer, respectively.

3. Method of Solution

The continuity equation (1) is satisfied by introducing the stream function ψ such that

$$u = \frac{\partial \psi}{\partial y}, \qquad v = -\frac{\partial \psi}{\partial x}. \tag{6}$$

Introducing the following nondimensional variables:

$$\xi = \frac{x}{L}, \quad \eta = \frac{\mathrm{Pe}^{1/2}}{L\xi^{1/2}} y, \quad \psi = \alpha \mathrm{Pe}^{1/2}\xi^{1/2} f(\xi, \eta)$$
$$T - T_\infty(x) = (T_w - T_{\infty,0})\,\theta(\xi, \eta),$$
$$C - C_\infty(x) = (C_w - C_{\infty,0})\,\phi(\xi, \eta). \tag{7}$$

Substituting (6) and (7) in (2), (3), and (4), we obtain

$$f'' + 2F_c f' f'' = \frac{\mathrm{Ra}}{\mathrm{Pe}}\left[\theta' + B\phi'\right], \tag{8}$$

$$\theta'' + \frac{1}{2} f\theta' - \epsilon_1 \xi f' + D_f \phi'' = \xi\left[f'\frac{\partial \theta}{\partial \xi} - \theta'\frac{\partial f}{\partial \xi}\right], \tag{9}$$

$$\frac{1}{\mathrm{Le}}\phi'' + \frac{1}{2} f\phi' - \epsilon_2 \xi f' + S_r \theta'' = \xi\left[f'\frac{\partial \phi}{\partial \xi} - \phi'\frac{\partial f}{\partial \xi}\right], \tag{10}$$

where the prime denotes differentiation with respect to η, $\mathrm{Ra} = Kg\beta_T(T_w - T_{\infty,0})L/\alpha\nu$ is the Rayleigh number, $\mathrm{Pr} = \nu/\alpha$ is the Prandtl number, $F_c = c\sqrt{K}\mathrm{Pe}/L\mathrm{Pr}$ is the Forchheimer number, $\mathrm{Le} = \alpha/D$ is the diffusivity ratio and $B = \beta_C(C_w - C_{\infty,0})/\beta_T(T_w - T_{\infty,0})$ is the buoyancy ratio, $\epsilon_1 = AL/(T_w - T_{\infty,0})$ and $\epsilon_2 = BL/(C_w - C_{\infty,0})$ are the thermal and solutal stratification parameters, respectively, $D_f = (DK_T/\alpha C_s C_p)((C_w - C_{\infty,0})/(T_w - T_{\infty,0}))$ is Dufour parameter, and $S_r = (DK_T/\alpha T_m)((T_w - T_{\infty,0})/(C_w - C_{\infty,0}))$ is Soret parameter.

The boundary conditions (5a) and (5b) in terms of f, θ, and ϕ becomes

$$f(\xi, 0) + 2\xi\left(\frac{\partial f}{\partial \xi}\right)_{\eta=0} = 0, \qquad \theta(\xi, 0) = 1 - \epsilon_1 \xi,$$
$$\phi(\xi, 0) = 1 - \epsilon_2 \xi, \tag{11a}$$

$$f'(\xi, \infty) = 1, \qquad \theta(\xi, \infty) = 0, \qquad \phi(\xi, \infty) = 0. \tag{11b}$$

Results of practical interest are both heat and mass transfer rates. The local Nusselt number Nu_ξ and the local Sherwood number Sh_ξ are, respectively, given by

$$\frac{\mathrm{Nu}_\xi}{\mathrm{Pe}^{1/2}} = -\xi^{1/2}\theta'(\xi, 0), \qquad \frac{\mathrm{Sh}_\xi}{\mathrm{Pe}^{1/2}} = -\xi^{1/2}\phi'(\xi, 0). \tag{12}$$

4. Results and Discussion

Equations (8)–(10) with the boundary conditions (11a) and (11b) constitute a nonlinear nonhomogeneous differential equations for which closed form solution cannot be obtained. Hence, these equations have been solved numerically using an implicit finite-difference method known as the Keller-box scheme [15]. This method has four main steps. The first step is converting (8) to (10) into a system of first-order equations. The second step is replacing partial derivatives by central finite difference approximation. The third step is linearizing the nonlinear algebraic equations by Newton's method and then casting as the matrix vector form. The last step is solving linearized system of equations using the block-tridiagonal-elimination technique. Here, the initial values for velocity temperature and concentration are arbitrarily chosen so that they satisfy the boundary conditions. The independence of the results at least up to the 4th decimal place on the mesh density was examined. A convergence criterion based on the relative difference between the current and previous iterations was used. When this difference reached 10^{-5}, the solutions were assumed to have converged and the iterative process was terminated. This method has been proven to be adequate and give accurate results for boundary layer equations. In the present study, the boundary conditions for η at ∞ are replaced by sufficiently large value of η, where the velocity, temperature, and concentration profiles approach to zero. We have taken $\eta_\infty = 8$ and a grid size of η of 0.01 and $\xi = 0.1$ is fixed. In order to study the effects of stratification parameters, ϵ_1 and ϵ_2 computations were carried out for the fixed values of $F_c = 0.5$, $\mathrm{Le} = 1.0$, $B = 0.5$, $D_f = 0.3$, $S_r = 0.2$, and $\mathrm{Ra}/\mathrm{Pe} = 1.0$ while ϵ_1 and ϵ_2 were varied over a range.

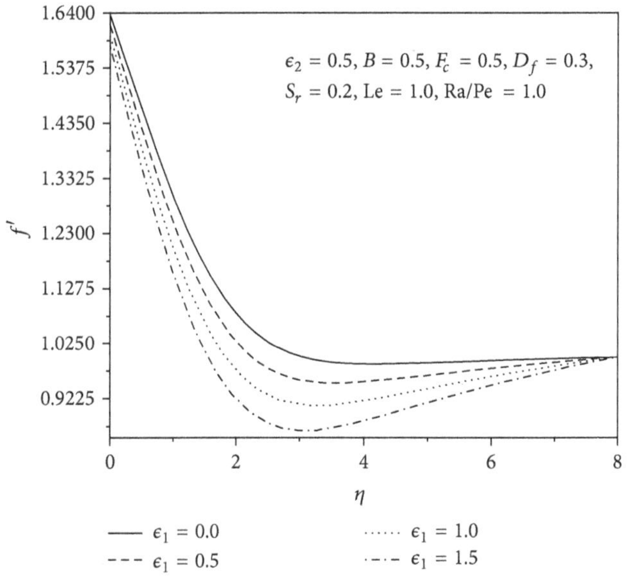

FIGURE 2: Variation of nondimensional velocity with thermal stratification parameter.

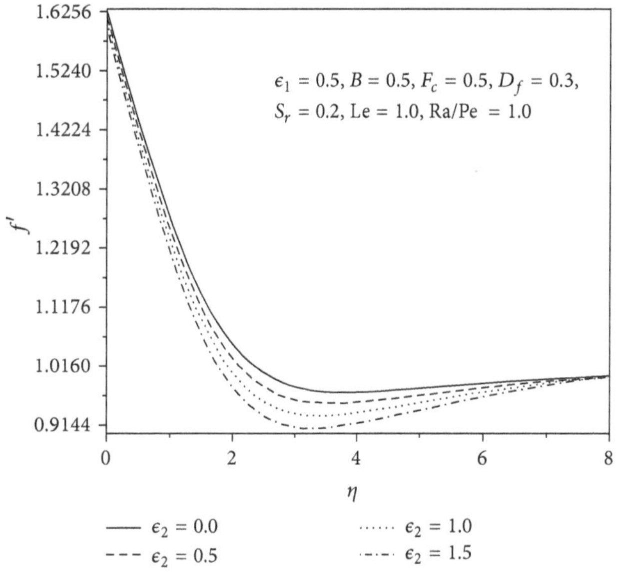

FIGURE 4: Variation of nondimensional velocity with solutal stratification parameter.

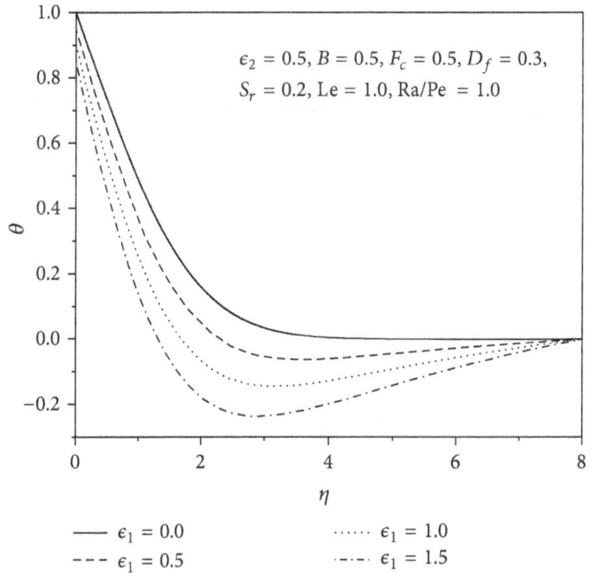

FIGURE 3: Variation of nondimensional temperature with thermal stratification parameter.

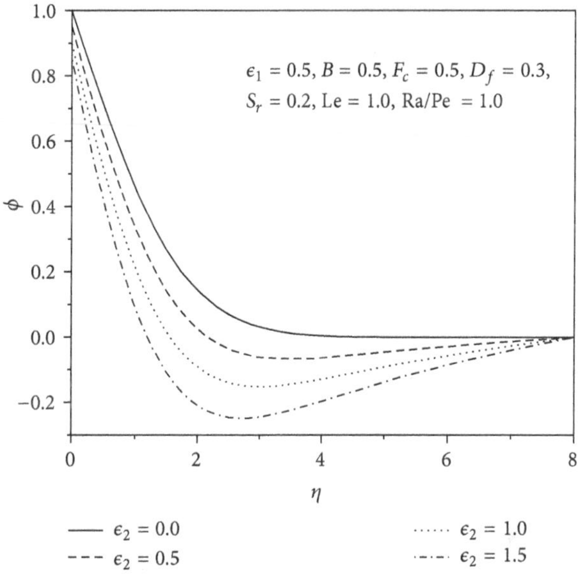

FIGURE 5: Variation of nondimensional concentration with solutal stratification parameter.

The variation of the nondimensional velocity and temperature profiles with η for different values of thermal stratification parameter ϵ_1 is illustrated in Figures 2 and 3. It is observed from Figure 2 that the velocity of the fluid decreases by nearly 11% with the increase of thermal stratification parameter ϵ_1 from 0 to 1.5. This is due to that thermal stratification reduces the effective convective potential between the heated plate and ambient fluid in the medium. Hence, the thermal stratification effect reduces the velocity in the boundary layer. From Figure 3, it is clear that the temperature of the fluid decreases by around 72% with the increase of thermal stratification parameter ϵ_1 from 0 to 1.5. When the thermal stratification is taken into consideration, the effective temperature difference between the plate and the ambient fluid will decrease; therefore, the thermal boundary layer is thickened and the temperature is reduced. The figure depicting the effect of thermal stratification parameter on nondimensional concentration is not included as the variation is very less.

Figures 4 and 5 depict the effect of solutal stratification parameter ϵ_2 on the nondimensional velocity, and concentration. It is noticed from Figure 4 that the velocity of the fluid decreases by about 5% with the increase of solutal stratification parameter ϵ_2 from 0 to 1.5. From Figure 5, it

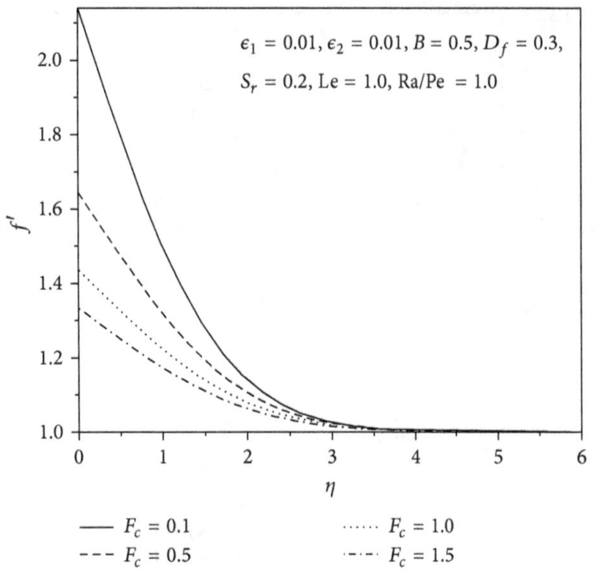

FIGURE 6: Variation of nondimensional velocity with Forchheimer parameter.

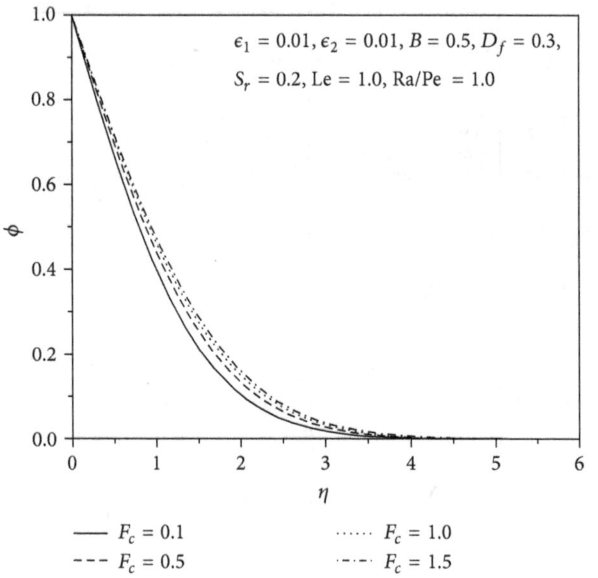

FIGURE 8: Variation of nondimensional concentration with Forchheimer parameter.

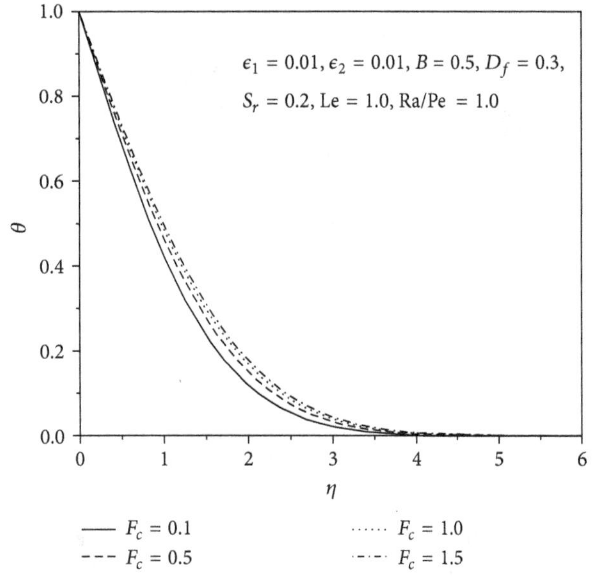

FIGURE 7: Variation of nondimensional temperature with Forchheimer parameter.

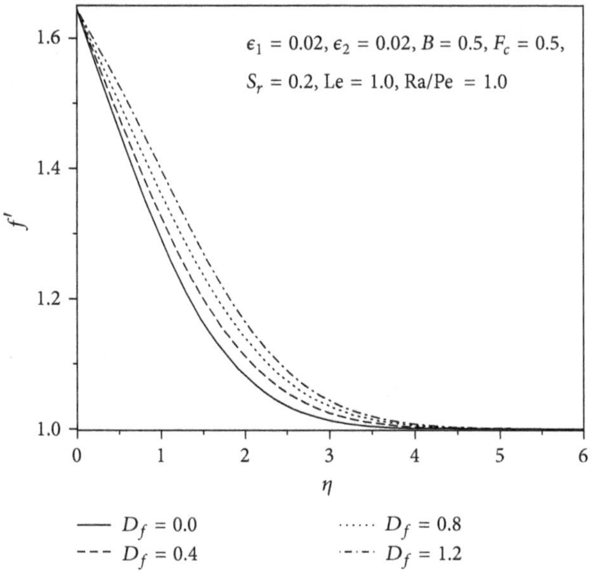

FIGURE 9: Variation of nondimensional velocity with Dufour parameter.

is observed that the concentration of the fluid decreases by about 80% with the increase of the solutal stratification parameter ϵ_2 from 0 to 1.5. The graph showing the influence of solutal stratification on nondimensional temperature is not included as the impact is very low. It is observed that the nondimensional temperature and concentration values are becoming negative inside the boundary layer for different values of the stratification parameters depending on the values of other parameters. This is in tune with the observation made in [16–19]. This is because the fluid near the plate can have temperature or concentration lower than the ambient medium.

Figures 6–8 depict the effect of Forchheimer number F_c on the nondimensional velocity, temperature, and concentration. It is observed from Figure 6 that the velocity of the fluid decreases by nearly 22% with the increase of Forchheimer number F_c from 0.1 to 1.5. Since F_c represents the inertial drag, thus an increase in the Forchheimer number increases the resistance to the flow and so a decrease in the fluid velocity ensues. Here $F_c = 0$ represents the case where the flow is Darcian. The velocity is maximum in this case due to the total absence of inertial drag. It is noticed from Figure 7 that the temperature of the fluid increases by about 18% with the increase of Forchheimer number F_c from

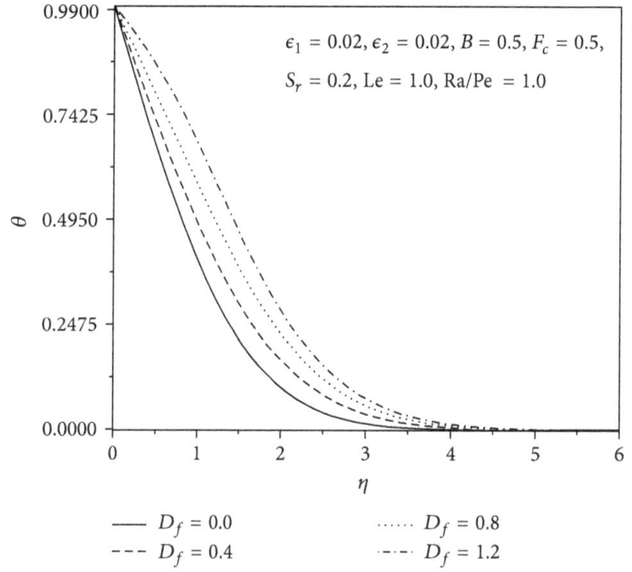

FIGURE 10: Variation of nondimensional temperature with Dufour parameter.

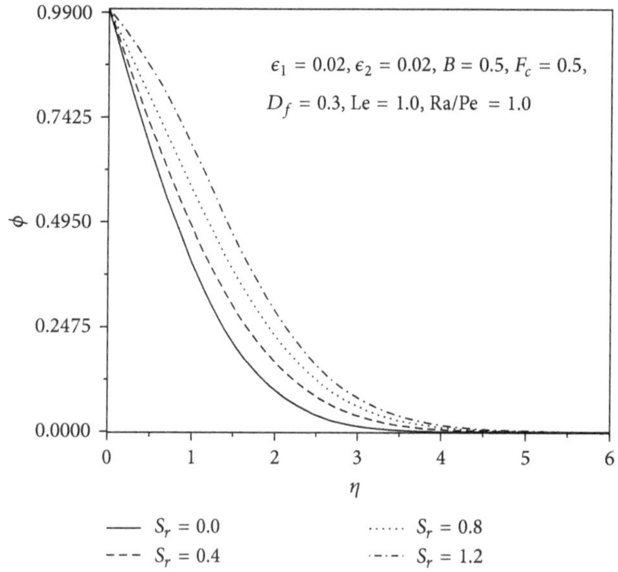

FIGURE 12: Variation of nondimensional concentration with Soret parameter.

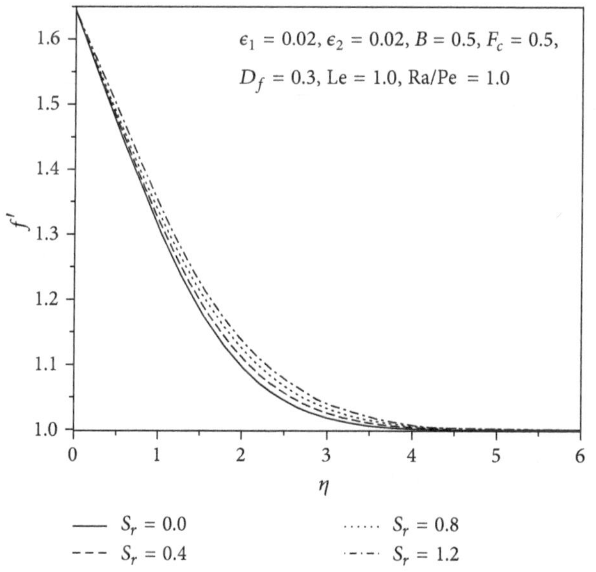

FIGURE 11: Variation of nondimensional velocity with Soret parameter.

0.1 to 1.5. An increase in F_c leads to rise temperature. As the fluid is decelerated, energy is dissipated as heat and it serves to enhance temperature in the boundary layer. From Figure 8, it is observed that the concentration of the fluid increases 19% with the increase of the Forchheimer number F_c from 0.1 to 1.5. As the Forchheimer number increases, the concentration boundary layer thickness increases. The increase in non-Darcy parameter reduces the intensity of the flow but enhances the thermal and concentration boundary layer thicknesses.

Figures 9 and 10 present the variation of nondimensional velocity and temperature with Dufour parameter D_f. It is seen from Figure 9 that the velocity of the fluid is enhanced by nearly 9% with the increase of Dufour parameter D_f from 0 to 1.2. The nondimensional temperature is enhanced by around 69% with the increase of Dufour parameter D_f from 0 to 1.2 as shown in Figure 10. The Dufour parameter denotes the contribution of the concentration gradients to the thermal energy flux in the flow. It can be seen that an increase in the Dufour parameter produces a significant increase in the velocity and temperature. The graph showing the effect of Dufour parameter on concentration is not included in the paper as the impact is very low.

Figures 11 and 12 depict the effect of Soret parameter S_r on nondimensional velocity and concentration. It is noticed from Figure 11 that the velocity of the fluid increased by nearly 4% with rise of Soret parameter S_r from 0 to 1.2. Soret parameter is the ratio of temperature difference to the concentration. Hence, the higher value of Soret parameter stands for a larger temperature difference and precipitous gradient. Thus the fluid velocity enhances due to greater thermal diffusion factor. Figure 12 shows that the nondimensional concentration is enhanced by about 70% with the raise of Soret parameter S_r from 0 to 1.2. The figure depicting the effect of Soret parameter on nondimensional temperature is not presented in this paper due to less variation in the temperature with varying Soret parameters.

The variation of local heat and mass transfer coefficients (Nusselt number Nu_ξ and Sherwood number Sh_ξ) with thermal and solutal stratification parameters is presented in Figures 13 and 14. It is found from Figure 13 that the local heat transfer rate enhances by nearly 54% with the increase in the value of thermal stratification parameter ϵ_1 from 1.0 to 2.0. Physically, positive values of the stratification parameter have the tendency to decrease the boundary layer thickness due to the reduction in the temperature difference between the plate and the free stream. This causes increase in the Nusselt

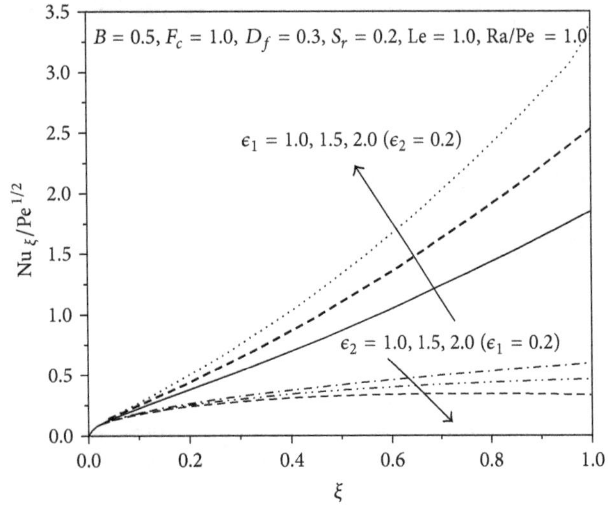

FIGURE 13: Variation of heat transfer rate with stratification parameters.

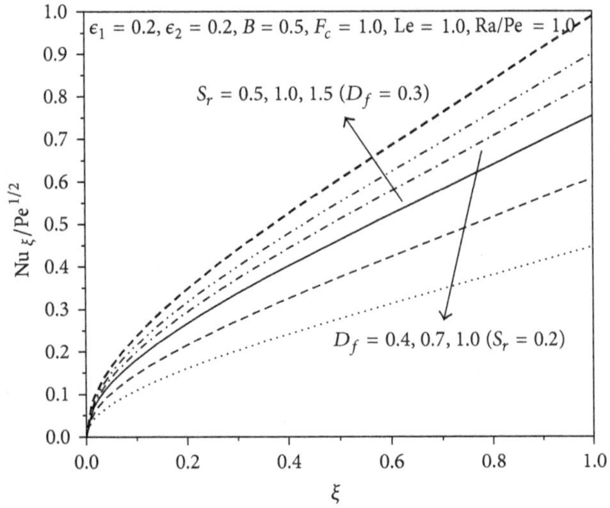

FIGURE 15: Variation of heat transfer rate with Dufour and Soret parameters.

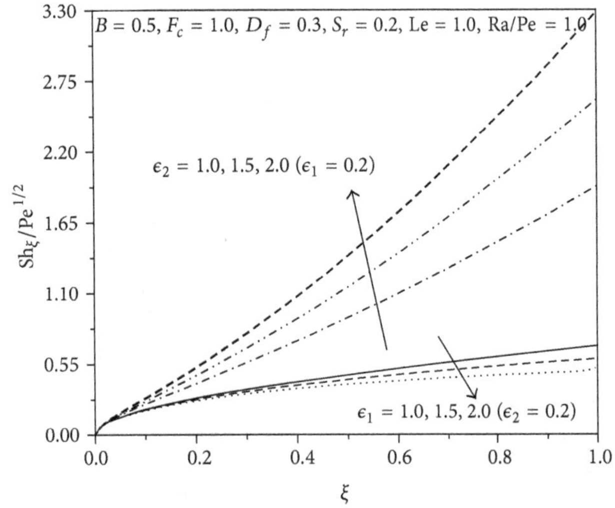

FIGURE 14: Variation of mass transfer rate with stratification parameters.

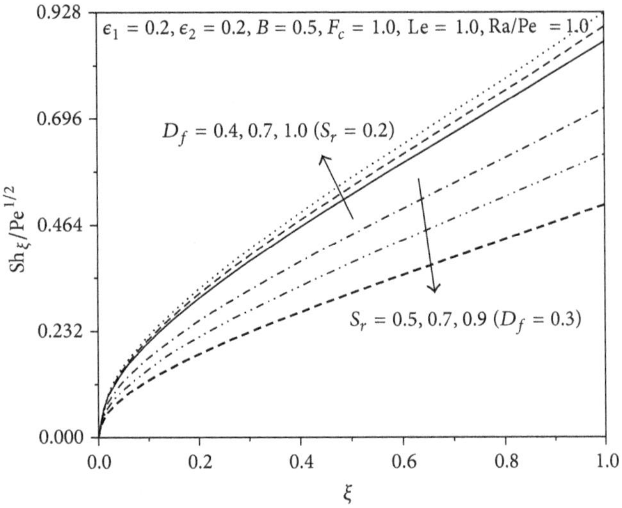

FIGURE 16: Variation of mass transfer rate with Dufour and Soret parameters.

number. It is also clear from the same graph that the local heat transfer rate decreases by around 22% with the raise of ϵ_2 from 1.0 to 2.0. Figure 14 illustrates that the local mass transfer coefficient decreases by closely 16% with the increase in the thermal stratification parameter ϵ_1 from 1.0 to 2.0. This is due to the effective mass transfer between the plate and the ambient medium decreases as the thermal stratification effect increases. It is seen from the same figure that the local mass transfer coefficient enhances by 53% with the increase of solutal stratification parameter ϵ_2 from 1.0 to 2.0.

The effect of Dufour and Soret parameters on local heat and mass transfer coefficients is exhibited in Figures 15 and 16. It is found from Figure 15 that the local heat transfer rate enhances by about 18% with the increase in Soret parameter S_r from 0.5 to 1.5 but decreases by 40% with the raise of Dufour parameter D_f from 0.4 to 1.0. Figure 16 reveals that the local mass transfer coefficient increases by nearly 7% with the increase in the Dufour parameter D_f from 0.4 to 1.0, and we notice that the local mass transfer coefficient enhances by about 29% with the decrease in the Soret parameter S_r from 0.9 to 0.5.

The variation of local heat and mass transfer coefficients with Forchheimer number and Lewis number is shown in Figures 17 and 18. It is found from Figure 17 that local heat transfer rate decreases by about 5% with the increase of Forchheimer number F_c from 0.5 to 0.9, and mass transfer rate decreases by 7% with the increase in Forchheimer number F_c from 0.1 to 0.5. Since F_c represents the inertial drag, thus an increase in the Forchheimer number increases the resistance to the flow. Figure 18 shows that the local heat transfer rate decreases by about 24% with the increase in Lewis number Le from 1.0 to 3.0, whereas the mass transfer

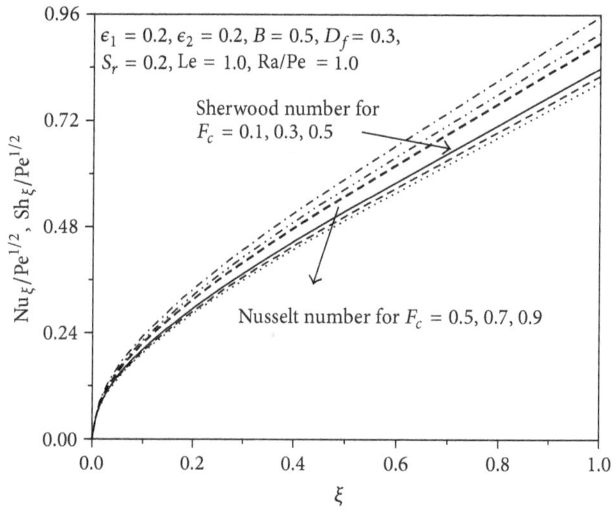

FIGURE 17: Variation of heat and mass transfer rates with Forchheimer parameter.

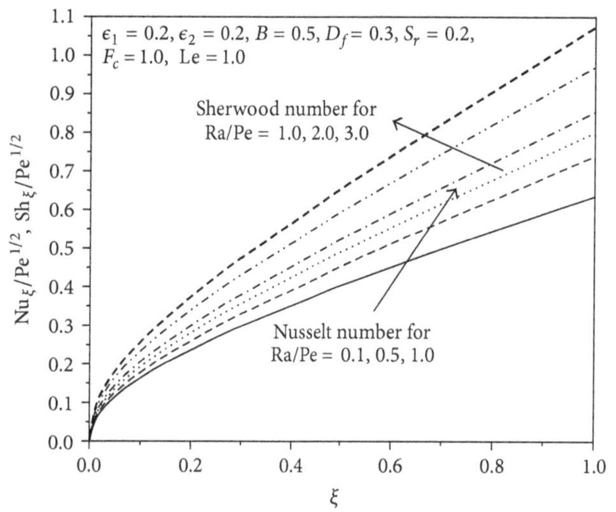

FIGURE 19: Variation of heat and mass transfer rates with mixed convection parameter.

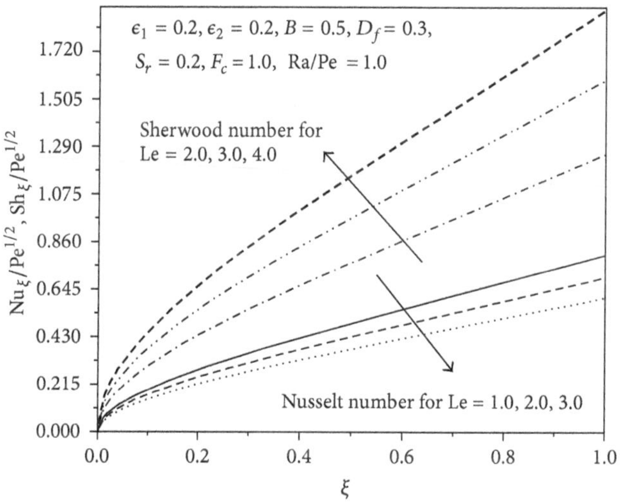

FIGURE 18: Variation of heat and mass transfer rates with Lewis number.

rate is enhanced by nearly 51% with the raise of Lewis number Le from 2.0 to 4.0.

The influence of mixed convection parameter on local heat and mass transfer coefficients is shown in Figure 19. The figure depicts that the local heat transfer rate increases by 22% with the increase of mixed convection parameter from 0.1 to 1.0, and mass transfer rate increase by about 25% with the increase of mixed convection parameters from 1.0 to 3.0.

5. Conclusions

Mixed convection heat and mass transfer from a vertical plate in a doubly stratified viscous fluid saturated non-Darcy porous medium in the presence of Soret and Dufour effects is studied. Numerically, nonsimilar solutions are obtained for different values of thermal stratification parameter, solutal stratification parameter, buoyancy parameter, Forchheimer number, Soret, and Dufour parameters. An increase in the thermal stratification parameter, ϵ_1, decreases the velocity, temperature, and local mass transfer coefficient but increases local heat transfer coefficient. The higher value of solutal stratification parameter ϵ_2 resulting in lower velocity, concentration, and local heat transfer coefficient but higher local mass transfer coefficient. The influence of Forchheimer number is to decrease the velocity and the local heat and mass transfer coefficients but to increase nondimensional temperature and concentration. The influence of Dufour parameter is to increase the nondimensional velocity and temperature. The higher value of Soret parameter results in the higher velocity and concentration. The presence of Soret parameter increases the local heat transfer rate but decreases the local mass transfer rate. The local heat transfer rate is decreased and local mass transfer rate is increased due to the presence of Dufour parameter. The local heat transfer rate is decreased whereas the local mass transfer rate is increased with the increase of Lewis number. The significance of mixed convection parameter is to increase both the local heat and mass transfer rates.

Conflict of Interests

The authors declare that there is no conflict of interests regarding the publication of this paper.

References

[1] D. A. Nield and A. Bejan, *Convection in Porous Media*, Springer, New York, NY, USA, 4th edition, 2013.

[2] I. Pop and D. B. Ingham, *Convective Heat Transfer Mathematical and Computational Modelling of Viscous Fluids and Porous Media*, Elsevier Science & Technology Books, Pergaman, UK, 2001.

[3] A. Bejan, *Convection Heat Transfer*, John Wiley, New York, NY, USA, 1994.

[4] S. Mukhopadhyay and A. Ishak, "Mixed convection flow along a stretching cylinder in a thermally stratified medium," *Journal of Applied Mathematics*, vol. 2012, Article ID 491695, 8 pages, 2012.

[5] I. A. Hassanien, A. Y. Bakier, and R. S. R. Gorla, "Effects of thermal dispersion and stratification on non-Darcy mixed convection from a vertical plate in a porous medium," *Heat and Mass Transfer*, vol. 34, no. 2-3, pp. 209–212, 1998.

[6] A. J. Chamkha and A.-R. A. Khaled, "Hydromagnetic simultaneous heat and mass transfer by mixed convection from a vertical plate embedded in a stratified porous medium with thermal dispersion effects," *Heat and Mass Transfer*, vol. 36, no. 1, pp. 63–70, 2000.

[7] A. Ishak, R. Nazar, and I. Pop, "Mixed convection boundary layer flow over a vertical surface embedded in a thermally stratified porous medium," *Physics Letters A*, vol. 372, no. 14, pp. 2355–2358, 2008.

[8] I. Muhaimin, R. Kandasamy, and A. B. Khamis, "Numerical investigation of variable viscosities and thermal stratification effects on MHD mixed convective heat and mass transfer past a porous wedge in the presence of a chemical reaction," *Applied Mathematics and Mechanics*, vol. 30, no. 11, pp. 1353–1364, 2009.

[9] B. V. Rathish Kumar and S. V. S. S. N. V. G. Krishna Murthy, "A finite element study of double diffusive mixed convection in a concentration stratified Darcian fluid saturated porous enclosure under injection/suction effect," *Journal of Applied Mathematics*, vol. 2012, Article ID 594701, 29 pages, 2012.

[10] E. R. G. Eckert and R. M. Drake, *Analysis of Heat and Mass Transfer*, McGraw Hill, New York, NY, USA, 1972.

[11] M. A. Seddeek, "Thermal-diffusion and diffusion-thermo effects on mixed free-forced convective flow and mass transfer over an accelerating surface with a heat source in the presence of suction and blowing in the case of variable viscosity," *Acta Mechanica*, vol. 172, no. 1-2, pp. 83–94, 2004.

[12] S. Shateyi, S. S. Motsa, and P. Sibanda, "The effects of thermal radiation, hall currents, soret, and dufour on MHD flow by mixed convection over a vertical surface in porous media," *Mathematical Problems in Engineering*, vol. 2010, Article ID 627475, 20 pages, 2010.

[13] D. Srinivasacharya and C. Ramreddy, "Soret and Dufour effects on mixed convection in a non-Darcy porous medium saturated with micropolar fluid," *Nonlinear Analysis: Modelling and Control*, vol. 16, no. 1, pp. 100–115, 2011.

[14] C.-Y. Cheng, "Soret and dufour effects on mixed convection heat and mass transfer from a vertical wedge in a porous medium with constant wall temperature and concentration," *Transport in Porous Media*, vol. 94, pp. 123–132, 2012.

[15] T. Cebeci and P. Bradshaw, *Physical and Computational Aspects of Convective Heat Transfer*, Springer, New York, NY, USA, 1984.

[16] P. A. L. Narayana and P. V. S. N. Murthy, "Soret and dufour effects on free convection heat and mass transfer in a doubly stratified darcy porous medium," *Journal of Porous Media*, vol. 10, no. 6, pp. 613–623, 2007.

[17] B. Gebhart, Y. Jaluria, R. Mahajan, and B. Sammakia, *Buoyancy Induced Flows and Transport*, Hemisphere, New York, NY, USA, 1998.

[18] L. Prandtl, *Essentials of Fluid Dynamics*, Blackie, London, UK, 1952.

[19] Y. Jaluria and K. Himasekhar, "Buoyancy-induced two-dimensional vertical flows in a thermally stratified environment," *Computers and Fluids*, vol. 11, no. 1, pp. 39–49, 1983.

10

Numerical Performance of Higher-Order Semicompact Scheme for Arbitrary Triangular Cavity Flow

Xiaofeng Wang[1,2] and Dongyang Shi[1]

[1] *School of Mathematics and Statistics, Zhengzhou University, Zhengzhou 450001, China*
[2] *School of Mathematical Sciences, Henan Institute of Science and Technology, Xinxiang 453003, China*

Correspondence should be addressed to Xiaofeng Wang; wangxiaofengl166@163.com

Academic Editor: Song Cen

An efficient fourth-order semicompact finite difference scheme has been developed to solve steady incompressible Navier-Stokes (N-S) equations in stream function and vorticity formulation in a triangular cavity of arbitrary geometry. The governing equations are transformed into curvilinear coordinates by a simple linear transformation to handle the nonregular geometry of the problem. The main feature of the new higher-order semicompact scheme is that it can calculate a triangle flow with arbitrary shape for high Reynolds numbers. It is found that the solutions obtained with the present scheme are in good agreement with the analytical results or with the existing results depending on the availability.

1. Introduction

Numerical simulations for the solution of steady incompressible viscous flow within a driven cavity are a complex and significant topic, and a lot of researchers contributed to this subject [1–4]. Although there are still some minor discrepancies in the results, the square cavity flow has been essentially characterized [5]. As pointed out in McQuain et al. [6], the results for the square cavity may not be applied to other important geometries such as a triangular cavity. Also the latter shapes are more common in practice.

The problem under consideration is the steady motion of an incompressible viscous flow in a triangular cavity of arbitrary geometry. This flow was studied numerically by McQuain et al. [6], Jyotsna and Vanka [7], Li and Tang [8], and Gaskell et al. [9]. McQuain et al. [6] applied the Batchelor's mean square law to triangular cavity flow and analytically obtained the inviscid core vorticity for infinite Reynolds number. Recent calculations of the steady problem in an equilateral triangular cavity have been given in Ribbens et al. [10] for $Re \leq 500$. Erturk and Gokcol [11] have presented high accurate, fine grid solutions of 2D steady incompressible flow in triangle cavities. The governing equations are solved up to very low residuals at various Reynolds numbers. The results of these studies, however, show some discrepancies. Furthermore, the scheme in [11] is only second-order spatial accuracy. This constitutes the main motivation of this study.

The main object of this study is the development of an accurate and efficient scheme for solving the N-S problems in triangular geometries. In the present paper, we have attempted an efficient fourth-order semicompact scheme (compactness of the scheme is relaxed for few terms of the governing equations) for stream function vorticity form of incompressible Navier-Stokes equations in curvilinear coordinates inside a triangular cavity. A geometric transformation handling triangles of arbitrary shape is also presented. The developed equations are solved numerically subject to appropriate boundary conditions by a fourth-order accurate finite difference method. Overall, besides opening up new possibilities, the method may be considered an efficient one for computation of flow for this physical configuration and the results thus obtained are not only useful for engineers in the process design; they also supplement the existing literature.

2. Governing Equations

A triangular driven cavity of general shape is given by locating its three vertices at $\widehat{A}(x_a, h)$, $\widehat{B}(x_b, h)$, and $\widehat{C}(x_c, y_c)$, with the upper side moving to the right via a constant velocity U_0; see Figure 1(a). The 2D steady incompressible flow inside a triangular cavity is governed by the N-S equations. We consider the N-S equations in streamfunction and vorticity formulation in Cartesian coordinates, such that [5]

$$\nabla_{xy}^2 \psi = -\omega, \quad \nabla_{xy}^2 \omega = \mathrm{Re}\left(\mathbf{U} \cdot \nabla_{xy} \omega\right), \tag{1}$$

where

$$\nabla_{xy}^2 = \frac{\partial^2}{\partial x^2} + \frac{\partial^2}{\partial y^2}, \qquad \nabla_{xy} = \left(\frac{\partial}{\partial x}, \frac{\partial}{\partial y}\right),$$
$$\mathbf{U} = (u, v), \qquad u = \frac{\partial \psi}{\partial y}, \qquad v = -\frac{\partial \psi}{\partial x}, \qquad \omega = \frac{\partial v}{\partial x} - \frac{\partial u}{\partial y}, \tag{2}$$

where ψ is the streamfunction, ω is the vorticity, u and v are the components of the velocity in x- and y-directions, respectively, and Re is Reynolds number. We note that these equations are non dimensional and a length scale of $(h-c_y)/3$ and a velocity scale of U_0, that is, the velocity of the lid, are used to nondimensionalize the parameters and Reynolds number is defined accordingly.

By a simple linear transformation

$$\xi = \frac{x - x_a}{x_b - x_a} + \frac{(x_a - x_c)(y - h)}{(x_b - x_a)(y_c - h)}, \quad \eta = \frac{\sqrt{(y_c - h)^2 + (x_c - x_a)^2}}{(x_b - x_a)(y_c - h)}(y - h), \tag{3}$$

the triangle $\triangle \widehat{A}\widehat{B}\widehat{C}$ is transformed to a right-angled triangle with vertices $A(0,0)$, $B(1,0)$, and $C(0, \sqrt{(y_c - h)^2 + (x_c - x_a)^2}/(x_b - x_a))$; see Figure 1(b). From these relations, we can calculate the transformation metrics as follows:

$$\frac{\partial \xi}{\partial x} = \frac{1}{x_b - x_a}, \qquad \frac{\partial \xi}{\partial y} = \frac{x_a - x_c}{(x_b - x_a)(y_c - h)},$$
$$\frac{\partial \eta}{\partial x} = 0, \qquad \frac{\partial \eta}{\partial y} = \frac{\sqrt{(y_c - h)^2 + (x_c - x_a)^2}}{(x_b - x_a)(y_c - h)}, \tag{4}$$

and also

$$\frac{\partial^2 \xi}{\partial x^2} = \frac{\partial^2 \xi}{\partial y^2} = \frac{\partial^2 \eta}{\partial x^2} = \frac{\partial^2 \eta}{\partial y^2} = 0. \tag{5}$$

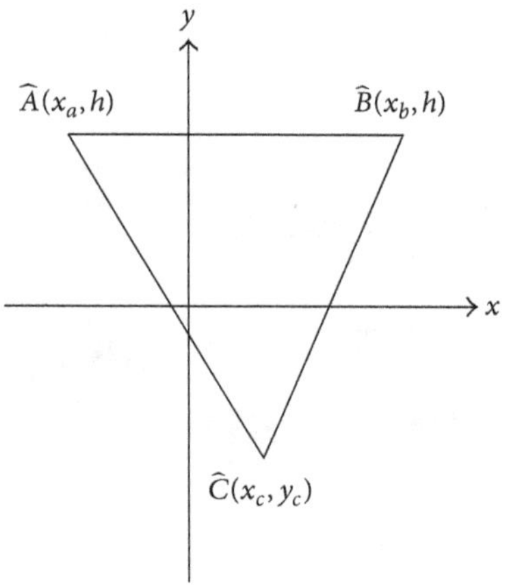

(a)

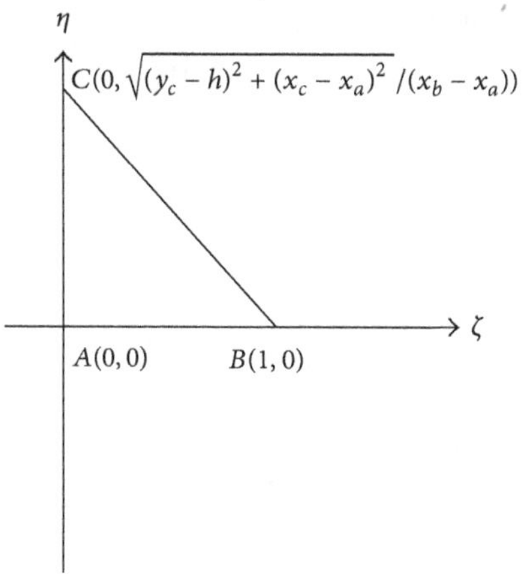

(b)

FIGURE 1: Geometric transformation of the triangular cavity: Physical domain (a) and Computational domain (b).

Using the chain rule, the governing N-S equations (1) in general curvilinear coordinates are as follows:

$$\begin{aligned}
&\left(\left(\frac{\partial \xi}{\partial x}\right)^2 + \left(\frac{\partial \xi}{\partial y}\right)^2\right)\frac{\partial^2 \psi}{\partial \xi^2} + \left(\left(\frac{\partial \eta}{\partial x}\right)^2 + \left(\frac{\partial \eta}{\partial y}\right)^2\right)\frac{\partial^2 \psi}{\partial \eta^2} \\
&\quad + \left(\frac{\partial^2 \xi}{\partial x^2} + \frac{\partial^2 \xi}{\partial y^2}\right)\frac{\partial \psi}{\partial \xi} + \left(\frac{\partial^2 \eta}{\partial x^2} + \frac{\partial^2 \eta}{\partial y^2}\right)\frac{\partial \psi}{\partial \eta} \\
&\quad + 2\left(\frac{\partial \xi}{\partial x}\frac{\partial \eta}{\partial x} + \frac{\partial \xi}{\partial y}\frac{\partial \eta}{\partial y}\right)\frac{\partial^2 \psi}{\partial \xi \partial \eta} = -\omega, \\
&\left(\left(\frac{\partial \xi}{\partial x}\right)^2 + \left(\frac{\partial \xi}{\partial y}\right)^2\right)\frac{\partial^2 \omega}{\partial \xi^2} + \left(\left(\frac{\partial \eta}{\partial x}\right)^2 + \left(\frac{\partial \eta}{\partial y}\right)^2\right)\frac{\partial^2 \omega}{\partial \eta^2} \\
&\quad + \left(\frac{\partial^2 \xi}{\partial x^2} + \frac{\partial^2 \xi}{\partial y^2}\right)\frac{\partial \omega}{\partial \xi} + \left(\frac{\partial^2 \eta}{\partial x^2} + \frac{\partial^2 \eta}{\partial y^2}\right)\frac{\partial \omega}{\partial \eta} \\
&\quad + 2\left(\frac{\partial \xi}{\partial x}\frac{\partial \eta}{\partial x} + \frac{\partial \xi}{\partial y}\frac{\partial \eta}{\partial y}\right)\frac{\partial^2 \omega}{\partial \xi \partial \eta} = \Omega,
\end{aligned} \tag{6}$$

where

$$\Omega = \mathrm{Re}\frac{\partial \xi}{\partial x}\frac{\partial \eta}{\partial y}\left(\frac{\partial \psi}{\partial \eta}\frac{\partial \omega}{\partial \xi} - \frac{\partial \psi}{\partial \xi}\frac{\partial \omega}{\partial \eta}\right). \tag{7}$$

Substituting for the transformation metrics, we obtain the equations that govern the flow in a triangular cavity shown in Figure 1(b) as follows:

$$\nabla^2_{\xi\eta}\psi = -\omega, \tag{8}$$

$$\nabla^2_{\xi\eta}\omega = J\left(\frac{\partial\psi}{\partial\eta}\frac{\partial\omega}{\partial\xi} - \frac{\partial\psi}{\partial\xi}\frac{\partial\omega}{\partial\eta}\right), \tag{9}$$

where

$$\nabla^2_{\xi\eta} = \alpha\frac{\partial^2}{\partial\xi^2} + \beta\frac{\partial^2}{\partial\eta^2} + \gamma\frac{\partial^2}{\partial\xi\partial\eta}, \qquad \alpha = \left(\frac{\partial\xi}{\partial x}\right)^2 + \left(\frac{\partial\xi}{\partial y}\right)^2,$$
$$\beta = \left(\frac{\partial\eta}{\partial y}\right)^2, \qquad \gamma = 2\frac{\partial\xi}{\partial y}\frac{\partial\eta}{\partial y}, \qquad J = \mathrm{Re}\frac{\partial\xi}{\partial x}\frac{\partial\eta}{\partial y}, \tag{10}$$

and we obtain

$$\alpha = \beta, \qquad u = \frac{\partial\psi}{\partial\xi}\frac{\partial\xi}{\partial y} + \frac{\partial\psi}{\partial\eta}\frac{\partial\eta}{\partial y}, \qquad v = -\frac{\partial\psi}{\partial\xi}\frac{\partial\xi}{\partial x}. \tag{11}$$

The boundary conditions in xy-plane are follows [11]:

$$\psi = 0 \quad \text{on all three sides}, \tag{12}$$

$$(u, v)\cdot\tau = \begin{cases} 1 & \text{on the top side,} \\ 0 & \text{on the other two sides,} \end{cases} \tag{13}$$

$$(u, v)\cdot n = 0 \quad \text{on all three sides}, \tag{14}$$

where n is the outside normal unit vector and τ is the tangential unit vector in clockwise direction. In the $\xi\eta$-plane condition (11) remains the same form, while (13)-(14) need to be converted as follows. On the top side $\widehat{AB}$, substituting (11) into (13)-(14), with $\tau = (1, 0)$ and $n = (0, 1)$ yields

$$\frac{\partial\psi}{\partial\xi} = 0,$$
$$\frac{\partial\psi}{\partial\eta} = \frac{\partial y}{\partial\eta} = \frac{(x_b - x_a)(y_c - h)}{\sqrt{(y_c - h)^2 + (x_c - x_a)^2}} \quad \text{on side } AB. \tag{15}$$

On side $\widehat{BC}$ we have

$$\tau = \frac{(x_c - x_b, y_c - h)}{\sqrt{(x_c - x_b)^2 + (y_c - h)^2}},$$
$$n = \frac{(h - y_c, x_c - x_b)}{\sqrt{(x_c - x_b)^2 + (y_c - h)^2}}. \tag{16}$$

Combining this with (11) we obtain

$$\frac{\partial\psi}{\partial\xi} = 0, \qquad \frac{\partial\psi}{\partial\eta} = 0 \quad \text{on side } BC. \tag{17}$$

Similarly on side $\widehat{CA}$ by using

$$\tau = \frac{(x_a - x_c, h - y_c)}{\sqrt{(x_a - x_c)^2 + (y_c - h)^2}},$$
$$n = \frac{(y_c - h, x_a - x_c)}{\sqrt{(x_a - x_c)^2 + (y_c - h)^2}}. \tag{18}$$

we obtain

$$\frac{\partial\psi}{\partial\xi} = 0, \qquad \frac{\partial\psi}{\partial\eta} = 0 \quad \text{on side } CA. \tag{19}$$

3. Higher-Order Semicompact Scheme

3.1. Discretization of the Stream Function Equation. It is known that the higher-order compact approximation for any Poisson type of operator acting on f is given by [12, 13]

$$D_x^2\left(1 + \frac{\Delta h^2}{12}D_y^2\right)f + D_y^2\left(1 + \frac{\Delta h^2}{12}D_x^2\right)f = \left(1 + \frac{\Delta h^2}{12}\left(D_x^2 + D_y^2\right)\right)g, \tag{20}$$

where g is any known function, Δh is the constant step length in x and y directions, and

$$D_x^2 f(x, y) = \frac{1}{\Delta h}\left(f(x - \Delta h, y) - 2f(x, y) + f(x + \Delta h, y)\right),$$
$$D_y^2 f(x, y) = \frac{1}{\Delta h}\left(f(x, y - \Delta h) - 2f(x, y) + f(x, y + \Delta h)\right). \tag{21}$$

If we define

$$\left(1 + \frac{\Delta h^2}{12}\left(D_\xi^2 + D_\eta^2\right)\right)\omega = \tilde{\omega}, \tag{22}$$

and then approximate the stream function (8) in the form of (25) we get

$$\alpha D_\xi^2\left(1 + \frac{\Delta h^2}{12}D_\eta^2\right)\psi + \beta D_\eta^2\left(1 + \frac{\Delta h^2}{12}D_\xi^2\right)\psi + \tilde{\gamma}\psi = -\tilde{\omega} + O\left(\Delta h^4\right), \tag{23}$$

where

$$\tilde{\gamma} = \gamma D_{\xi\eta}^2\left(1 - \frac{\Delta h^2}{12}\left(D_\xi^2 + D_\eta^2\right)\right),$$
$$D_\xi f(\xi, \eta) = \frac{1}{2\Delta h}\left(f(\xi - \Delta h, \eta) - f(\xi + \Delta h, \eta)\right), \tag{24}$$
$$D_\eta f(\xi, \eta) = \frac{1}{2\Delta h}\left(f(\xi, \eta - \Delta h) - f(\xi, \eta + \Delta h)\right).$$

A 2D finite difference scheme is compact if it only involves at most the 8 nearest neighboring grid points (of the center point) in the approximation formula [14]. Then in (23), first and second terms on the left side of the equation can be discretized on the compact nine point stencil as in any other compact scheme. However, discretization of the third term $\tilde{\gamma}\psi$ needs some special attention. We refer this term as *TERM*1 and the discretization of this term is considered after the discussion of the vorticity equation.

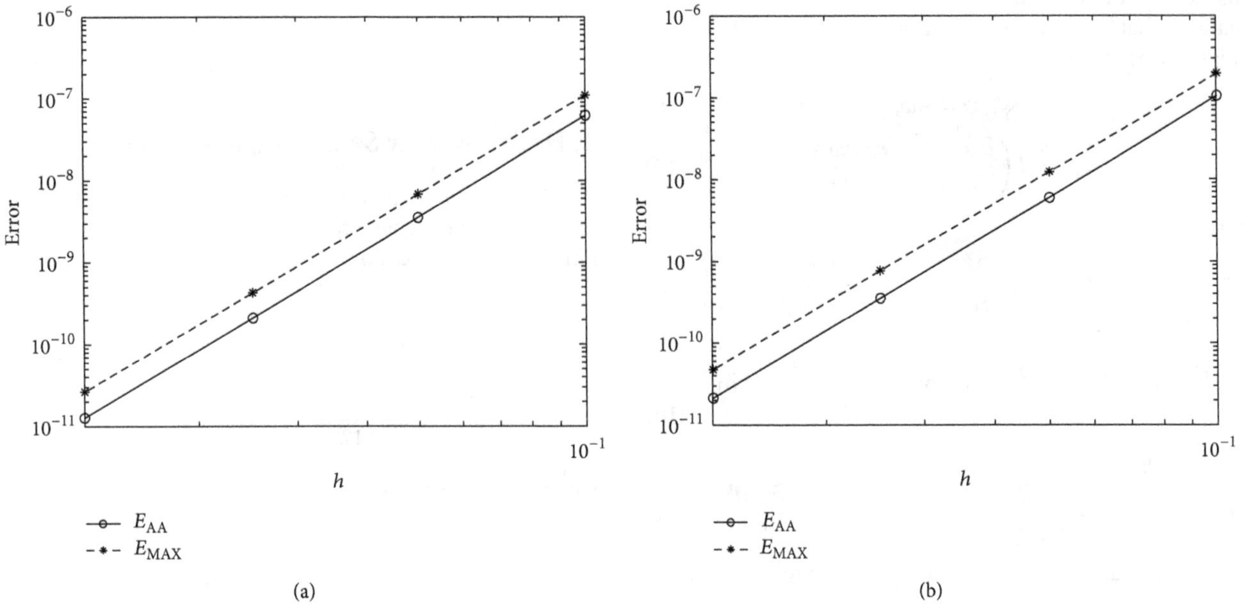

FIGURE 2: The average absolute error E_{AA} and the maximum norm error E_{MAX} in streamfunction and vorticity variables with respect to grid spacing. (a) streamfunction and (b) vorticity.

TABLE 1: Properties of the center of the primary eddy, located at (x, y) with streamfunction value and vorticity for equilateral triangle.

Re	Source	ψ	ω	(x, y)
	Present	−0.2348	−1.4335	(0.0271, 0.4844)
	Li and Tang [8]	−0.2340	−1.3990	(0.0000, 0.4750)
1	Erturk and Gokcol [11]	−0.2329	−1.3788	(0.0101, 0.4668)
	Gaskell et al. [9]	−0.2320	−1.3750	(0.0100, 0.4670)
	McQuain et al. [6]	−0.2330	−1.3630	(0.0170, 0.4600)
	Present	−0.2377	−1.4533	(0.3248, 0.4375)
	Li and Tang [8]	−0.2350	−1.4380	(0.3680, 0.4380)
50	Erturk and Gokcol [11]	−0.2369	−1.4689	(0.3484, 0.4434)
	Gaskell et al. [9]	−0.2360	−1.4690	(0.3500, 0.4420)
	McQuain et al. [6]	−0.2370	−1.4640	(0.3460, 0.4450)
	Present	−0.2489	−1.3795	(0.3291, 0.3700)
	Li and Tang [8]	−0.2440	−1.2640	(0.3680, 0.3630)
100	Erturk and Gokcol [11]	−0.2482	−1.3669	(0.3315, 0.3555)
	Gaskell et al. [9]	−0.2490	−1.3650	(0.3330, 0.3560)
	McQuain et al. [6]	−0.2470	−1.3730	(0.3290, 0.3550)
	Present	−0.2663	−1.2388	(0.1624, 0.2500)
	Li and Tang [8]	−0.2740	−1.1530	(0.1520, 0.2130)
350	Erturk and Gokcol [11]	−0.2724	−1.1985	(0.1556, 0.2383)
	Gaskell et al. [9]	−0.2740	−1.2010	(0.1490, 0.2340)
	McQuain et al. [6]	−0.2680	−1.2320	(0.1730, 0.2650)
	Present	−0.2748	−1.1992	(0.1299, 0.2350)
	Li and Tang [8]	−0.2780	−1.1240	(0.1080, 0.2130)
500	Erturk and Gokcol [11]	−0.2774	−1.1791	(0.1319, 0.2207)
	Gaskell et al. [9]	−0.2810	−1.1870	(0.1280, 0.2170)
	McQuain et al. [6]	−0.2690	−1.2500	(0.1730, 0.2650)
	Present	−0.2902	−1.2067	(0.0866, 0.2200)
1000	Li and Tang [8]	−0.2790	−1.0480	(0.1080, 0.1380)
	Erturk and Gokcol [11]	−0.2844	−1.1629	(0.1116, 0.1973)
	Present	−0.2962	−1.2282	(0.0664, 0.2250)
1500	Li and Tang [8]	−0.2770	−0.9980	(0.1080, 0.1380)
	Erturk and Gokcol [11]	−0.2873	−1.1639	(0.1015, 0.1914)
	Present	−0.2951	−1.1827	(0.0693, 0.2029)
2000	Li and Tang [8]	—	—	—
	Erturk and Gokcol [11]	—	—	—

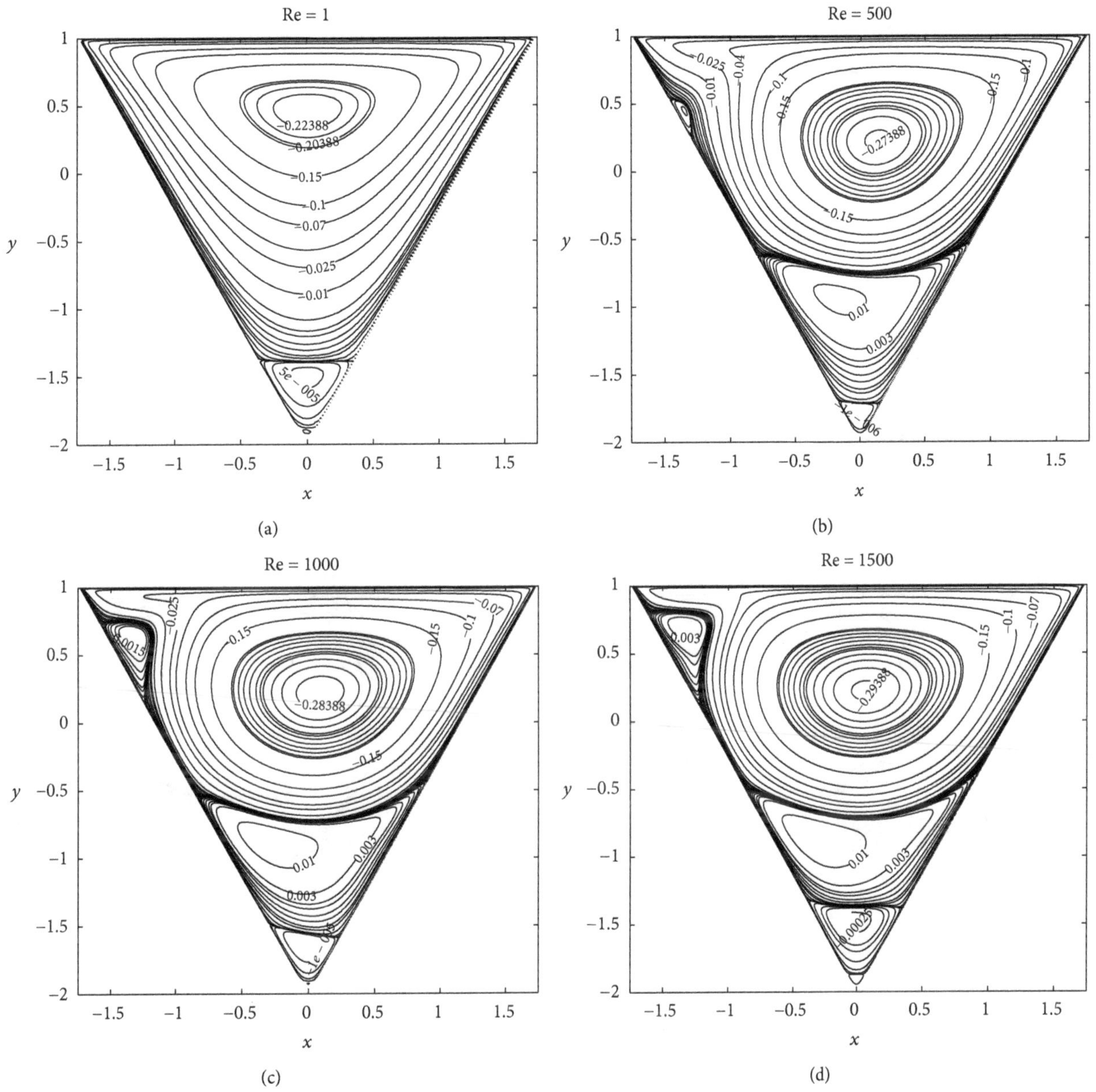

FIGURE 3: Streamline contours of triangular cavity flow at various Reynolds numbers.

3.2. Discretization of the Vorticity Equation. Equation (9) with fourth-order approximation can be written as

$$\alpha D_\xi^2\left(1+\frac{\Delta h^2}{12}D_\eta^2\right)\omega+\beta D_\eta^2\left(1+\frac{\Delta h^2}{12}D_\xi^2\right)\omega+\tilde{\gamma}\omega$$
$$=J\left(1+\frac{\Delta h^2}{12}\left(D_\xi^2+D_\eta^2\right)\right)\left(D_\eta\psi D_\xi\omega-D_\xi\psi D_\eta\omega\right)$$
$$+O\left(\Delta h^4\right). \tag{25}$$

The first and second terms on the left hand side of (25) can be discretized within the compact stencil. We call the third term $\tilde{\gamma}\omega$ as $TERM2$ which needs a larger stencil to approximate to fourth-order. The term on the right hand side of (25) contains

$$\left(D_\xi^2+D_\eta^2\right)\left(D_\eta\psi D_\xi\omega-D_\xi\psi D_\eta\omega\right)$$
$$=2D_{\xi\eta}^2\omega\left(D_\eta^2-D_\xi^2\right)\psi+2D_{\xi\eta}^2\psi\left(D_\xi^2-D_\eta^2\right)\omega$$
$$-D_\xi\left(D_\xi^2+D_\eta^2\right)\psi D_\eta\omega-D_\xi\psi D_\eta\left(D_\xi^2+D_\eta^2\right)\omega$$
$$+D_\eta\left(D_\xi^2+D_\eta^2\right)\psi D_\xi\omega+D_\eta\psi D_\xi\left(D_\xi^2+D_\eta^2\right)\omega. \tag{26}$$

The first two terms on the right hand side of (26) again can be discretized within the compact stencil to the fourth-order;

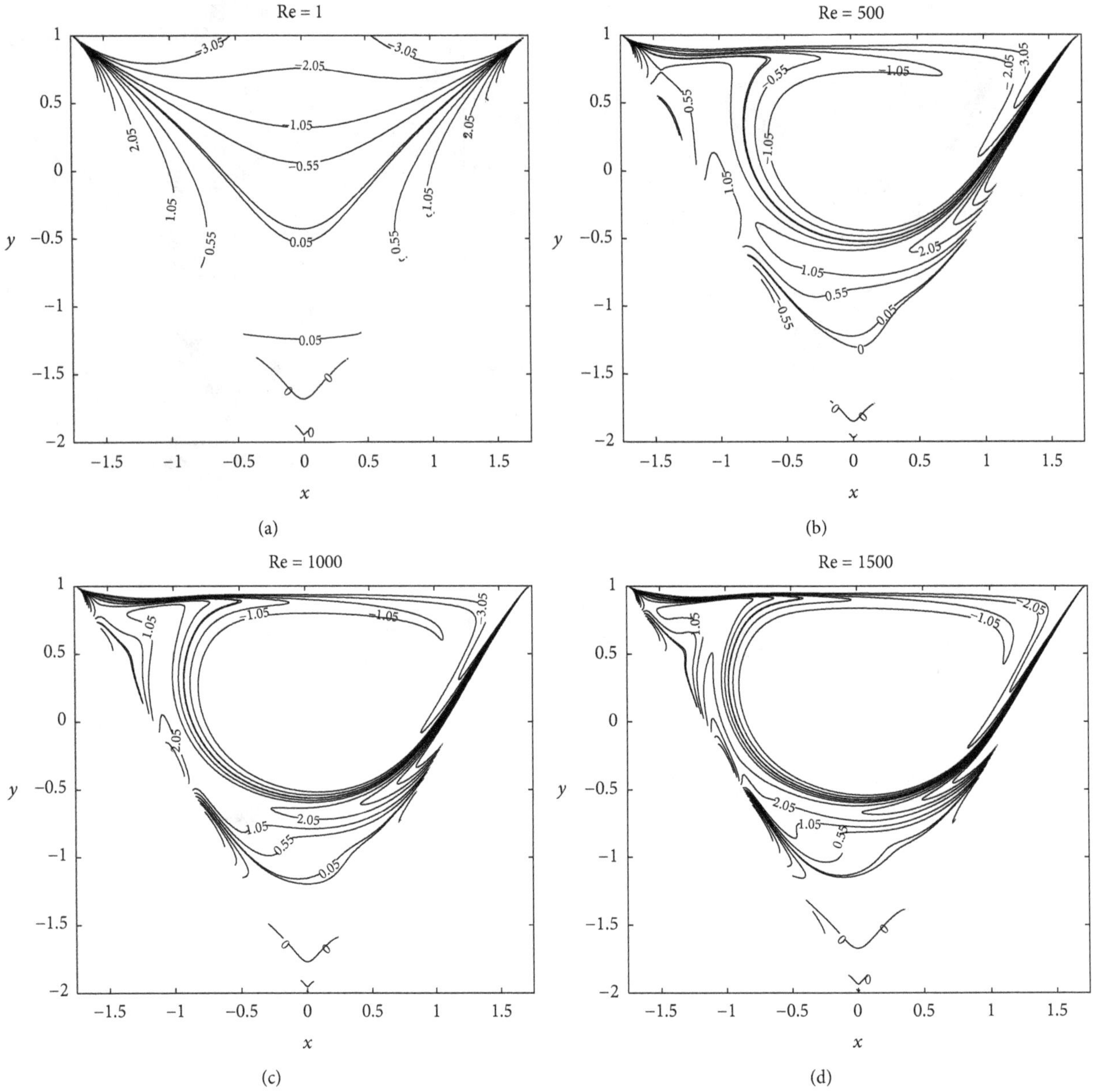

FIGURE 4: Vorticity contours of triangular cavity flow at various Reynolds numbers.

however, the last four terms, call them as *TERM*3, *TERM*4, *TERM*5, and *TERM*6, need larger stencil.

3.3. Numerical Treatment of the Terms TERM1–TERM6. We obtain from (8) that

$$\frac{\partial^2 \psi}{\partial \xi^2} + \frac{\partial^2 \psi}{\partial \eta^2} = -\frac{\omega}{\alpha} - \frac{\gamma}{\alpha}\frac{\partial^2 \psi}{\partial \xi \partial \eta}. \tag{27}$$

Therefore,

$$D_{\xi\eta}^2 \left(D_\xi^2 + D_\eta^2\right) \psi = -\frac{1}{\alpha} D_{\xi\eta}^2 \omega - \frac{\gamma}{\alpha} D_{\xi\xi\eta\eta}^4 \psi. \tag{28}$$

Similarly, from (9) we obtain

$$\begin{aligned} D_{\xi\eta}^2 \left(D_\xi^2 + D_\eta^2\right) \omega &= -\frac{\gamma}{\alpha} D_{\xi\xi\eta\eta}^4 \omega + \frac{J}{\alpha}\left(D_{\xi\eta\eta}^3 \psi D_\xi \omega + D_\eta \psi D_{\xi\xi\eta}^3 \omega \right. \\ &\quad \left. + D_\eta^2 \psi D_\xi^2 \omega\right) \\ &\quad - \frac{J}{\alpha}\left(D_{\xi\xi\eta}^3 \psi D_\eta \omega + D_\xi \psi D_{\xi\eta\eta}^3 \omega + D_\xi^2 \psi D_\eta^2 \omega\right). \end{aligned} \tag{29}$$

In (29), we have third derivatives such as $D_\xi(D_\xi^2 + D_\eta^2)\psi$ and $D_\eta(D_\xi^2 + D_\eta^2)\omega$ which present some problems to the fourth-order compact formulation. In order to find an expression

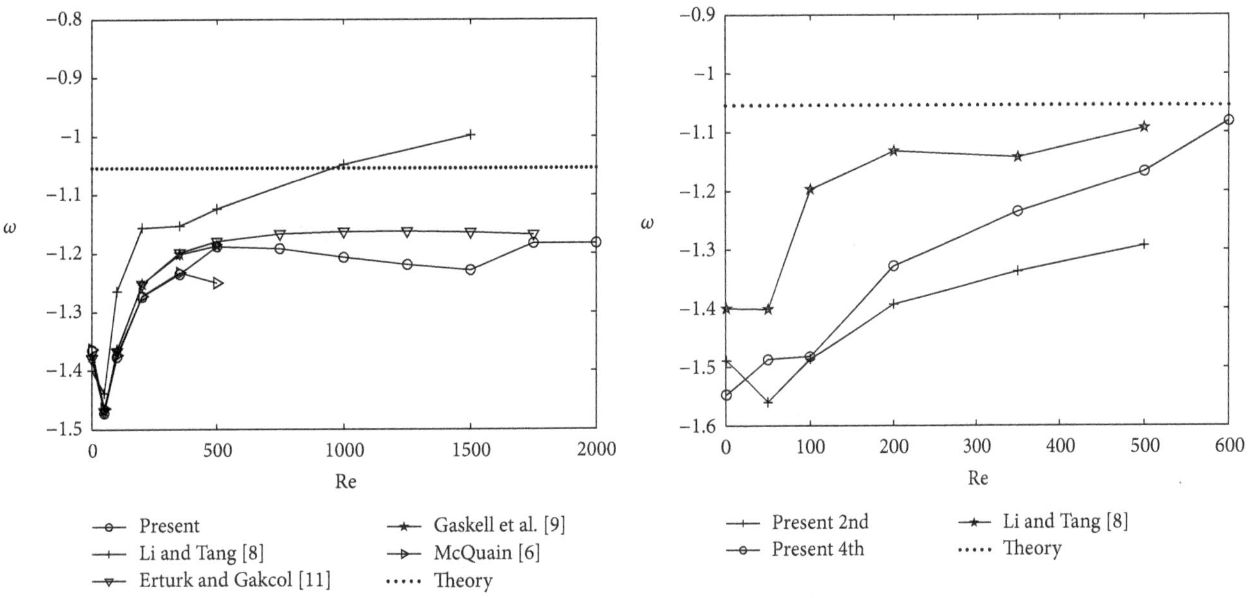

(a) (b)

FIGURE 5: Comparison of the vorticity at the center of the primary eddy in the equilateral triangular cavity (a) and right-oriented triangular cavity (b).

for these derivatives, we use (8)-(9) to obtain the following equations:

$$D_\kappa\left(D_\xi^2 + D_\eta^2\right)\psi = -\frac{1}{\alpha}D_\kappa\omega - \frac{\gamma}{\alpha}D^3_{\xi\kappa\eta}\psi, \quad \kappa = \xi, \eta, \tag{30}$$

$$D_\kappa\left(D_\xi^2 + D_\eta^2\right)\omega = -\frac{\gamma}{\alpha}D^3_{\xi\kappa\eta}\omega + \frac{J}{\alpha}\left(D_\xi\omega D^2_{\kappa\eta}\psi + D^2_{\xi\kappa}\omega D_\eta\psi\right) - \frac{J}{\alpha}\left(D^2_{\xi\kappa}\psi D_\eta\omega + D_\xi\psi D^2_{\kappa\eta}\omega\right), \quad \kappa = \xi, \eta. \tag{31}$$

3.4. Numerical Boundary Conditions. The numerical implementation of the boundary conditions for u, v, and ψ are straightforward. On side AB boundary, we have

$$\omega_{i,1} = -\beta\frac{\partial^2\psi_{i,1}}{\partial\eta^2}, \quad \alpha\frac{\partial^2\omega_{i,1}}{\partial\xi^2} + \beta\frac{\partial^2\omega_{i,1}}{\partial\eta^2} + \gamma\frac{\partial^2\omega_{i,1}}{\partial\xi\partial\eta} = JK\frac{\partial\omega_{i,1}}{\partial\xi}, \tag{32}$$

where $i = 2, 3, \ldots, N$ and

$$K = \frac{\partial\psi_{i,1}}{\partial\eta} = \frac{(x_b - x_a)(y_c - h)}{\sqrt{(y_c - h)^2 + (x_c - x_a)^2}}. \tag{33}$$

Employing Taylor series expansion, we get

$$K = \frac{\partial\psi_{i,1}}{\partial\eta} = D^+_\eta\psi_{i,1} - \frac{\Delta h}{2}\frac{\partial^2\psi_{i,1}}{\partial\eta^2} - \frac{\Delta h^2}{6}\frac{\partial^3\psi_{i,1}}{\partial\eta^3} - \frac{\Delta h^3}{24}\frac{\partial^4\psi_{i,1}}{\partial\eta^4} + O\left(\Delta h^4\right), \tag{34}$$

where D^+_η is the first-order forward difference operator in the η-direction and Using (32) in (34), we get the fourth-order accuracy $O(\Delta h^4)$ expression as follows:

$$K = D^+_\eta\psi_{i,1} + \frac{\Delta h}{2\beta}\omega_{i,1} + \frac{\Delta h^2}{6\beta}D^+_\eta\omega_{i,1} - \frac{\Delta h^3}{24\beta^2}\left(JKD^+_\xi - \alpha D^2_\xi - \gamma\delta^2_{\xi\eta}\right)\omega_{i,1}, \tag{35}$$

where D^+_ξ is the first-order forward difference operator in the ξ-direction, $\delta^2_{\xi\eta}$ is the first-order cross difference operator, that is

$$\delta^2_{\xi\eta}\omega_{i,1} = \frac{1}{\Delta h^2}\left(\omega_{i,2} - \omega_{i,1} - \omega_{i-1,2} + \omega_{i-1,1}\right) + O(\Delta h). \tag{36}$$

Substituting (36) into (35), we get

$$\omega_{i,1} = \frac{\Theta - (JK\Delta h - \alpha)\,\omega_{i+1,1} + (\alpha + \gamma)\,\omega_{i-1,1} + \gamma\omega_{i-1,2}}{\gamma - JK\Delta h + 2\alpha - 8\beta}, \tag{37}$$

where $\Theta = \theta(\psi_{i,2} - K\Delta h) + (4\beta + \gamma)\omega_{i,2}$, $\theta = 24\beta^2/\Delta h^2$.

Similarly on side AC the vorticity is equal to

$$\omega_{1,j} = \frac{\theta\psi_{2,j} + (4\beta + \gamma)\,\omega_{2,j} + \alpha\omega_{1,j+1} + (\alpha + \gamma)\,\omega_{1,j-1} + \gamma\omega_{2,j-1}}{\gamma + 2\alpha - 8\beta}, \tag{38}$$

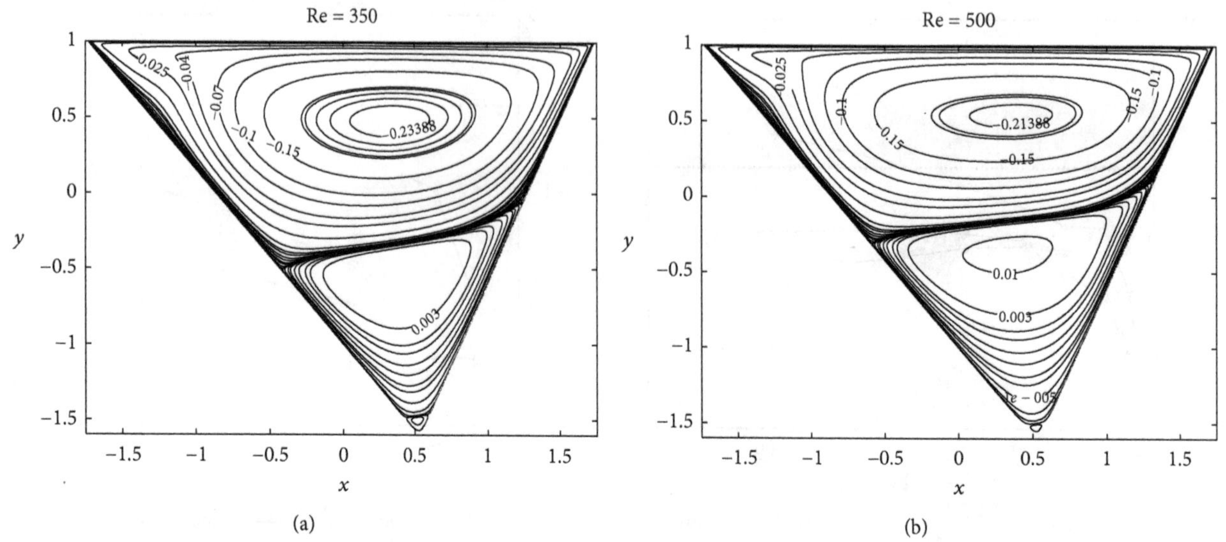

FIGURE 6: Streamlines distributions for a right-oriented triangular cavity at different Reynolds numbers.

where $j = 2, 3, \ldots, N$. On side BC, it is equal to

$$\omega_{i,k} = \frac{\theta\psi_{i-1,k} + (4\beta + \gamma)\,\omega_{i-1,k} + \alpha\omega_{i,k-1} + (\alpha + \gamma)\,\omega_{i,k+1} + \gamma\omega_{i-1,k+1}}{\gamma + 2\alpha - 8\beta}, \tag{39}$$

where $k = N + 2 - i$, $i = 2, 3, \ldots, N$. It is found that the point $(i, k+1)$ is outside the 9-point domain being required and the value of $\omega_{i,k+1}$ is not available; we replace $\omega_{i,k+1}$ by using the following equation:

$$\omega_{i,k+1} = 2\omega_{i,k} - \omega_{i,k-1} + \omega_{i-1,k+1} - 2\omega_{i-1,k} + \omega_{i-1,k-1} + O\left(\Delta h^4\right). \tag{40}$$

At the three singular corners, the values of vorticity need some special attention in the process of discretization. We use the average of vorticity on the two adjacent points with the singular corners.

4. Results and Discussion

Following the numerical procedure described in the previous section, the new schemes (23) and (25) are fourth-order accurate; therefore, they need to be solved in an iterative manner such as SOR iteration with the relaxation parameter $\varpi = 2/3$. In all of the cases considered, we start the iterations from a homogeneous initial guess and continue until a certain condition of convergence is satisfied. As convergence criteria we decided to use the difference of the ψ-ω variables between two steps normalized by the previous value of the corresponding variable, such that

$$\max\left(\left|\psi^{n+1} - \psi^{n}\right|\right) < 10^{-8}, \quad \max\left(\left|\omega^{n+1} - \omega^{n}\right|\right) < 10^{-8}. \tag{41}$$

At this convergence level, this would indicate that the variables ψ and ω are changing less than 0.000001% of their value between two iterations at every grid point in the mesh. These residuals indicate the degree to which the numerical solution has converged to steady state.

We considered different triangle geometries with different Reynolds numbers. Numerical tests for a variety of triangular geometries have been investigated, with Re up to 2000 for an equilateral cavity and 1500 for scalene cavities. It should be pointed out that Reynolds number is based on h, which is consistent with the definition in the case of the equilateral cavity. We solved the flow in this triangle cavity at various Reynolds numbers ranging between 1 and 2000. We note that if the length of one side of the triangle $2\sqrt{3}$ was used in nondimensionalization, as it was used by Erturk and Gokcol [11], then our Reynolds number of 2000 would $2\sqrt{3}$-fold such that it would correspond to a Reynolds number of 6928.

4.1. Equilateral Triangular Cavity. Let us first consider a nondimensional equilateral triangle with coordinates of corner points

$$x_a = -\sqrt{3}, \quad x_b = \sqrt{3}, \quad h = 1, \quad x_c = 0, \quad y_c = -2, \tag{42}$$

which was also considered by McQuain et al. [6], Jyotsna and Vanka [7], Li and Tang [8], Gaskell et al. [9], and Ribbens et al. [10]. We solved the flow in this triangle cavity at various Reynolds numbers ranging between 1 and 2000.

In order to verify the accuracy of the present numerical study, we consider arbitrary triangular cavity flow with the following analytical solutions [15]:

$$\psi = \frac{y - x}{\mathrm{Re}} - e^{x+y}, \qquad \omega = 2e^{x+y}. \tag{43}$$

For this model problem, as the boundary conditions we decided to use the analytical solutions defined in (43) at

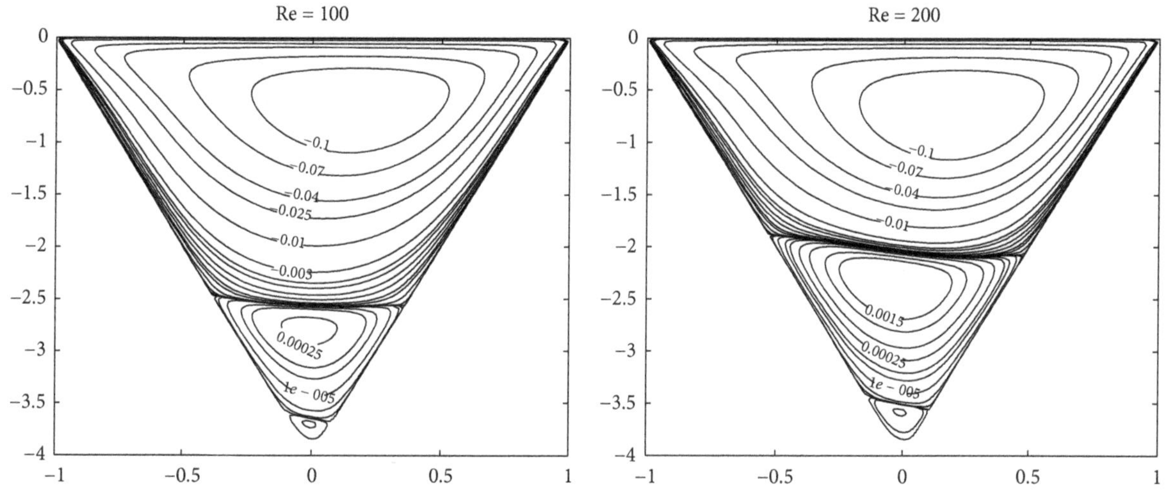

(a) (b)

FIGURE 7: Contour figures of isosceles triangle.

the grid points on the boundaries. This way we would be able to avoid any effect of numerical boundary condition approximations on the numerical solution and concentrate on the accuracy of the solution of both formulations in the interior domain for given analytical values at the boundaries. We note that in (43), the vorticity ω is independent of Re and the solution of the streamfunction ψ at different Reynolds numbers looks almost the same in a contour plot. We solve this model problem using different grid meshes, $11 \times 11/2$, $21 \times 21/2$, $41 \times 41/2$, $81 \times 81/2$, respectively. In these higher-order solutions, the average absolute (E_{AA}) error and the the maximum norm (E_{MAX}) error between the exact solution Exa_k given in (43) and the numerical solution Sol_k obtained from (23) and (25) are defined as

$$E_{AA} = \frac{1}{S}\sum_k \left|\text{Exa}_k - \text{Sol}_k\right|, \qquad E_{\text{MAX}} = \max_k \left|\text{Exa}_k - \text{Sol}_k\right|, \tag{44}$$

where S being the total number of grid points. Since the grid size is decreased by a factor of 2, we can calculate the convergence rate m using the following formula [16]:

$$m = \frac{\log\left(E_{\Delta h}/E_{\Delta h/2}\right)}{\log 2}. \tag{45}$$

As observed by Wan and Zhou [17], the accuracy of the present results is getting better as the grid number increases, and we can see that when the grid spacing is decreased progressively by half, the scheme maintains fourth-order of spatial accurate; the convergence rate is very close to $m \approx 4$. In Figure 2, the average absolute error E_{AA} and the the maximum norm error E_{MAX} are plotted with respect to the grid spacing in a log-log scale [16]. From Figure 2, we can again clearly see that the fully HOC formulation indeed provide fourth-order accurate solutions; the slope between $\log E$ and $\log h$ is close to $m \approx 4$.

Figures 3 and 4 show the streamline and vorticity contours of the triangular cavity flow for a variety of Reynolds numbers obtained with using $(128 \times 128)/2$ grids points. These figures show the streams and vortices that as the Reynolds number increases. From these contour figures, we conclude that the fourth-order compact formulation provides very smooth solutions and it is seen that as Re becomes large, the location of the center of the primary eddy and its streamfunction value seem to have converged. In terms of quantitative analysis, Table 1 tabulates the center of the primary eddy and the streamfunction and vorticity values at the core, together with results found in the literature. Our results are in good agreement with the results in [6] and [9] up to the maximum Reynolds number (Re = 500).

In Figure 5(a), we plot the vorticity values at the center of the primary vortex tabulated in Table 1, with respect to Reynolds number. The results of Li and Tang [8] start to behave differently starting from Re = 100 from the rest of the results. Also in Figure 5(a), the results of McQuain et al. [6] start to deviate from present results and Gaskell et al.'s [9] results after Re = 200. Having a different behavior, the vorticity value of McQuain et al. [6] shows an increase such that their vorticity value at Re = 500 is greater than the vorticity value at Re = 350, whereas the present results and Erturk et al.'s [11] results show a continuous decrease until Re $\leq$ 1500. We believe that these behaviors are due to the coarse grids used in those studies and in order to resolve these behaviors we decided to solve the same flow problem using several coarse grid meshes. For a given grid mesh we have solved the flow for increasing Reynolds number until a particular Re, where the solution was not convergent but oscillating. For this particular Re when the number of grid points was increased, the convergence was recovered and we were able to obtain a solution.

According to the mean square law [18], the value of vorticity is approximately constant at the primary vortex. For an equilateral cavity with length of side $2\sqrt{3}$ this constant is

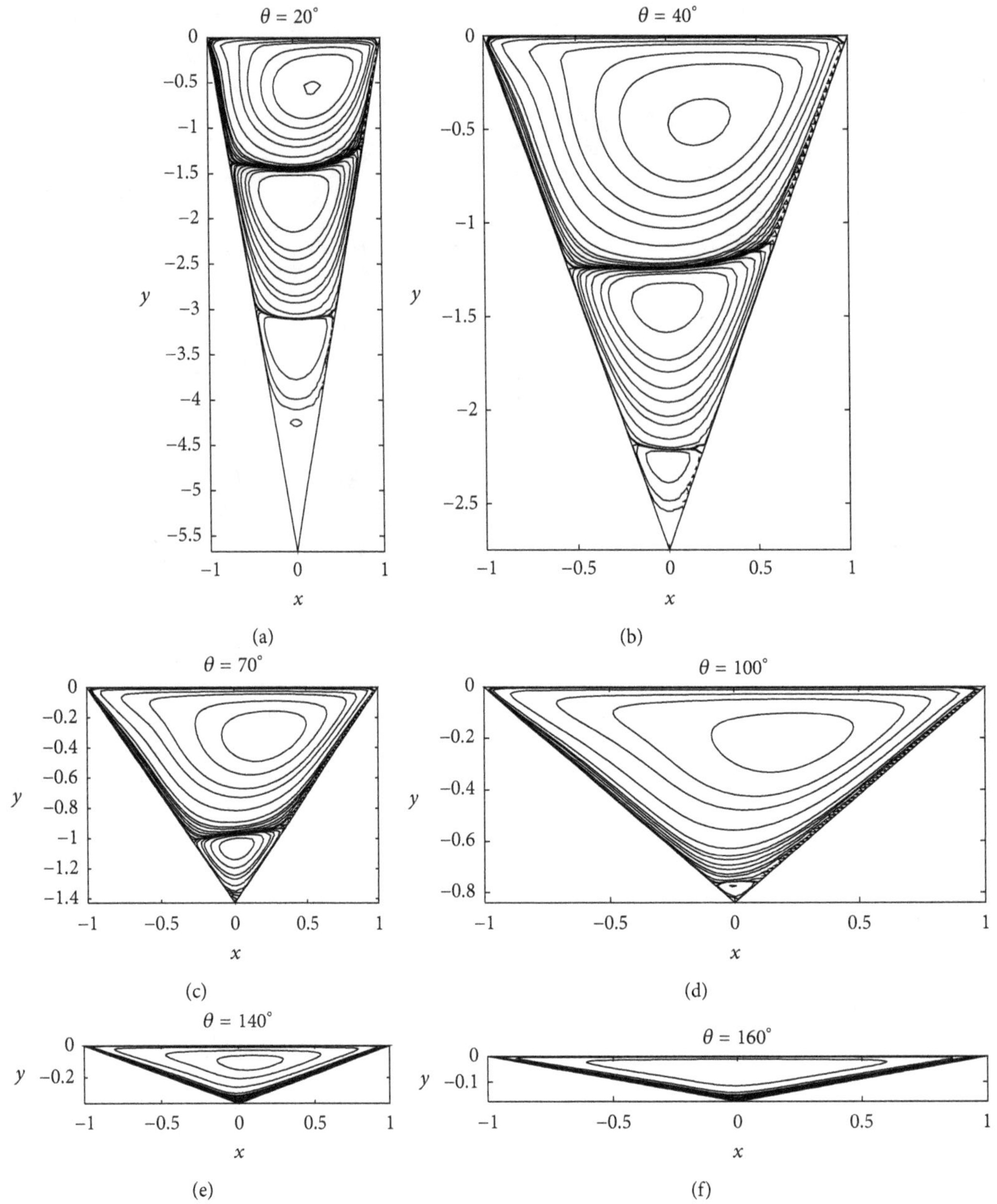

FIGURE 8: Contour figures of isosceles triangle with various corner angle.

found to be 1.054 by McQuain et al. [6]. This infinite Reynolds number value of the vorticity of 1.054 is shown with the dotted line in Figure 5(a). We note that especially the eddies at the bottom corner occupy some portion of the corner as Reynolds number increases. However, the increase in the size of the portion of the bottom corner eddies almost stops after Re = 500 and the size of the primary eddy remains almost constant beyond and so is the value of the vorticity at the core of the primary eddy. It looks like, for an equilateral triangular cavity flow at high Reynolds numbers, the mean square law predicts the strength of the primary eddy within an error due to the effect of the secondary eddies. For circular or elliptic boundaries Ribbens et al. [10], for square cavity flow Erturk and Gokcol [11], and for rectangle cavities McQuain et al. [6] have shown that the mean square law is approximately valid and successful in predicting the strength of the primary eddy at high Reynolds numbers.

4.2. Scalene Triangle Cavity. Owing to the unsymmetric geometry, flow motion in a scalene triangular cavity is very difficult to simulate. It was conjectured in McQuain et al. [6] and Ribbens et al. [10] that it might be unavoidably ill-conditioned. Then we considered a scalene triangle with coordinates of corner points

$$x_a = -\sqrt{3}, \qquad x_b = \sqrt{3}, \qquad h = 1,$$
$$x_c = \frac{\sqrt{3}}{2}, \qquad y_c = -2. \tag{46}$$

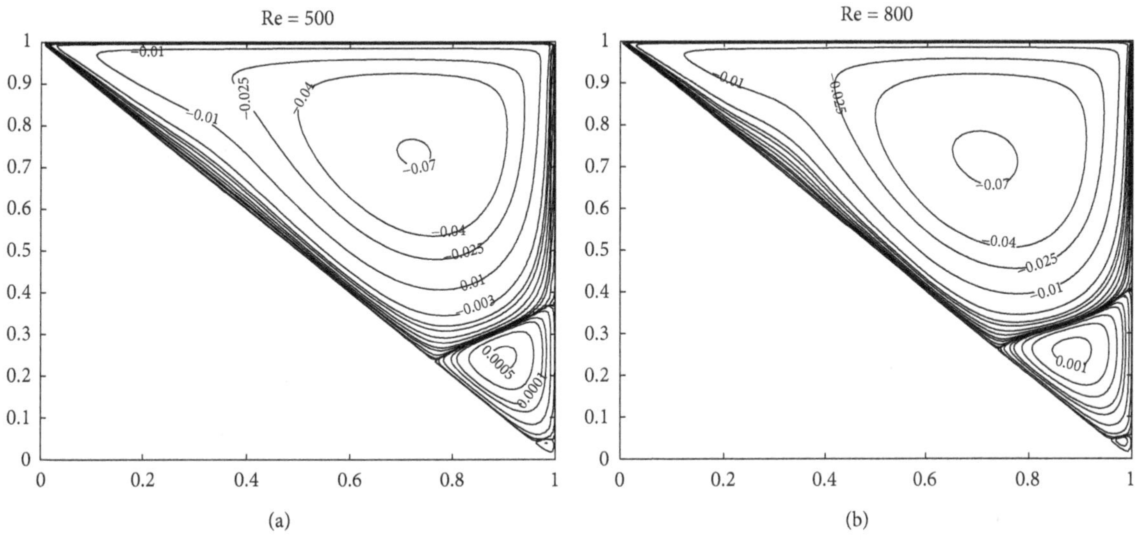

FIGURE 9: Contour figures of right hand side aligned right triangle at different Reynolds numbers.

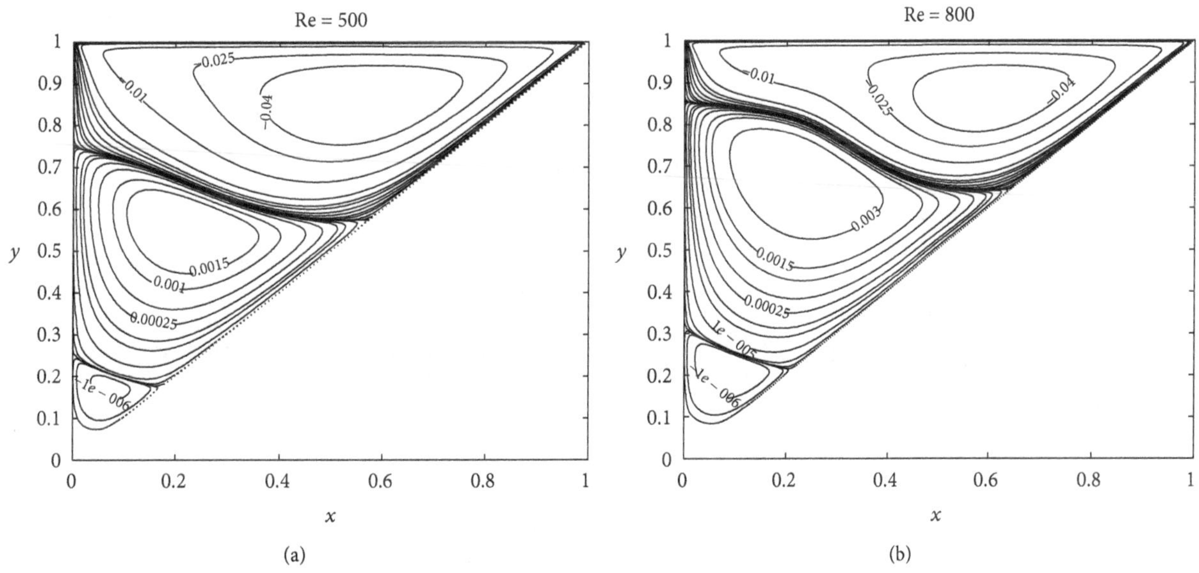

FIGURE 10: Contour figures of left hand side aligned right triangle at various Reynolds numbers.

Using this triangle geometry, we could carry out the calculations for Reynolds numbers as high as 800. In Figure 5(b), we plot the vorticity values at the center of the primary vortex obtained with present study and Li and Tang. [8]. We can see from Figure 5(b) that the vorticity values are getting more and more close to the theoretical value as the Reynolds number increases. In Figure 6, we plot the streamlines and vorticity contours for Re = 350 and 500. Again it is seen that as Re becomes large, the location of the center of the primary eddy and its streamfunction value seem to have converged.

4.3. Isosceles Triangle Cavity. We first considered an isosceles triangle which was also considered by Jyotsna and Vanka [7], Gaskell et al. [9], and Erturk and Gokcol [11], where

$$x_a = -1, \quad x_b = 1, \quad h = 0, \quad x_c = 0, \quad y_c = -4. \qquad (47)$$

We note that our definition of Reynolds number is equivalent to one fourth of the Reynolds number definition used by Jyotsna and Vanka [7]. Figure 7 shows the streamline contours of the flow in this triangle at various Reynolds numbers. Comparing the location of the primary eddy center with that of Jyotsna and Vanka [7], Gaskell et al. [9], and Erturk and Gokcol [11], we believe that our results are more accurate.

Then we considered an isosceles triangle which was also considered by Gaskell et al. [9] and Erturk and Gokcol. [11], with

$$x_a = -1, \quad x_b = 1, \quad h = 0, \quad x_c = 0, \quad y_c = -\cot\frac{\theta}{2}, \qquad (48)$$

where $\theta = 5^\circ, 10^\circ, 20^\circ, 40^\circ, 70^\circ, 100^\circ, 140^\circ, 160^\circ$, and 170°. Figure 8 qualitatively shows the change in the flow with Re = 200 in an isosceles triangle as the corner angle, θ, changes.

We also considered an isosceles right triangle with the 90° corner being at the top right corner, such that with corner points [11]

$$x_a = 0, \quad x_b = 1, \quad h = 1, \quad x_c = 1, \quad y_c = 0. \qquad (49)$$

Figure 9 shows the flow topology as a function of Reynolds numbers and at last, we also considered an isosceles right triangle with the 90° corner being at the top left corner, such that with corner points [11]

$$x_a = 0, \quad x_b = 1, \quad h = 1, \quad x_c = 0, \quad y_c = 0. \qquad (50)$$

Using this triangle geometry, we were able to obtain the solutions and Figure 10 shows the flow topology as a function of Reynolds numbers. For the considered triangle geometry, we can see that the eddy is closest to the moving lid. As it is obvious from these figures, in both cases, the flow behaves very differently as Reynolds number increases, which shows that the flow structures in a triangle cavity are greatly affected by the triangle geometry.

5. Conclusion

In this work we have developed a new semicompact fourth-order scheme for the time-independent ψ-ω form of the 2D, incompressible N-S equations governing the fluid flow in a triangular driven cavity. Our numerical scheme has been proved robust for a wide range of Reynolds numbers and applicable to triangular cavities with arbitrary shape. The key point with the present scheme is that it allows direct iteration for low-to-high Reynolds numbers. It is the success of the current method with the wide range of Re and mesh sizes that indicates the potential of this method as an accurate and stable numerical method applicable to a wide range of problems. We have tested the present method for both the equilateral triangular cavity problem and the scalene triangular cavity problem, and excellent agreement is found in all the cases, both qualitatively and quantitatively.

Conflict of Interests

The authors declare that there is no conflict of interests regarding the publication of this paper.

Acknowledgment

This work is supported by the National Natural Science Foundation of China (no. U1304106).

References

[1] E. Barragy and G. F. Carey, "Stream function-vorticity driven cavity solution using p finite elements," *Computers and Fluids*, vol. 26, no. 5, pp. 453–468, 1997.

[2] Z. F. Tian and Y. B. Ge, "A fourth-order compact finite difference scheme for the steady stream function-vorticity formulation of the Navier-Stokes/Boussinesq equations," *International Journal for Numerical Methods in Fluids*, vol. 41, no. 5, pp. 495–518, 2003.

[3] J. C. Kalita, A. K. Dass, and D. C. Dalal, "A transformation-free HOC scheme for steady convection-diffusion on non-uniform grids," *International Journal for Numerical Methods in Fluids*, vol. 44, no. 1, pp. 33–53, 2004.

[4] X. H. Wu, J. Z. Wu, and J. M. Wu, "Effective vorticity-velocity formulations for three-dimensional incompressible viscous flows," *Journal of Computational Physics*, vol. 122, no. 1, pp. 68–82, 1995.

[5] M. Li, *Numerical solutions for the incompressible Navier-Stokes equations [Ph.D. thesis]*, Simon Fraser University, British Columbia, Canada, 1998.

[6] W. D. McQuain, C. J. Ribbens, C. -Y. Wang, and L. T. Watson, "Steady viscous flow in a trapezoidal cavity," *Computers & Fluids*, vol. 112, no. 1, pp. 173–181, 1994.

[7] R. Jyotsna and S. P. Vanka, "Multigrid calculation of steady, viscous flow in a triangular cavity," *Journal of Computational Physics*, vol. 122, no. 1, pp. 107–117, 1995.

[8] M. Li and T. Tang, "Steady viscous flow in a triangular cavity by efficient numerical techniques," *Applied Mathematics and Computation*, vol. 32, pp. 55–65, 1996.

[9] P. H. Gaskell, H. M. Thompson, and M. D. Savage, "A finite element analysis of steady viscous flow in triangular cavities," *Journal of Mechanical Engineering Science*, vol. 213, no. 3, pp. 263–276, 1999.

[10] C. J. Ribbens, L. T. Watson, and C. -Y. Wang, "Steady viscous flow in a triangular cavity," *Journal of Computational Physics*, vol. 112, no. 1, pp. 173–181, 1994.

[11] E. Erturk and O. Gokcol, "Fine grid numerical solutions of triangular cavity flow," *Applied Physics*, vol. 38, no. 1, pp. 97–105, 2007.

[12] G. D. Simth, *Numerical Solution of Partial Differential Equations*, Oxford University Press, Oxford, UK, 2nd edition, 1984.

[13] Y. V. S. S. Sanyasiraju and V. Manjula, "Higher order semi compact scheme to solve transient incompressible Navier-Stokes equations," *Computational Mechanics*, vol. 35, no. 6, pp. 441–448, 2005.

[14] W. F. Spotz and G. F. Carey, "High-order compact scheme for the steady stream-function vorticity equations," *International Journal for Numerical Methods in Engineering*, vol. 38, no. 20, pp. 3497–3512, 1995.

[15] C. W. Richards and C. M. Crane, "The accuracy of finite difference schemes for the numerical solution of the Navier-Stokes equations," *Applied Mathematical Modelling*, vol. 3, no. 3, pp. 205–211, 1979.

[16] J. Breil and P.-H. Maire, "A cell-centered diffusion scheme on two-dimensional unstructured meshes," *Journal of Computational Physics*, vol. 224, no. 2, pp. 785–823, 2007.

[17] D. C. Wan and Y. C. Zhou, "Numerical solution of incompressible flows by discrete singular," *International Journal for Numerical Methods in Fluids*, vol. 38, pp. 789–810, 2002.

[18] G. K. Batchelor, *An Introduction to Fluid Dynamics*, Cambridge University Press, Cambridge, UK, 1967.

11

Cubic Hermite Collocation Method for Solving Boundary Value Problems with Dirichlet, Neumann, and Robin Conditions

Ishfaq Ahmad Ganaie,[1] Shelly Arora,[2] and V. K. Kukreja[1]

[1] *Department of Mathematics, SLIET, Longowal, Punjab 148106, India*
[2] *Department of Mathematics, Punjabi University, Patiala, Punjab 147002, India*

Correspondence should be addressed to V. K. Kukreja; vkkukreja@gmail.com

Academic Editor: Viktor Popov

Cubic Hermite collocation method is proposed to solve two point linear and nonlinear boundary value problems subject to Dirichlet, Neumann, and Robin conditions. Using several examples, it is shown that the scheme achieves the order of convergence as four, which is superior to various well known methods like finite difference method, finite volume method, orthogonal collocation method, and polynomial and nonpolynomial splines and B-spline method. Numerical results for both linear and nonlinear cases are presented to demonstrate the effectiveness of the scheme.

1. Introduction

Since time immemorial mathematicians are ardently pursuing the solution of linear or nonlinear two point boundary value problems (BVPs) of the type:

$$\frac{d^2y}{dx^2} + \alpha_1(x)\frac{dy}{dx} + \alpha_2(x)y = f(x), \quad x \in [a,b] \tag{1}$$

subject to Dirichlet, Neumann, and Robin's boundary conditions.

Such BVPs have wide application in astronomy, biology, boundary layer theory, deflection in cables, diffusion process, electromagnetism, heat transfer, and other topics. It is well known that closed form analytical solution of such problems cannot be obtained in many cases; therefore, numerical techniques such as collocation method [1, 2], B-spline interpolation [3], Hermite cubic collocation [4–6], finite difference method [7–9], nonlinear shooting method [10], geometric Hermite interpolation [11], quintic B-spline collocation method [12], polynomial and nonpolynomial spline approaches [13–15], quartic spline solution [16], cubic spline collocation method [17], and finite volume element method [18] are frequently used.

In this study, cubic Hermite collocation method (CHCM) involves cubic Hermite basis function to reduce mathematical complexity. Different linear and nonlinear differential equations are solved subject to Dirichlet, Neumann, and Robin boundary conditions using the present method. Moreover, the decoupling technique [4] used to solve elliptic problems with Neumann and Dirichlet conditions is a particular case of present technique. In this paper, linear and nonlinear boundary value problems reported in recent papers [2, 3, 7, 9, 13, 14] are solved using CHCM. It is worth mentioning that CHCM is giving better results than finite difference method, finite element method, finite volume method, B-spline method, and polynomial and nonpolynomial spline approach with fourth order of convergence.

The paper comprises five sections. Section 1 deals with general introduction of the problem. Section 2 gives brief description of cubic Hermite collocation method. In Section 3, symbolic solution of (1) is presented. Seven numerical examples are discussed in Section 4 and finally overall conclusions are given in Section 5.

2. Proposed Technique

In the present method, the domain is divided into finite elements and then orthogonal collocation method with cubic Hermite as basis function is applied within each element.

The cubic Hermite interpolant of the function f relative to partition $a = x_1 < x_2 < \cdots < x_{N+1} = b$ is a function s that satisfies the following:

(1) on each subinterval $[x_k, x_{k+1}]$, s coincides with a cubic polynomial $s_k(x)$,

(2) s interpolates f and $f^{(1)}$ at $x_1, x_2, \ldots, x_{N+1}$,

(3) s and $s^{(1)}$ are continuous on $[a, b]$.

The cubic Hermite interpolant of f and its first derivative at $x = x_k$ requires that

$$s_k(x_k) = f(x_k), \qquad s_k^{(1)}(x_k) = f^{(1)}(x_k). \tag{2}$$

Combining the continuity of s and $s^{(1)}$ at $x = x_{k+1}$ with interpolation of f and its first derivative at $x = x_{k+1}$, one gets

$$\begin{gathered} s_k(x_{k+1}) = s_{k+1}(x_{k+1}) = f(x_{k+1}), \\ s_k^{(1)}(x_{k+1}) = s_{k+1}^{(1)}(x_{k+1}) = f^{(1)}(x_{k+1}). \end{gathered} \tag{3}$$

Hence, $s_k(x)$ is a third degree polynomial that interpolates both f and $f^{(1)}$ at $x = x_k$ and at $x = x_{k+1}$. Therefore, $s_k(x)$ can be written as

$$\begin{aligned} s_k(x) = {} & H_1^k(x) f(x_k) + H_2^k(x) f^{(1)}(x_k) \\ & + H_3^k(x) f^{(1)}(x_{k+1}) + H_4^k(x) f(x_{k+1}), \end{aligned} \tag{4}$$

where

$$H_{2p-1}^k(x) = \begin{cases} \left(\dfrac{x - x_{k-1}}{h_{k-1}}\right)^2 \left(3 - \dfrac{2(x - x_{k-1})}{h_{k-1}}\right); & x \in [x_{k-1}, x_k] \\ \left(1 - \dfrac{x - x_k}{h_k}\right)^2 \left(1 - \dfrac{2(x - x_k)}{h_k}\right); & x \in [x_k, x_{k+1}] \\ 0; & \text{otherwise,} \end{cases}$$

$$H_{2p}^k(x) = \begin{cases} -h_{k-1}\left(\dfrac{x - x_{k-1}}{h_{k-1}}\right)^2 \left(1 - \dfrac{x - x_{k-1}}{h_{k-1}}\right); & x \in [x_{k-1}, x_k] \\ h_j\left(1 - \dfrac{x - x_k}{h_k}\right)^2 \left(\dfrac{x - x_k}{h_k}\right); & x \in [x_k, x_{k+1}] \\ 0; & \text{otherwise.} \end{cases} \tag{5}$$

Here, $p = 1, 2$ and $k = 1, 2, \ldots, N$.

The grid points, x_k, are often called the "knots" of the piecewise polynomial since they are points where polynomials are "tied together." The Hermite polynomials do not require the subsidiary condition to make first derivative continuous. This fact reduces the number of equations by $(N - 1)$, where N is the number of elements.

The global variable x varies in the kth element, where $k = 1, 2, \ldots, N$. A new variable $u = (x - x_k)/h_k$ is introduced in kth element in such a way that as x varies from x_k to x_{k+1}, u varies from 0 to 1. Orthogonal collocation is applied on local variable u.

Approximation of function $y(u)$ in the kth element is given as [6]

$$\overline{y}(u) = \sum_{i=1}^{4} a_{i+2k-2} H_i(u). \tag{6}$$

To apply the collocation method, one must evaluate the trial function (6) and its derivatives at two internal collocation points $u = u_j$ $(j = 1, 2)$. These are given by

$$\begin{gathered} \overline{y}(u_j) = \sum_{i=1}^{4} a_{i+2k-2} H_i(u_j), \\ \frac{d\overline{y}}{du}(u_j) = \frac{1}{h_k}\sum_{i=1}^{4} a_{i+2k-2} A_{ji}, \\ \frac{d^2\overline{y}}{du^2}(u_j) = \frac{1}{h_k^2}\sum_{i=1}^{4} a_{i+2k-2} B_{ji}, \end{gathered} \tag{7}$$

where the Hermite polynomials and their first and second derivatives are defined as

$$\begin{aligned} & H_1(u_j) = (1 + 2u_j)(1 - u_j)^2, && H_2(u_j) = u_j(1 - u_j)^2 h_k, \\ & H_3(u_j) = u_j^2(3 - 2u_j), && H_4(u_j) = u_j^2(u_j - 1) h_k, \\ & A_{j1}(u_j) = 6u_j^2 - 6u_j, && A_{j2}(u_j) = (1 - 4u_j + 3u_j^2) h_k, \\ & A_{j3}(u_j) = 6u_j - 6u_j^2, && A_{j4}(u_j) = (3u_j^2 - 2u_j) h_k, \\ & B_{j1}(u_j) = 12u_j - 6, && B_{j2}(u_j) = (6u_j - 4) h_k, \\ & B_{j3}(u_j) = 6 - 12u_j, && B_{j4}(u_j) = (6u_j - 2) h_k, \end{aligned} \tag{8}$$

where u_j's are the zeros of shifted orthogonal Legendre polynomial $P_2^{(0,0)}(u)$ with $u_1 = 0.2113248654$ and $u_2 = 0.7886751346$, as shown in Figure 1.

3. Symbolic Solution

Equation (1) can be discretized using (7) as follows:

$$\begin{aligned} & \frac{1}{h_k^2}\sum_{i=1}^{4} a_{i+2k-2} B_{ji} + \beta_1(u) \frac{1}{h_k}\sum_{i=1}^{4} a_{i+2k-2} A_{ji} + \beta_2(u) \\ & \quad \times \sum_{i=1}^{4} a_{i+2k-2} H_{ji} = f(u), \end{aligned} \tag{9}$$

where, in the 1st element, a_i's vary from 1 to 4, 2nd element varies from 3 to 6, and so on. Symbolic solution is given subject to Neumann boundary conditions, for $a = 0$ and $b = 1$:

$$y^{(1)}(0) = c_0, \tag{10}$$

$$y^{(1)}(1) = c_L, \tag{11}$$

where c_0 and c_L are finite real constants.

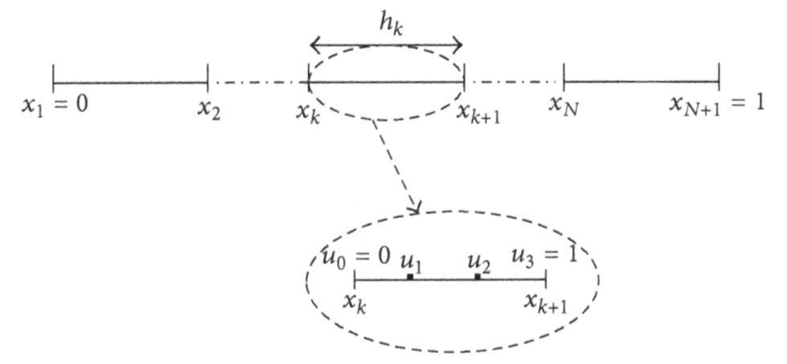

FIGURE 1: Subdivision of mesh points on the global domain. The four coefficients, in each k element, are estimated by using four collocation points u_0, u_1, u_2, and u_3.

In the first element, $u_j = 0 \Rightarrow A_{j1} = A_{j3} = A_{j4} = 0$; therefore, (10) becomes

$$\frac{1}{h_k}\sum_{i=1}^{4} a_i A_{ji}\left(u_j\right) = c_0 \Longrightarrow a_2 = c_0. \tag{12}$$

In the last element, $u_j = 1 \Rightarrow A_{j1} = A_{j2} = A_{j3} = 0$; therefore, (11) becomes

$$\frac{1}{h_k}\sum_{i=1}^{4} a_{i+2N-2} A_{ji}\left(u_j\right) = c_L \Longrightarrow a_{2N+1} = c_L. \tag{13}$$

A system of $(2N + 2)$ equations is obtained from (9) to (13). It includes all parameters of the system and the dependent variables at the boundaries. The support of each Hermite cubic basis function spans at most two subintervals; therefore, a band matrix is obtained with bandwidth two (Figure 2). Of these, two unknowns are found using boundary conditions and rest $2N$ are from discretized system of (9), using Mathematica. After substituting the appropriate values of a's in (7), the result can be obtained for any element.

3.1. Error Analysis. Suppose the function $f(x)$ and its four derivatives are continuous on $[a, b]$ and there is a positive constant which satisfies $|f^4(x)| \le M$ for all $x \in [a, b]$. If $H(x)$ is the cubic Hermite interpolant of f at a and b, then according to de Boor [19], $|f(x) - H(x)| \le \varepsilon h^4$, where $h = b - a$, $\varepsilon = M/384$.

Also the placement of the collocation points plays a critical role in obtaining the $O(h^4)$ estimate [20, 21]. For Gauss Legendre roots, as collocation points, an error estimate of $O(h^4)$ is obtained whereas for other choices of the collocation points, only second-order accuracy is obtained.

3.2. Algorithm of the Method

Step 1. Divide the domain $0 \le x \le 1$ into a mesh $0 = x_1 < x_2 < \cdots < x_{N+1} = 1$.

Step 2. Transform the global variable x into local variable $u = (x - x_k)/h_k$.

Step 3. Approximate the solution at $u = u_j$.

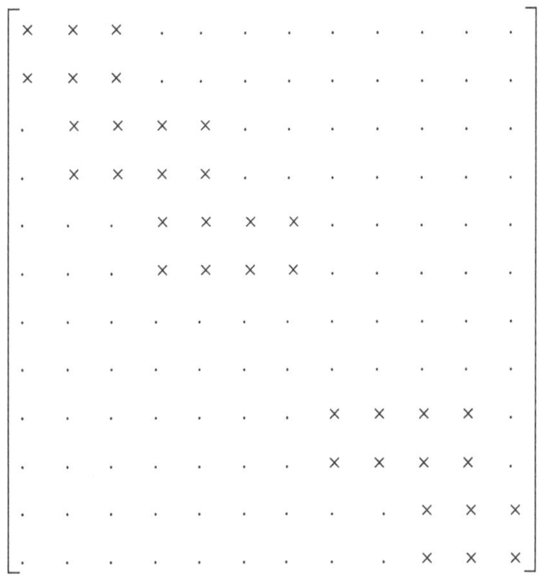

FIGURE 2: Pattern of nonzero elements of banded matrix arising from CHCM for $N = 6$, where ×'s represent nonzero elements and 0's are represented by dots.

Step 4. Obtain the trial function $\overline{y}(u_j)$ and the derivatives $(d\overline{y}/du)(u_j)$, $(d^2\overline{y}/du^2)(u_j)$.

Step 5. Carry out discretization of the model using Step 4.

Step 6. Evaluate Step 5 at collocation points $u_1 = 0.2113248654$ and $u_2 = 0.7886751346$.

Step 7. The obtained system in Step 6 is solved using any software.

4. Numerical Examples and Discussion

In this section, seven examples demonstrate the efficiency and accuracy of the method. Following formulae are used for estimation of error in this study.

Rate of convergence of CHCM is calculated using $\rho = (\ln(L^{\infty}(N2)/L^{\infty}(N1)))/(\ln(N2/N1))$.

Relative error is obtained by $(y_{\text{ex}} - y_{\text{nm}})/y_{\text{ex}}$.

Max norm is found by $L^{\infty} = \max_{i=1}^{n}|y(u_i) - \overline{y}(u_i)|$.

Example 1. Solve (1) for

$$\alpha_1(x) = 0, \qquad \alpha_2(x) = -H^2, \qquad f(x) = 0, \tag{14}$$

subject to Robin's boundary conditions

$$y(0) = 1, \qquad y^{(1)}(1) = 0, \tag{15}$$

which has exact solution [9], for planer geometry

$$y = \frac{\cosh H(1 - x)}{\cosh H}. \tag{16}$$

TABLE 1: Relative error for different values of parameter H for Example 1.

X	$H = 0$ $N = 10$	$H = 0.5$ $N = 10$	$H = 1$ $N = 10$	$H = 2$ $N = 20$	$H = 3$ $N = 20$	$H = 5$ $N = 30$	$H = 10$ $N = 40$
0.0	0	0	0	0	0	0	0
0.1	$1.0000E-10$	$5.963E-11$	$2.708E-9$	$5.1850E-9$	$3.6328E-8$	$8.9622E-8$	$9.0871E-7$
0.2	$1.0000E-10$	$1.145E-10$	$5.337E-9$	$1.0504E-8$	$7.3297E-8$	$1.7941E-7$	$1.8174E-6$
0.3	$1.0000E-10$	$1.640E-10$	$7.842E-9$	$1.5952E-8$	$1.1120E-7$	$2.6957E-7$	$2.7261E-6$
0.4	$1.0000E-10$	$2.079E-10$	$1.017E-8$	$2.1492E-8$	$1.5038E-7$	$3.6059E-7$	$3.6349E-6$
0.5	$1.0000E-10$	$2.457E-10$	$1.229E-8$	$2.7024E-8$	$1.9106E-7$	$4.5343E-7$	$4.5440E-6$
0.6	$1.0000E-10$	$2.773E-10$	$1.413E-8$	$3.2361E-8$	$2.3301E-7$	$5.4982E-7$	$5.4547E-6$
0.7	$1.0000E-10$	$3.020E-10$	$1.561E-8$	$3.7203E-8$	$2.7477E-7$	$6.5192E-7$	$6.3745E-6$
0.8	$1.0000E-10$	$3.199E-10$	$1.673E-8$	$4.1147E-8$	$3.1262E-7$	$7.5863E-7$	$7.3351E-6$
0.9	$1.0000E-10$	$3.307E-10$	$1.741E-8$	$4.3753E-8$	$3.4019E-7$	$8.5360E-7$	$8.3951E-6$
1.0	$1.0000E-10$	$3.344E-10$	$1.766E-8$	$4.4668E-8$	$3.5045E-7$	$4.2434E-7$	$9.0872E-6$

TABLE 2: Relative error between CHCM and exact values for Example 2.

x	Exact solution	CHCM (100 elements)	Relative error
0.0	0	0	0
0.1	0.059343034025940	0.059343034024546	$1.3944E-12$
0.2	0.110134207176555	0.110134207173893	$2.6620E-12$
0.3	0.151024408862577	0.151024408858821	$3.7560E-12$
0.4	0.182725813258852	0.182725813254164	$4.6880E-12$
0.5	0.197560538965947	0.197560538960736	$5.2110E-12$
0.6	0.196995306556119	0.196995306550776	$5.3430E-12$
0.7	0.178732867019218	0.178732867014236	$4.9820E-12$
0.8	0.145015397537614	0.145015397533471	$4.1430E-12$
0.9	0.085646323767636	0.085646323765122	$2.5130E-12$
1.0	0	0	0

TABLE 3: Max norm of errors for five methods with respect to exact solution.

Methods	h	Max norm/h^2
FDM	0.1	$8.24E-3$
	0.01	$8.31E-3$
FEM	0.1	$6.35E-3$
	0.01	$6.36E-3$
FVM	0.1	$3.18E-3$
	0.01	$3.18E-3$
B-spline	0.1	$2.9E-4$
	0.01	$2.89E-6$
CHCM	0.1	$7.240E-4$
	0.01	$5.352E-8$

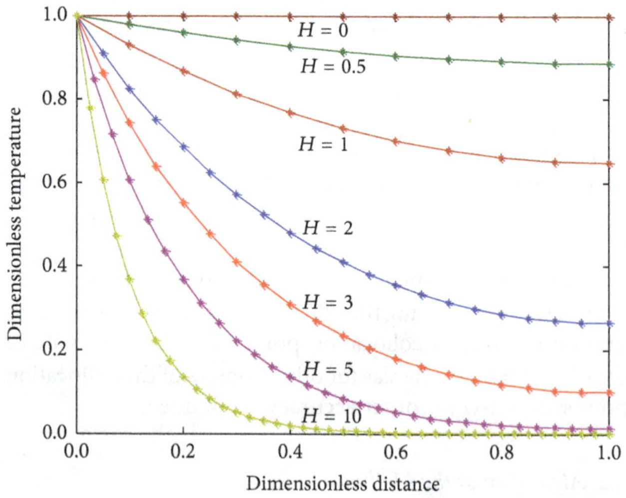

FIGURE 3: Temperature profiles in a rectangular fin for dimensionless heat transfer coefficient H.

The problem is solved for different values of dimensionless parameter H by taking 10 to 40 elements. The exact and numeric results are plotted in Figure 3. The results reported by [9] for 10 elements using finite difference method are matching with exact solution up to 3 decimal places, whereas using CHCM the results are matching up to 9 decimal places. This shows the superiority of cubic Hermite collocation method over the finite difference method. Relative error between CHCM and exact values is presented in Table 1.

TABLE 4: Relative error between CHCM and exact values for Example 3.

x	Exact solution	CHCM (60 elements)	Relative error
0.1	0.100042548871756	0.100042549016137	$1.4400E-10$
0.2	0.110172999814179	0.110173000081150	$2.6700E-10$
0.3	0.131412868154409	0.131412868530861	$3.7600E-10$
0.4	0.165903899721862	0.165903900198008	$4.7600E-10$
0.5	0.217124036179286	0.217124036742860	$5.6400E-10$
0.6	0.290238117151979	0.290238117780922	$6.2900E-10$
0.7	0.392618682595762	0.392618683248041	$6.5200E-10$
0.8	0.534589391091750	0.534589391690484	$5.9900E-10$
0.9	0.730466017481367	0.730466017892698	$4.1100E-10$
1.0	1	1	0

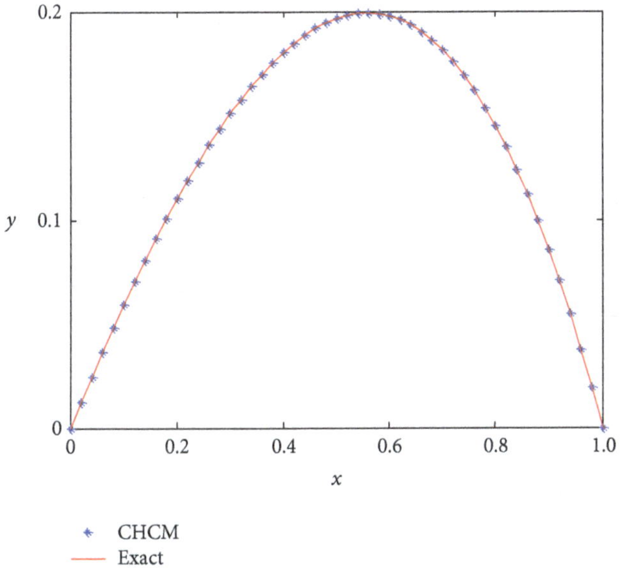

FIGURE 4: Comparison of CHCM with exact result for Example 2.

Example 2. Solve (1) for

$$\alpha_1(x) = -1, \qquad \alpha_2(x) = 0, \qquad f(x) = -e^{x-1} - 1, \quad (17)$$

subject to Dirichlet boundary conditions

$$y(0) = y(1) = 0, \quad 0 < x < 1, \quad (18)$$

which has exact solution [3, 7]

$$y(x) = x\left(1 - e^{x-1}\right). \quad (19)$$

The exact and numeric results are plotted in Figure 4. The CHCM results are matching up to 11 decimal places with the exact ones in Table 2. On comparing present results with the results of [3, 7], shown in Table 3, a big difference of errors between CHCM with finite difference method, finite element method, finite volume method, and B-spline method is observed. This indicates the supremacy of the present method.

TABLE 5: Rate of convergence of CHCM for Example 3.

N_k	$E(N_k)$	ρ
0.0666667	$1.6720E-07$	—
0.0333333	$1.0447E-08$	4.000345
0.0166667	$6.8968E-10$	3.921095

Example 3. Solve (1) for

$$\alpha_1(x) = 0, \qquad \alpha_2(x) = -\phi^2, \qquad f(x) = 0, \quad (20)$$

subject to Dirichlet boundary conditions

$$y(0) = 0.1, \qquad y(1) = 1, \quad (21)$$

which has exact solution [2], for $\phi^2 = 10$ and

$$y = \frac{g \sinh \phi x + f \sinh \phi (1-x)}{\sinh \phi}. \quad (22)$$

The exact and numeric results are plotted in Figure 5. The results are matching up to 9 decimal places as shown in Table 4. From Table 5, order of convergence is found to be 4.

Example 4. Solve (1) for

$$\alpha_1(x) = 0, \qquad \alpha_2(x) = 1, \qquad f(x) = -1, \quad (23)$$

subject to Neumann boundary conditions

$$y^{(1)}(0) = \frac{1-\cos(1)}{\sin(1)}, \qquad y^{(1)}(1) = -\frac{1-\cos(1)}{\sin(1)}, \quad (24)$$

which has exact solution [13, 14] as

$$y(x) = \cos x + \frac{1-\cos(1)}{\sin(1)} \sin x - 1. \quad (25)$$

Example 5. Solve (1) for

$$\alpha_1(x) = 0, \qquad \alpha_2(x) = x,$$
$$f(x) = \left(3 - x - x^2 + x^3\right) \sin x + 4 \cos x, \quad (26)$$

TABLE 6: Maximum absolute errors obtained by different methods in Example 4.

Methods	$N = 8$	$N = 16$	$N = 32$	$N = 64$	$N = 128$
Quadratic spline method [14]	$7.70E-04$	$1.93E-04$	$4.83E-05$	$1.21E-05$	$3.02E-06$
Cubic spline method [14]	$7.13E-04$	$1.78E-04$	$4.45E-05$	$1.11E-05$	$2.78E-06$
Nonpolynomial spline method [14]	$1.75E-04$	$2.16E-05$	$2.68E-06$	$3.33E-07$	$4.15E-08$
Polynomial spline approach [13]	$2.68E-05$	$4.53E-07$	$8.42E-09$	$2.21E-10$	$6.41E-12$
Present approach	$1.53E-07$	$9.35E-09$	$6.11E-10$	$3.00E-11$	$9.99E-13$

TABLE 7: Maximum absolute errors obtained by different methods in Example 5.

Methods	$N = 8$	$N = 16$	$N = 32$	$N = 64$	$N = 128$
Quadratic spline method [14]	$4.50E-02$	$3.08E-03$	$7.70E-04$	$1.93E-04$	$4.80E-05$
Cubic spline method [14]	$1.15E-02$	$2.88E-03$	$7.21E-04$	$1.80E-04$	$4.50E-05$
Nonpolynomial spline method [14]	$2.67E-03$	$3.24E-04$	$3.99E-05$	$4.94E-06$	$6.16E-07$
Polynomial spline approach [13]	$2.22E-04$	$5.05E-06$	$1.63E-07$	$5.58E-09$	$1.89E-10$
Present approach	$2.34E-06$	$1.43E-07$	$9.99E-09$	$8.00E-10$	$6.00E-11$

TABLE 8: Rate of convergence of CHCM for Examples 4 and 5.

N_k	Example 4		Example 5	
	$E(N_k)$	ρ	$E(N_k)$	ρ
1/8	$1.53E-07$	—	$2.34E-06$	—
1/16	$9.35E-09$	4.03	$1.43E-07$	4.04
1/32	$6.11E-10$	3.93	$9.99E-09$	3.84
1/64	$3.00E-11$	4.35	$8.00E-10$	3.64

TABLE 9: Rate of convergence of CHCM for Examples 6 and 7.

N_k	Example 6		Example 7	
	$E(N_k)$	ρ	$E(N_k)$	ρ
1/25	$2.4810E-03$	—	$3.3211E-03$	—
1/50	$1.8600E-04$	3.74	$1.9900E-04$	4.07
1/100	$1.1000E-05$	4.08	$1.2000E-05$	4.06

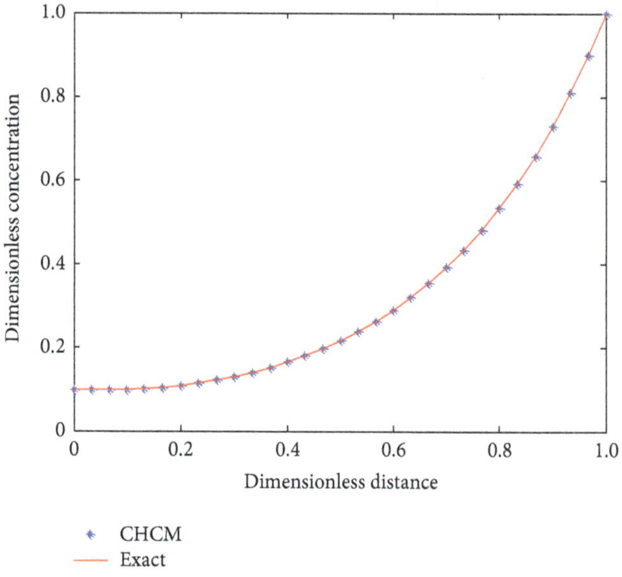

FIGURE 5: Comparison of CHCM with exact result for Example 3.

subject to Neumann boundary conditions

$$y^{(1)}(0) = -1, \qquad y^{(1)}(1) = 2\sin(1), \tag{27}$$

which has exact solution [13, 14] as

$$y(x) = \left(x^2 - 1\right)\sin x. \tag{28}$$

For different number of elements $N = 8, 16, 32, 64$, and 128, Examples 4 and 5 are solved using the present method. Maximum absolute errors of the numerical solutions are calculated and compared with those reported by [13, 14], for different methods, in Tables 6 and 7. The present approach is giving much more accurate results than the others. From Table 8, the order of convergence is found to be 4.

Example 6. Consider a nonlinear problem, that is, solving (1) for

$$\alpha_1(x) = 0, \qquad \alpha_2(x) = -y, \\ f(x) = 2\pi^2\cos(2\pi x) - \sin^4(\pi x), \tag{29}$$

subject to Neumann boundary conditions

$$y^{(1)}(0) = 0, \qquad y^{(1)}(1) = 0, \tag{30}$$

which has exact solution [13]

$$y(x) = \sin^2(\pi x). \tag{31}$$

The exact and CHCM results are showing good agreement in Figure 6 for $N = 50$.

Example 7. Consider a nonlinear problem, that is, solving (1) for

$$\alpha_1(x) = 0, \qquad \alpha_2(x) = \frac{e^{-2y}}{y}, \qquad f(x) = 0, \tag{32}$$

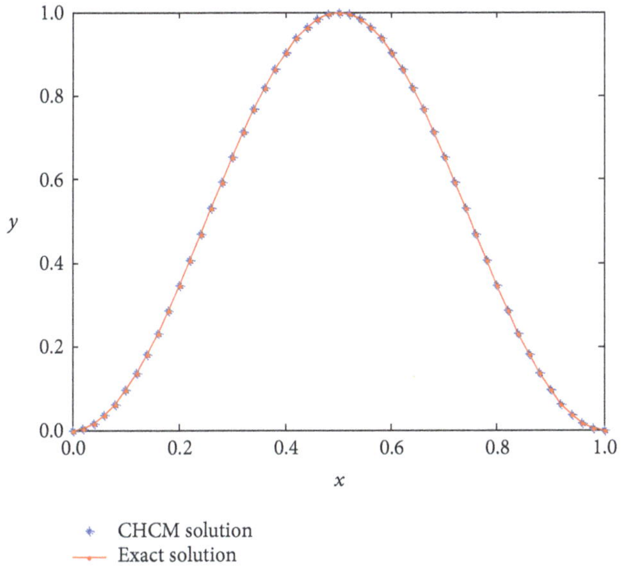

FIGURE 6: Comparison of CHCM with exact result for Example 6.

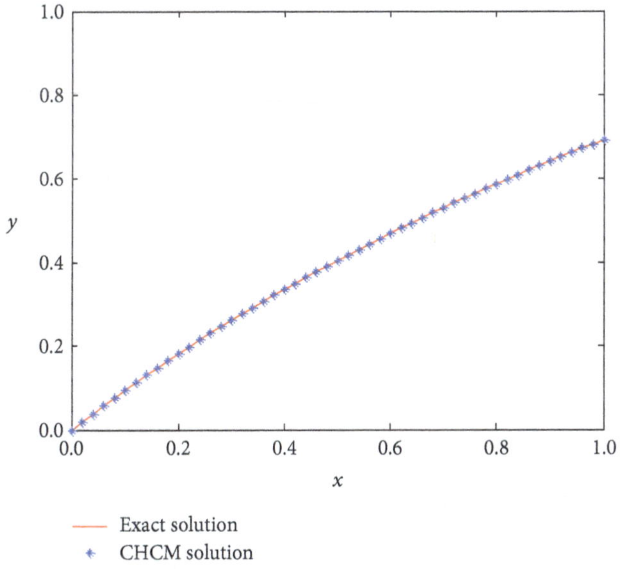

FIGURE 7: Comparison of CHCM with exact result for Example 7.

subject to Neumann boundary conditions

$$y^{(1)}(0) = 1, \qquad y^{(1)}(1) = \frac{1}{2}, \tag{33}$$

which has exact solution [13]

$$y(x) = \ln(1 + x). \tag{34}$$

An excellent matching is found between the exact and CHCM results in Figure 7 for $N = 50$. In Table 9, the order of convergence is again found to be 4 for Examples 6 and 7.

5. Conclusion

In this paper, cubic Hermite collocation method is tested for seven problems. The numerical results obtained are quite satisfactory and comparable with the existing solution available in the literature. The superiority over the finite difference method, finite element method, finite volume method, B-spline method, and polynomial and nonpolynomial spline approach shows the strength of this method. The convergence of the CHCM technique is of order 4.

Conflict of Interests

The authors declare that there is no conflict of interests regarding the publication of this paper.

Acknowledgments

This work is supported by NBHM, Mumbai, and UGC, New Delhi, India, in the form of Research Projects 2/48(14)/2009/R&D-II/2806 and 41-786/2012(SR), respectively. The JRF provided to Mr. I. A. Ganaie by NBHM, Mumbai, is thankfully acknowledged.

References

[1] J. H. Ahlberg and T. Ito, "A collocation method for two point boundary value problems," *Mathematics of Computation*, vol. 29, no. 131, pp. 761–776, 1975.

[2] M. A. Soliman and A. A. Ibrahim, "Studies on the method of orthogonal collocation III: the use of Jacobi orthogonal polynomials for the solution of boundary value problems," *Journal of King Saud University*, vol. 11, no. 2, pp. 191–202, 1999.

[3] H. Caglar, N. Caglar, and K. Elfaituri, "B-spline interpolation compared with finite difference, finite element and finite volume methods which applied to two-point boundary value problems," *Applied Mathematics and Computation*, vol. 175, no. 1, pp. 72–79, 2006.

[4] W. R. Dyksen and R. E. Lynch, "A new decoupling technique for the Hermite cubic collocation equations arising from boundary value problems," *Mathematics and Computers in Simulation*, vol. 54, no. 4-5, pp. 359–372, 2000.

[5] D. J. Higham, "Monotonic piecewise cubic interpolation, with applications to ODE plotting," *Journal of Computational and Applied Mathematics*, vol. 39, no. 3, pp. 287–294, 1992.

[6] P. M. Prenter, *Splines and Variational Methods*, Wiley interscience, New York, NY, USA, 1975.

[7] Q. Fang, T. Tsuchiya, and T. Yamamoto, "Finite difference, finite element and finite volume methods applied to two-point boundary value problems," *Journal of Computational and Applied Mathematics*, vol. 139, no. 1, pp. 9–19, 2002.

[8] M. Kumar, "A second order finite difference method and its convergence for a class of singular two-point boundary value problems," *Applied Mathematics and Computation*, vol. 146, no. 2-3, pp. 873–878, 2003.

[9] V. R. Subramanian and R. E. White, "Symbolic solutions for boundary value problems using Maple," *Computers and Chemical Engineering*, vol. 24, no. 11, pp. 2405–2416, 2000.

[10] S. N. Ha, "A nonlinear shooting method for two-point boundary value problems," *Computers and Mathematics with Applications*, vol. 42, no. 10-11, pp. 1411–1420, 2001.

[11] M. Krajnc, "Geometric Hermite interpolation by cubic G1 splines," *Nonlinear Analysis: Theo y, Methods and Applications*, vol. 70, no. 7, pp. 2614–2626, 2009.

[12] F.-G. Lang and X.-P. Xu, "Quintic B-spline collocation method for second order mixed boundary value problem," *Computer Physics Communications*, vol. 183, no. 4, pp. 913–921, 2012.

[13] L.-B. Liu, H.-W. Liu, and Y. Chen, "Polynomial spline approach for solving second-order boundary-value problems with Neumann conditions," *Applied Mathematics and Computation*, vol. 217, no. 16, pp. 6872–6882, 2011.

[14] M. A. Ramadan, I. F. Lashien, and W. K. Zahra, "Polynomial and nonpolynomial spline approaches to the numerical solution of second order boundary value problems," *Applied Mathematics and Computation*, vol. 184, no. 2, pp. 476–484, 2007.

[15] S. S. Siddiqi and G. Akram, "Solution of fifth order boundary value problems using nonpolynomial spline technique," *Applied Mathematics and Computation*, vol. 175, no. 2, pp. 1574–1581, 2006.

[16] S. S. Siddiqi, G. Akram, and A. Elahi, "Quartic spline solution of linear fifth order boundary value problems," *Applied Mathematics and Computation*, vol. 196, no. 1, pp. 214–220, 2008.

[17] W. Sun, "Hermite cubic spline collocation methods with upwind features," *ANZIAM Journal*, vol. 42, pp. 1379–1397, 2000.

[18] Z. Xiong and Y. Chen, "Finite volume element method with interpolated coefficients for two-point boundary value problem of semilinear differential equations," *Computer Methods in Applied Mechanics and Engineering*, vol. 196, no. 37–40, pp. 3798–3804, 2007.

[19] C. de Boor, *A Practical Guide To Splines*, Springer, New York, NY, USA, 2001.

[20] J. Douglas and T. Dupont, "A finite element collocation method for quasi-linear parabolic equations," *Mathematics of Computation*, vol. 121, pp. 17–28, 1973.

[21] I. A. Ganaie, B. Gupta, N. Parumasur, P. Singh, and V. K. Kukreja, "Asymptotic convergence of cubic Hermite collocation method for axial dispersion model," *Applied Mathematics and Computation*, vol. 220, no. 1, pp. 560–567, 2013.

12

Axially Symmetric Vibrations of Composite Poroelastic Spherical Shell

Rajitha Gurijala and Malla Reddy Perati

Department of Mathematics, Kakatiya University, Andhra Pradesh 506009, Warangal, India

Correspondence should be addressed to Malla Reddy Perati; mperati@yahoo.com

Academic Editor: Z.X. Guo

This paper deals with axially symmetric vibrations of composite poroelastic spherical shell consisting of two spherical shells (inner one and outer one), each of which retains its own distinctive properties. The frequency equations for pervious and impervious surfaces are obtained within the framework of Biot's theory of wave propagation in poroelastic solids. Nondimensional frequency against the ratio of outer and inner radii is computed for two types of sandstone spherical shells and the results are presented graphically. From the graphs, nondimensional frequency values are periodic in nature, but in the case of ring modes, frequency values increase with the increase of the ratio. The nondimensional phase velocity as a function of wave number is also computed for two types of sandstone spherical shells and for the spherical bone implanted with titanium. In the case of sandstone shells, the trend is periodic and distinct from the case of bone. In the case of bone, when the wave number lies between 2 and 3, the phase velocity values are periodic, and when the wave number lies between 0.1 and 1, the phase velocity values decrease.

1. Introduction

In day-to-day problems, composite structures play an important role. The term composite is applied to materials that are created by mechanically bonding two or more different elastic materials together. On the other hand, in structural engineering, spherical shell shape trims the internal volume and minimizes the surface area that saves material cost. Spherical shell forms an important class of structural configurations in aerospace as well as ground structures as they offer high strength-to-weight and stiffness-to-weight ratios. Besides some of manmade structures, the skull and the bones at shoulder joint and ankle joint are approximately in the shape of spherical shells. The said composite spherical structures are poroelastic in nature.

Kumar [1] studied the axially symmetric vibrations of fluid-filled spherical shells employing three-dimensional equations of linear elasticity. For torsional vibrations of solid prolate spheroids and thick prolate spheroidal shells, frequency equations and mode shapes are presented in analytic form [2]. Paul [3] studied the radial vibrations of poroelastic spherical shells. Employing Biot's theory [4], Shah and Tajuddin [5, 6] discussed torsional vibrations of poroelastic spheroidal shells and axially symmetric vibrations of fluid-filled poroelastic spherical shell. In the paper in [5], they derived frequency equations for poroelastic thin spherical shell, thick spherical shell, and poroelastic solid sphere and concluded that the frequency is the same for all the three cases. In the paper in [6], radial and rotatory vibrations of fluid-filled and empty poroelastic spherical shells are investigated. Vibration analysis of a poroelastic composite hollow sphere is discussed by Shanker et al. [7]. They derived frequency equations for poroelastic composite hollow sphere and a poroelastic composite hollow sphere with rigid core. Some structures may be far from their center of curvature and thickness might be very small when compared to radii of curvature. In this case, we have ring modes [8]. To the best of authors knowledge, poroelastic composite spherical shell and its ring modes are not yet investigated. Hence the same are warranted. In the present paper, we investigate the axially symmetric (independent of azimuthal coordinate) vibrations of composite poroelastic spherical shell in the framework of Biot's theory. Frequency equations are obtained for both pervious and impervious surfaces. Also, frequency against the ratio of outer and inner radii and the phase velocity against the wave number are computed. Comparative study is made between the modes of composite spherical shell and its ring modes.

The rest of the paper is organized as follows. In Section 2, basic governing equations, formulation, and solution of the problem are given. In Section 3, frequency equations are derived for both pervious and impervious surfaces. Particular case is derived in Section 4, while numerical results are presented in Section 5. Finally, conclusion is given in Section 6.

2. Governing Equations and Solution of the Problem

The equations of motion of a homogeneous, isotropic poroelastic solid [4] in the presence of dissipation b are

$$\begin{aligned} N\nabla^2\vec{u} + (A+N)\nabla e + Q\nabla\varepsilon &= \frac{\partial^2}{\partial t^2}\left(\rho_{11}\vec{u} + \rho_{12}\vec{U}\right) + b\frac{\partial}{\partial t}\left(\vec{u} - \vec{U}\right), \\ Q\nabla e + R\nabla\varepsilon &= \frac{\partial^2}{\partial t^2}\left(\rho_{12}\vec{u} + \rho_{22}\vec{U}\right) - b\frac{\partial}{\partial t}\left(\vec{u} - \vec{U}\right), \end{aligned} \tag{1}$$

where ∇^2 is the Laplace operator, $\vec{u}(u, v, 0)$ and $\vec{U}(U, V, 0)$ are solid and fluid displacements, e and ε are the dilatations of solid and fluid, A, N, Q, R are all poroelastic constants, b is the dissipative coefficient, and ρ_{ij} are mass coefficients. The relevant solid stresses σ_{ij} and fluid pressure s are

$$\begin{aligned} \sigma_{ij} &= 2Ne_{ij} + (Ae + Q\varepsilon)\,\delta_{ij} \quad (i, j = 1, 2, 3), \\ s &= Qe + R\varepsilon. \end{aligned} \tag{2}$$

In (2), δ_{ij} is the well-known Kronecker delta function.

Let (r, θ, ϕ) be the spherical polar coordinates. Consider a composite isotropic poroelastic spherical shell with outer and inner radii r_2 and r_1, respectively, made up of two different materials and the inner one is solid spherical shell whereas the outer one is a thick walled hollow spherical shell having thickness $h = (r_2 - r_1 > 0)$. Poroelastic constants of outer shell and inner shell are ${}_2P$, ${}_2N$, ${}_2Q$, ${}_2R$ and ${}_1P$, ${}_1N$, ${}_1Q$, ${}_1R$, respectively. We introduce the displacement potentials ϕ's and ψ's which are the functions of r, θ, and t as follows:

$$\begin{aligned} u &= \frac{\partial\phi_1}{\partial r} + \frac{1}{r}\frac{\partial^2\psi_1}{\partial\theta^2} + \frac{\cot\theta}{r}\frac{\partial\psi_1}{\partial\theta}, \\ v &= \frac{1}{r}\frac{\partial\phi_1}{\partial\theta} - \frac{\partial^2\psi_1}{\partial r\partial\theta} - \frac{1}{r}\frac{\partial\psi_1}{\partial\theta}, \\ U &= \frac{\partial\phi_2}{\partial r} + \frac{1}{r}\frac{\partial^2\psi_2}{\partial\theta^2} + \frac{\cot\theta}{r}\frac{\partial\psi_2}{\partial\theta}, \\ V &= \frac{1}{r}\frac{\partial\phi_2}{\partial\theta} - \frac{\partial^2\psi_2}{\partial r\partial\theta} - \frac{1}{r}\frac{\partial\psi_2}{\partial\theta}. \end{aligned} \tag{3}$$

For free harmonic vibrations, the potential functions ϕ_1, ϕ_2, ψ_1, and ψ_2 are expressed as follows:

$$\begin{aligned} \phi_1 &= f_1(r)\, p_l^m(\cos\theta)\, e^{i\omega t}, \\ \phi_2 &= f_2(r)\, p_l^m(\cos\theta)\, e^{i\omega t}, \\ \psi_1 &= g_1(r)\, p_l^m(\cos\theta)\, e^{i\omega t}, \\ \psi_2 &= g_2(r)\, p_l^m(\cos\theta)\, e^{i\omega t}, \end{aligned} \tag{4}$$

where ω is the frequency of wave, $p_l^m(\cos\theta)$ is the associated Legendre polynomial, where l is the order of spherical harmonic and $m = 0, 1, \ldots l$, i is the complex unity, and t is time. Equations (1) and (4), after a long calculation, yield

$$\begin{aligned} \phi_1 &= \left(A_1 J_n\left({}_2\xi_1 r\right) + B_1 Y_n\left({}_2\xi_1 r\right) + A_2 J_n\left({}_2\xi_2 r\right) + B_2 Y_n\left({}_2\xi_2 r\right)\right) p_l^m(\cos\theta)\, e^{i\omega t}, \\ \phi_2 &= \left(A_1\left({}_2\delta_1^2\right) J_n\left({}_2\xi_1 r\right) + B_1\left({}_2\delta_1^2\right) Y_n\left({}_2\xi_1 r\right) + A_2\left({}_2\delta_2^2\right) J_n\left({}_2\xi_2 r\right) + B_2\left({}_2\delta_2^2\right) Y_n\left({}_2\xi_2 r\right)\right) \\ &\quad \times p_l^m(\cos\theta)\, e^{i\omega t}, \\ \psi_1 &= \left(A_3 J_n\left({}_2\xi_3 r\right) + B_3 Y_n\left({}_2\xi_3 r\right)\right) p_l^m(\cos\theta)\, e^{i\omega t}, \\ \psi_2 &= -\frac{{}_2M_{12}}{{}_2M_{22}}\left(A_3 J_n\left({}_2\xi_3 r\right) + B_3 Y_n\left({}_2\xi_3 r\right)\right) \times p_l^m(\cos\theta)\, e^{i\omega t}. \end{aligned} \tag{5}$$

In (5), $J_n(x)$, $Y_n(x)$ are spherical Bessel functions of first and second kinds of order n, respectively, and

$$n = -\frac{1}{2} + \frac{1}{2p_l^m(\cos\theta)} \times \left(\left(p_l^m(\cos\theta)\right)^2 - 4p_l^m(\cos\theta) \times \left(p_l^{m''}(\cos\theta)\sin^2\theta - 2p_l^{m'}(\cos\theta)\sin\theta\right)\right)^{1/2}. \tag{6}$$

The displacement components of outer part and inner part can readily be evaluated from (3) and are given by

$$\begin{aligned} {}_2u &= \left(A_1\,{}_2D_{11}(r) + B_1\,{}_2D_{12}(r) + A_2\,{}_2D_{13}(r) + B_2\,{}_2D_{14}(r) + A_3\,{}_2D_{15}(r) + B_3\,{}_2D_{16}(r)\right) \times p_l^m(\cos\theta)\, e^{i\omega t}, \\ {}_2v &= \left(A_1\,{}_2D_{21}(r) + B_1\,{}_2D_{22}(r) + A_2\,{}_2D_{23}(r) + B_2\,{}_2D_{24}(r) + A_3\,{}_2D_{25}(r) + B_3\,{}_2D_{26}(r)\right) \end{aligned}$$

$$\times p_l^{m'}(\cos\theta)\sin\theta e^{i\omega t},$$

$$_1u = (C_1\,{}_1D_{11}(r) + C_2\,{}_1D_{13}(r) + C_3\,{}_1D_{15}(r)) \times p_l^{m}(\cos\theta)e^{i\omega t},$$

$$_1v = (C_1\,{}_1D_{21}(r) + C_2\,{}_1D_{23}(r) + C_3\,{}_1D_{25}(r)) \times p_l^{m'}(\cos\theta)\sin\theta e^{i\omega t}. \quad (7)$$

In (7),

$$\begin{aligned}
{}_2D_{11}(r) &= \frac{n}{r}J_n({}_2\xi_1 r) - {}_2\xi_1 J_{n+1}({}_2\xi_1 r),\\
{}_2D_{12}(r) &= \frac{n}{r}Y_n({}_2\xi_1 r) - {}_2\xi_1 Y_{n+1}({}_2\xi_1 r),\\
{}_2D_{13}(r) &= \frac{n}{r}J_n({}_2\xi_2 r) - {}_2\xi_2 J_{n+1}({}_2\xi_2 r),\\
{}_2D_{14}(r) &= \frac{n}{r}Y_n({}_2\xi_2 r) - {}_2\xi_2 Y_{n+1}({}_2\xi_2 r),\\
{}_2D_{15}(r) &= \left(p_l^{m''}(\cos\theta)\sin^2\theta - (1+\cot\theta)\right.\\
&\quad \left.\times p_l^{m'}(\cos\theta)\sin\theta\right)\frac{1}{r}J_n({}_2\xi_3 r),\\
{}_2D_{16}(r) &= \left(p_l^{m''}(\cos\theta)\sin^2\theta - (1+\cot\theta)\,p_l^{m'}\right.\\
&\quad \left.\times(\cos\theta)\sin\theta\right)\frac{1}{r}Y_n({}_2\xi_3 r),\\
{}_2D_{21}(r) &= -\frac{1}{r}J_n({}_2\xi_1 r),\\
{}_2D_{22}(r) &= -\frac{1}{r}Y_n({}_2\xi_1 r),\\
{}_2D_{23}(r) &= -\frac{1}{r}J_n({}_2\xi_2 r),\\
{}_2D_{24}(r) &= -\frac{1}{r}Y_n({}_2\xi_2 r),\\
{}_2D_{25}(r) &= \left(\frac{n+1}{r}\right)J_n({}_2\xi_3 r) - {}_2\xi_3 J_{n+1}({}_2\xi_3 r),\\
{}_2D_{26}(r) &= \left(\frac{n+1}{r}\right)Y_n({}_2\xi_3 r) - {}_2\xi_3 Y_{n+1}({}_2\xi_3 r).
\end{aligned} \quad (8)$$

The notations ${}_1D_{ij}$ $(i = 1,2\ j = 1,3,5)$ are the same as ${}_2D_{ij}$ $(i = 1,2\ j = 1,3,5)$ and ${}_p\xi_q = (\omega/{}_pV_q)$ $(q = 1,2,3,\ p = 1,2)$. The notations ${}_pV_1$, ${}_pV_2$, and ${}_pV_3$ are dilatational wave velocities of first and second kinds and shear wave velocity, respectively. The notations A_1, B_1, A_2, B_2, A_3, B_3, C_1, C_2, C_3 are all arbitrary constants. By substituting the displacements in (2), the relevant stresses and fluid pressure pertaining to outer and inner parts are the following:

$$\begin{aligned}
{}_2\sigma_{rr} + {}_2s &= (A_1\,{}_2M_{11}(r) + B_1\,{}_2M_{12}(r)\\
&\quad + A_2\,{}_2M_{13}(r) + B_2\,{}_2M_{14}(r)\\
&\quad + A_3\,{}_2M_{15}(r) + B_3\,{}_2M_{16}(r))\\
&\quad \times p_l^{m}(\cos\theta)e^{i\omega t},\\
{}_2\sigma_{r\theta} &= (A_1\,{}_2M_{21}(r) + B_1\,{}_2M_{22}(r)\\
&\quad + A_2\,{}_2M_{23}(r) + B_2\,{}_2M_{24}(r)\\
&\quad + A_3\,{}_2M_{25}(r) + B_3\,{}_2M_{26}(r))\\
&\quad \times p_l^{m'}(\cos\theta)\sin\theta e^{i\omega t},\\
{}_2s &= (A_1\,{}_2M_{31}(r) + B_1\,{}_2M_{32}(r)\\
&\quad + A_2\,{}_2M_{33}(r) + B_2\,{}_2M_{34}(r))\\
&\quad \times p_l^{m}(\cos\theta)e^{i\omega t},\\
\frac{\partial({}_2s)}{\partial r} &= (A_1\,{}_2N_{41}(r) + B_1\,{}_2N_{42}(r)\\
&\quad + A_2\,{}_2N_{43}(r) + B_2\,{}_2N_{44}(r))\\
&\quad \times p_l^{m}(\cos\theta)e^{i\omega t},\\
{}_1\sigma_{rr} + {}_1s &= (C_1\,{}_1M_{11}(r) + C_2\,{}_1M_{13}(r)\\
&\quad + C_3\,{}_1M_{15}(r))\,p_l^{m}(\cos\theta)e^{i\omega t},\\
{}_1\sigma_{r\theta} &= (C_1\,{}_1M_{21}(r) + C_2\,{}_1M_{23}(r)\\
&\quad + C_3\,{}_1M_{25}(r))\,p_l^{m'}(\cos\theta)\sin\theta e^{i\omega t},\\
{}_1s &= (C_1\,{}_1M_{31}(r) + C_2\,{}_1M_{33}(r))\,p_l^{m}(\cos\theta)e^{i\omega t},\\
\frac{\partial({}_1s)}{\partial r} &= (C_1\,{}_1N_{31}(r) + C_2\,{}_1N_{33}(r))\,p_l^{m}(\cos\theta)e^{i\omega t},
\end{aligned} \quad (9)$$

where

$$\begin{aligned}
{}_pM_{11}(r) &= \Bigg(\left(({}_pP + {}_pQ) + ({}_pQ + {}_pR)\,{}_p\delta_1^2\right)\\
&\quad \times\left(\frac{n(n-1)}{r^2} - {}_p\xi_1^2\right)J_n({}_p\xi_1 r)\\
&\quad + \frac{2}{{}_p\xi_1 r^3}J_{n+1}({}_p\xi_1 r)\Bigg)\\
&\quad + \left(({}_pA + {}_pQ) + ({}_pQ + {}_pR)\,{}_p\delta_1^2\right)
\end{aligned}$$

$$\times\left(\frac{2n+L}{r^2}J_n\left({}_p\xi_1 r\right)-\frac{2\,{}_p\xi_1}{r}J_{n+1}\left({}_p\xi_1 r\right)\right),$$

$$\begin{aligned}
{}_pM_{15}(r) &= 2\,{}_pN\left(\frac{(n-1)L_2}{r^2}J_n\left({}_p\xi_3 r\right)\right.\\
&\quad\left.-\frac{{}_p\xi_3 L_2}{r}J_{n+1}\left({}_p\xi_3 r\right)\right),\\
{}_pM_{21}(r) &= 2\,{}_pN\left(\frac{(1-n)}{r^2}J_n\left({}_p\xi_1 r\right)+\frac{{}_p\xi_1}{r}J_{n+1}\right.\\
&\quad\left.\times\left({}_p\xi_1 r\right)\right),\\
{}_pM_{25}(r) &= {}_pN\left(\left(\frac{L_1+n(n+1)}{r^2}-{}_p\xi_3^2\right)J_n\left({}_p\xi_3 r\right)\right.\\
&\quad\left.+\left(\frac{{}_p\xi_3-2\,{}_p\xi_3(n+1)}{r}\right)J_{n+1}\left({}_p\xi_3 r\right)\right),\\
{}_pM_{31}(r) &= \left({}_pQ+{}_pR\,{}_p\delta_1^2\right)\\
&\quad\times\left(\left(\frac{n(n+1)+L}{r^2}-{}_p\xi_1^2\right)J_n\left({}_p\xi_1 r\right)\right.\\
&\quad\left.+\left(\frac{2}{{}_p\xi_1 r^3}-\frac{2\,{}_p\xi_1}{r}\right)J_{n+1}\left({}_p\xi_1 r\right)\right),\\
{}_pN_{41}(r) &= \left({}_pQ+{}_pR\,{}_p\delta_1^2\right)\\
&\quad\times\left(\left(n^2+n+L\right)\right.\\
&\quad\times\left(\frac{n}{r}J_n\left({}_p\xi_1 r\right)-{}_p\xi_1 J_{n+1}\left({}_p\xi_1 r\right)\right)\\
&\quad+\left(\frac{2}{r}-(4+n)\,{}_p\xi_1^2 r\right)J_n\left({}_p\xi_1 r\right)\\
&\quad+\left({}_p\xi_1^3 r^2-\frac{2(n+1)}{{}_p\xi_1 r^2}+2\,{}_p\xi_1(n+1)\right.\\
&\quad\left.-\frac{2}{{}_p\xi_1 r^2}-2\,{}_p\xi_1\right)\\
&\quad\left.\times J_{n+1}\left({}_p\xi_1 r\right)\right).
\end{aligned}\tag{10}$$

In all the above, ${}_pM_{12}(r)$, ${}_pM_{16}(r)$, ${}_pM_{22}(r)$, ${}_pM_{26}(r)$, ${}_pM_{32}(r)$, and ${}_pN_{32}(r)$ are similar expressions as in ${}_pM_{11}(r)$, ${}_pM_{15}(r)$, ${}_pM_{21}(r)$, ${}_pM_{25}(r)$, ${}_pM_{31}(r)$, and ${}_pN_{31}(r)$ with J_n, J_{n+1} replaced by Y_n, Y_{n+1} , respectively; ${}_pM_{13}(r)$, ${}_pM_{23}(r)$, ${}_pM_{33}(r)$, and ${}_pN_{33}(r)$ are similar expressions to ${}_pM_{11}(r)$, ${}_pM_{21}(r)$, ${}_pM_{31}(r)$, and ${}_pN_{31}(r)$ with ${}_p\xi_1$ and ${}_p\delta_1$ replaced by ${}_p\xi_2$ and ${}_p\delta_2$, respectively; ${}_pM_{11}(r)$, ${}_pM_{21}(r)$, ${}_pM_{14}(r)$, ${}_pM_{24}(r)$, ${}_pM_{34}(r)$, and ${}_pN_{34}(r)$ are similar expressions to ${}_pM_{31}(r)$ and ${}_pN_{31}(r)$ with ${}_p\xi_1$, ${}_p\delta_1$, J_n, and J_{n+1} replaced by ${}_p\xi_2$, ${}_p\delta_2$, Y_n, and Y_{n+1}, respectively.

In (10),

$$\begin{aligned}
{}_p\delta_q^2 &= \frac{-\left({}_pP\,{}_pR-{}_pQ^2\right)\,{}_1V_q^{-2}}{{}_pR\,{}_pM_{12}-{}_pQ\,{}_pM_{22}}\\
&\quad\times\left(\frac{1}{r^2}\left(n(n+1)-r^2\,{}_p\xi_q^2\right.\right.\\
&\quad-\left(\left(p_l^{m''}(\cos\theta)\sin^2\theta\right.\right.\\
&\quad\left.-2p_l^{m'}(\cos\theta)\sin\theta\right)\\
&\quad\left.\left.\left.\times\left(p_l^m(\cos\theta)\right)^{-1}\right)\right)\right)\\
&\quad-\frac{{}_pR\,{}_pM_{11}-{}_pQ\,{}_pM_{12}}{{}_pR\,{}_pM_{12}-{}_pQ\,{}_pM_{22}},\\
&\qquad p=1,2,\ q=1,2,\\
{}_pP &= {}_pA+2\,{}_pN,\\
{}_pM_{11} &= {}_p\rho_{11}-\frac{i\,{}_pb}{\omega},\\
{}_pM_{12} &= {}_p\rho_{12}+\frac{i\,{}_pb}{\omega},\\
{}_pM_{22} &= {}_p\rho_{22}-\frac{i\,{}_pb}{\omega},\\
L &= \frac{1}{p_l^m(\cos\theta)}\left(p_l^{m''}(\cos\theta)\sin^2\theta\right.\\
&\quad\left.-2p_l^{m'}(\cos\theta)\sin\theta\right),\\
L_1 &= \frac{1}{p_l^{m'}(\cos\theta)\sin\theta}\\
&\quad\times\left(-p_l^{m'''}(\cos\theta)\sin^3\theta+p_l^{m''}(\cos\theta)\sin\theta\right.\\
&\quad\left.\times(3+\sin\theta)+p_l^{m'}(\cos\theta)(\sin\theta-\cos\theta)\right),\\
L_2 &= \frac{1}{p_l^m(\cos\theta)}\left(p_l^{m''}(\cos\theta)\sin^2\theta-(1+\cot\theta)\right.\\
&\quad\left.\times p_l^{m'}(\cos\theta)\sin\theta\right).
\end{aligned}\tag{11}$$

3. Boundary Conditions and Frequency Equations

The boundary conditions for the stress-free outer surface and for the perfect bonding between the outer and the inner parts are the following:

$$\begin{aligned}&({}_2\sigma_{rr} + {}_2s) - ({}_1\sigma_{rr} + {}_1s) = 0,\\ &{}_2\sigma_{r\theta} - {}_1\sigma_{r\theta} = 0,\\ &{}_2s = 0,\\ &{}_1s = 0,\\ &{}_2u - {}_1u = 0,\\ &{}_2v - {}_1v = 0,\\ &\text{at } r = r_1,\end{aligned} \tag{12}$$

$$\begin{aligned}&{}_2\sigma_{rr} + {}_2s = 0,\\ &{}_2\sigma_{r\theta} = 0,\\ &{}_2s = 0,\\ &\text{at } r = r_2.\end{aligned} \tag{13}$$

Equations (12) and (13) pertain to a pervious surface; in the case of impervious surface, the boundary conditions are the same as those of the pervious surface except the third and fourth equations of (12) and the third equation of (13) on fluid pressure; instead, here we have

$$\frac{\partial({}_2s)}{\partial r} = 0, \quad \frac{\partial({}_1s)}{\partial r} = 0 \quad \text{at } r = r_1. \tag{14}$$

$$\frac{\partial({}_2s)}{\partial r} = 0 \quad \text{at } r = r_2. \tag{15}$$

Equations (12) and (13) result in a system of nine homogeneous equations in nine arbitrary constants: $A_1, B_1, A_2, B_2, A_3, B_3, C_1, C_2, C_3$. For a nontrivial solution, determinant of coefficients is zero. Accordingly, we obtain the following frequency equation for a pervious surface:

$$\begin{vmatrix} {}_2M_{11}(r_1) & {}_2M_{12}(r_1) & {}_2M_{13}(r_1) & {}_2M_{14}(r_1) & {}_2M_{15}(r_1) & {}_2M_{16}(r_1) & {}_1M_{11}(r_1) & {}_1M_{13}(r_1) & {}_1M_{15}(r_1)\\ {}_2M_{21}(r_1) & {}_2M_{22}(r_1) & {}_2M_{23}(r_1) & {}_2M_{24}(r_1) & {}_2M_{25}(r_1) & {}_2M_{26}(r_1) & {}_1M_{21}(r_1) & {}_1M_{23}(r_1) & {}_1M_{25}(r_1)\\ {}_2M_{31}(r_1) & {}_2M_{32}(r_1) & {}_2M_{33}(r_1) & {}_2M_{34}(r_1) & 0 & 0 & 0 & 0 & 0\\ {}_1M_{31}(r_1) & {}_1M_{33}(r_1) & 0 & 0 & 0 & 0 & 0 & 0 & 0\\ {}_2D_{11}(r_1) & {}_2D_{12}(r_1) & {}_2D_{13}(r_1) & {}_2D_{14}(r_1) & {}_2D_{15}(r_1) & {}_2D_{16}(r_1) & {}_1D_{11}(r_1) & {}_1D_{13}(r_1) & {}_1D_{15}(r_1)\\ {}_2D_{21}(r_1) & {}_2D_{22}(r_1) & {}_2D_{23}(r_1) & {}_2D_{24}(r_1) & {}_2D_{25}(r_1) & {}_2D_{26}(r_1) & {}_1D_{21}(r_1) & {}_1D_{23}(r_1) & {}_1D_{25}(r_1)\\ {}_2M_{11}(r_2) & {}_2M_{12}(r_2) & {}_2M_{13}(r_2) & {}_2M_{14}(r_2) & {}_2M_{15}(r_2) & {}_2M_{16}(r_2) & 0 & 0 & 0\\ {}_2M_{21}(r_2) & {}_2M_{22}(r_2) & {}_2M_{23}(r_2) & {}_2M_{24}(r_2) & {}_2M_{25}(r_2) & {}_2M_{26}(r_2) & 0 & 0 & 0\\ {}_2M_{31}(r_2) & {}_2M_{32}(r_2) & {}_2M_{33}(r_2) & {}_2M_{34}(r_2) & 0 & 0 & 0 & 0 & 0 \end{vmatrix} = 0. \tag{16}$$

In the case of impervious surface, the frequency equation is

$$\begin{vmatrix} {}_2M_{11}(r_1) & {}_2M_{12}(r_1) & {}_2M_{13}(r_1) & {}_2M_{14}(r_1) & {}_2M_{15}(r_1) & {}_2M_{16}(r_1) & {}_1M_{11}(r_1) & {}_1M_{13}(r_1) & {}_1M_{15}(r_1)\\ {}_2M_{21}(r_1) & {}_2M_{22}(r_1) & {}_2M_{23}(r_1) & {}_2M_{24}(r_1) & {}_2M_{25}(r_1) & {}_2M_{26}(r_1) & {}_1M_{21}(r_1) & {}_1M_{23}(r_1) & {}_1M_{25}(r_1)\\ {}_2N_{31}(r_1) & {}_2N_{32}(r_1) & {}_2N_{33}(r_1) & {}_2N_{34}(r_1) & 0 & 0 & 0 & 0 & 0\\ {}_1N_{31}(r_1) & {}_1N_{33}(r_1) & 0 & 0 & 0 & 0 & 0 & 0 & 0\\ {}_2D_{11}(r_1) & {}_2D_{12}(r_1) & {}_2D_{13}(r_1) & {}_2D_{14}(r_1) & {}_2D_{15}(r_1) & {}_2D_{16}(r_1) & {}_1D_{11}(r_1) & {}_1D_{13}(r_1) & {}_1D_{15}(r_1)\\ {}_2D_{21}(r_1) & {}_2D_{22}(r_1) & {}_2D_{23}(r_1) & {}_2D_{24}(r_1) & {}_2D_{25}(r_1) & {}_2D_{26}(r_1) & {}_1D_{21}(r_1) & {}_1D_{23}(r_1) & {}_1D_{25}(r_1)\\ {}_2M_{11}(r_2) & {}_2M_{12}(r_2) & {}_2M_{13}(r_2) & {}_2M_{14}(r_2) & {}_2M_{15}(r_2) & {}_2M_{16}(r_2) & 0 & 0 & 0\\ {}_2M_{21}(r_2) & {}_2M_{22}(r_2) & {}_2M_{23}(r_2) & {}_2M_{24}(r_2) & {}_2M_{25}(r_2) & {}_2M_{26}(r_2) & 0 & 0 & 0\\ {}_2N_{31}(r_2) & {}_2N_{32}(r_2) & {}_2N_{33}(r_2) & {}_2N_{34}(r_2) & 0 & 0 & 0 & 0 & 0 \end{vmatrix} = 0. \tag{17}$$

4. Poroelastic Thick Walled Hollow Spherical Shell: A Particular Case

The composite spherical shell will reduce to the poroelastic thick walled hollow spherical shell, under some special substitutions, that is discussed next.

Consider the case where ${}_2A = A$, ${}_2N = N$, ${}_2Q = Q$, ${}_2R = R$, ${}_2M_{ij} = M_{ij}$ and ${}_1A = 0$, ${}_1N = 0$, ${}_1Q = 0$, ${}_1R = 0$ so that ${}_1M_{ij} = 0$ in (16). Then, composite spherical shell will become a thick walled spherical shell in the case of pervious surface and its frequency equation is given by

$$|M_{ij}| = 0 \quad (i = 1, \ldots, 6,\ j = 1, \ldots, 6). \tag{18}$$

In (18), the elements are similar to those of (10) without left subscript. Now, we consider the case where $\xi_i r_1, \xi_i r_2 \to \infty$ $(i = 1, 2, 3)$, that is, the case of $(h/r_1) \to 0$. In the region of small (h/r_1), for $n \neq 0$, pertinent modes are essentially ring-extensional and ring-flexural ones [8]. Considering the determinant in (18) as a function D of $\xi_i r_1$ $(i = 1, 2, 3)$ and h/r_1, we obtain

$$\begin{aligned} D\left(\xi_i r_1, \frac{h}{r_1}\right) &= D(\xi_i r_1, 0) + \frac{h}{r_1}\frac{\partial}{\partial(h/r_1)}D(\xi_i r_1, 0)\\ &\quad + \frac{1}{2}\left(\frac{h}{r_1}\right)^2 \frac{\partial^2}{\partial(h/r_1)^2}D(\xi_i r_1, 0) + \cdots . \end{aligned} \tag{19}$$

It is found that

$$D(\xi_i r_1, 0) = \frac{\partial}{\partial (h/r_1)} D(\xi_i r_1, 0) = 0. \qquad (20)$$

Hence, for a small h/r_1, we have

$$D\left(\xi_i r_1, \frac{h}{r_1}\right) \approx \frac{1}{2}\left(\frac{h}{r_1}\right)^2 \frac{\partial^2}{\partial (h/r_1)^2} D(\xi_i r_1, 0). \qquad (21)$$

Further,

$$\frac{\partial^2}{\partial (h/r_1)^2} D(\xi_i r_1, 0) = r_1^3 \begin{vmatrix} M_{11}(r_1) & M_{12}(r_1) & M_{13}(r_1) & M_{14}(r_1) & M_{15}(r_1) & M_{16}(r_1) \\ M_{21}(r_1) & M_{22}(r_1) & M_{23}(r_1) & M_{24}(r_1) & M_{25}(r_1) & M_{26}(r_1) \\ M_{31}(r_1) & M_{32}(r_1) & M_{33}(r_1) & M_{34}(r_1) & 0 & 0 \\ M'_{11}(r_1)+M^*_{11}(r_1) & M'_{12}(r_1)+M^*_{12}(r_1) & M'_{13}(r_1)+M^*_{13}(r_1) & M'_{14}(r_1)+M^*_{14}(r_1) & M_{15}(r_1)+M^*_{15}(r_1) & M_{16}(r_1)+M^*_{16}(r_1) \\ M'_{21}(r_1)+M^*_{21}(r_1) & M'_{22}(r_1)+M^*_{22}(r_1) & M'_{23}(r_1)+M^*_{23}(r_1) & M'_{24}(r_1)+M^*_{24}(r_1) & M'_{25}(r_1)+M^*_{25}(r_1) & M'_{26}(r_1)+M^*_{26}(r_1) \\ M'_{31}(r_1)+M^*_{31}(r_1) & M'_{32}(r_1)+M^*_{32}(r_1) & M'_{33}(r_1)+M^*_{33}(r_1) & M'_{34}(r_1)+M^*_{34}(r_1) & 0 & 0 \end{vmatrix}, \qquad (22)$$

where M_{ij} $(i = 1,\ldots,6,\ j = 1,\ldots,6)$ are given by (18) and primes denote differentiation with respect to r_1 and M^*_{ij} $(i = 1,\ldots,6,\ j = 1,\ldots,6)$ that are given in the Appendix. For a nontrivial solution, determinant of coefficient is zero, that is, $D(\xi_i r_1, h/r_1) = 0$; accordingly, we get the frequency equation for the ring modes.

5. Numerical Results

Due to dissipative nature of the medium, waves are attenuated. Attenuation presents some difficulty in the definition of phase velocity. If dissipative coefficient b is nonzero, then the densities will be complex numbers that make the implicit frequency equations complex valued which cannot be solved so easily. Therefore, the case $b = 0$ is to be considered in what follows. Albeit the problem is poroelastic in nature, the only thing is that attenuation is not considered for the said reason. The following nondimensional parameters are introduced to investigate the frequency equations:

$$a_1 = \frac{{}_2P}{{}_1H}, \quad a_2 = \frac{{}_2Q}{{}_1H}, \quad a_3 = \frac{{}_2R}{{}_1H}, \quad a_4 = \frac{{}_2N}{{}_1H},$$

$$d_1 = \frac{{}_2\rho_{11}}{{}_1\rho}, \quad d_2 = \frac{{}_2\rho_{12}}{{}_1\rho}, \quad d_3 = \frac{{}_2\rho_{22}}{{}_1\rho},$$

$$_2\tilde{x} = \left(\frac{{}_2V_0}{{}_2V_1}\right)^2, \quad _1\tilde{y} = \left(\frac{{}_2V_0}{{}_2V_2}\right)^2, \quad _1\tilde{z} = \left(\frac{{}_2V_0}{{}_2V_3}\right)^2,$$

$$b_1 = \frac{{}_1P}{{}_1H}, \quad b_2 = \frac{{}_1Q}{{}_1H}, \quad b_3 = \frac{{}_1R}{{}_1H}, \quad b_4 = \frac{{}_1N}{{}_1H},$$

$$g_1 = \frac{{}_1\rho_{11}}{{}_1\rho}, \quad g_2 = \frac{{}_1\rho_{12}}{{}_1\rho}, \quad g_3 = \frac{{}_1\rho_{22}}{{}_1\rho},$$

$$_1\tilde{x} = \left(\frac{{}_1V_0}{{}_1V_1}\right)^2, \quad _1\tilde{y} = \left(\frac{{}_1V_0}{{}_1V_2}\right)^2, \quad _1\tilde{z} = \left(\frac{{}_1V_0}{{}_1V_3}\right)^2,$$

$$\Omega = \frac{\omega h}{{}_1c_0}, \quad m_1 = \frac{c}{{}_1c_0}, \quad c = \frac{\omega}{k}. \qquad (23)$$

In (23), Ω is nondimensional frequency, c is phase velocity, m_1 is nondimensional phase velocity, k is the wave number, and ${}_1H = {}_1P + 2\,{}_1Q + {}_1R$, ${}_1\rho = {}_1\rho_{11} + 2\,{}_1\rho_{12} + {}_1\rho_{22}$; also ${}_1c_0$ and ${}_1V_0$ are reference velocities and are given by ${}_1c_0^2 = {}_1N/{}_1\rho$, ${}_1V_0^2 = {}_1H/{}_1\rho$, and h is the thickness of the poroelastic spherical shell. Let $g = r_2/r_1$, so that $h/r_1 = g - 1$, $h/r_2 = (g-1)/g$. Employing these nondimensional quantities in the frequency equations, we will get two implicit relations; one is between the nondimensional frequency (Ω) and the ratio of outer and inner radii (g), and another is the relation between phase velocity (m_1) and the nondimensional wave number (kr_2). The numerical results are presented for the following cases.

5.1. Sandstone Composite Shells. Nondimensional frequency (Ω) and phase velocity (m_1) are computed for two types of composite spherical shells, namely, composite spherical shell 1 and composite spherical shell 2 using the numerical process performed in MATLAB. In composite spherical shell 1, outer shell is made up of sandstone saturated with water [9] and inner shell is made up of sandstone saturated with kerosene [10]. In composite spherical shell 2, the roles of materials are reversed. The physical parameters of these composite spherical shells following (19) are given in Table 1. The value of θ is taken to be 30° arbitrarily. The value of m is taken to be 1 and the value of l is taken to be 2, following [2]. The velocities of the dilatational waves and shear wave are computed using Biot's theory [4]. The numerical values are depicted in Figures 1 and 7.

TABLE 1: Material parameters.

Material parameters	Composite spherical shell 1	Composite spherical shell 2
a_1	0.445	1.819
a_2	0.034	0.011
a_3	0.015	0.054
a_4	0.123	0.780
d_1	0.887	0.891
d_2	−0.001	0
d_3	0.099	0.125
${}_2\tilde{x}$	1.863	0.489
${}_2\tilde{y}$	8.884	2.330
${}_2\tilde{z}$	7.183	1.142
b_1	0.96	0.843
b_2	0.006	0.065
b_3	0.028	0.028
b_4	0.412	0.234
g_1	0.887	0.901
g_2	0	−0.001
g_3	0.123	0.101
${}_1\tilde{x}$	0.913	0.999
${}_1\tilde{y}$	4.347	4.763
${}_1\tilde{z}$	1.129	3.851

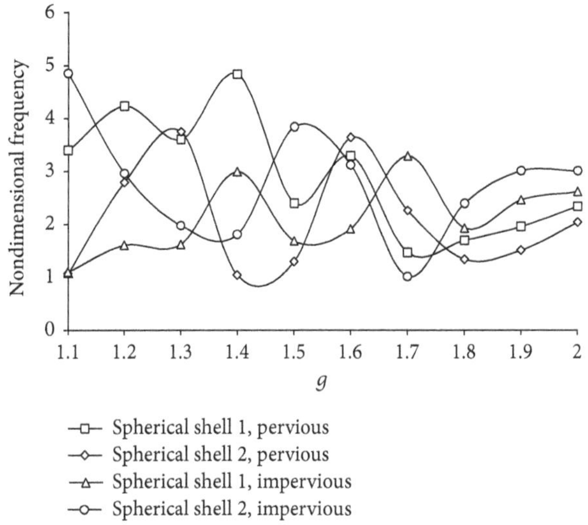

FIGURE 1: Variation of nondimensional frequency with ratio of outer and inner radii (g).

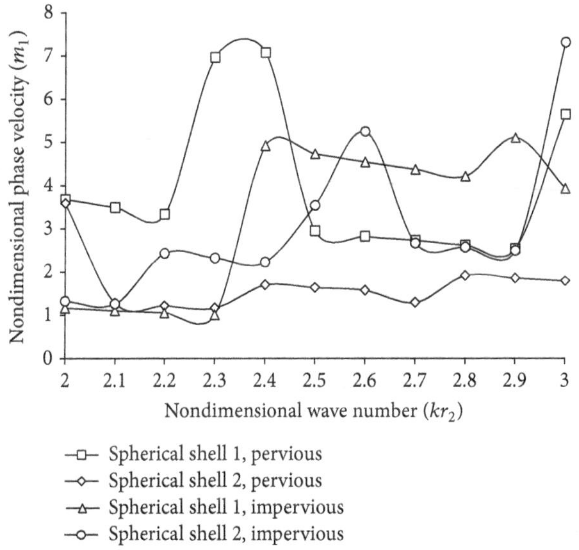

FIGURE 2: Variation of nondimensional phase velocity with wave number when $g = 2$.

Figure 1 depicts the nondimensional frequency (Ω) against the ratio of outer and inner radii (g) for poroelastic composite spherical shells 1 and 2, in the case of both pervious and impervious surfaces. From the figure, it is observed that the frequency values of spherical shell 1 are, in general, less than those of shell 2 for both pervious and impervious surfaces. Also it is found that the frequency values of pervious surface are, in general, less than those of the impervious surface in the case of spherical shell 1 and greater in shell 2.

Figure 2 shows the nondimensional phase velocity (m_1) against the nondimensional wave number (kr_2) in the case of both pervious and impervious surfaces. The phase velocity values of spherical shell 1 are, in general, greater than those of shell 2 in the case of pervious surface and the trend is reversed in the case of impervious surface. From this figure it is also found that the phase velocity values of pervious surface are, in general, greater than those of impervious surface for both spherical shells 1 and 2.

5.2. Spherical Bone Implanted with Titanium. If the spherical bone is implanted with titanium, then we obtain a composite spherical shell consisting of two different solids; one is bone and the other is titanium. The natural selection of titanium is obvious for its favorable characteristics including immunity to corrosion, biocompatibility, and the capacity for joining with bone, which is Osseo integration. Its density, Young's modulus, and Poisson ratio are $0.0004215\,\text{lb sec}^2/\text{inch}^4$, 105 GPA, and 0.32, respectively. Lame's constants and thereby dilatational wave velocity and shear wave velocity are computed. The values of bone poroelastic parameters and its mass coefficients are computed by using the inputs [11]. The values of Young's modulus and Poisson ratio are taken to be $3 \times 10^6\,\text{lb/inch}^2$ and 0.28, respectively [11]. Mass coefficients of solid part and fluid part are taken to be $1.65 \times 10^{-4}\,\text{lb sec}^2/\text{inch}^4$ and $0.14 \times 10^{-4}\,\text{lb sec}^2/\text{inch}^4$, respectively, [11]. These values are close to those of the experimental results. The values in the said study are detected at micrometer level [12]. These computations are based on the $V(z)$-curve method, which involves surface acoustic waves (SAW) that are propagating along the surface of a specimen. The dilatational wave velocities and shear wave velocity are computed, which are $V_1 = 2.016 \times 10^5\,\text{inch/sec}$, $V_2 =$

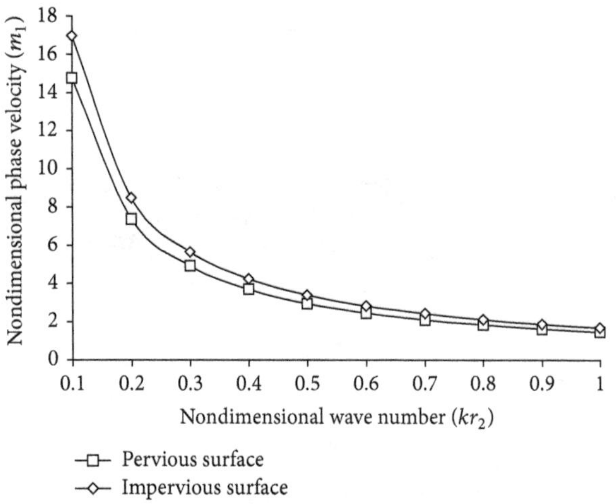

FIGURE 3: Variation of nondimensional phase velocity with wave number of spherical bone implanted with titanium for fixed $g = 2$.

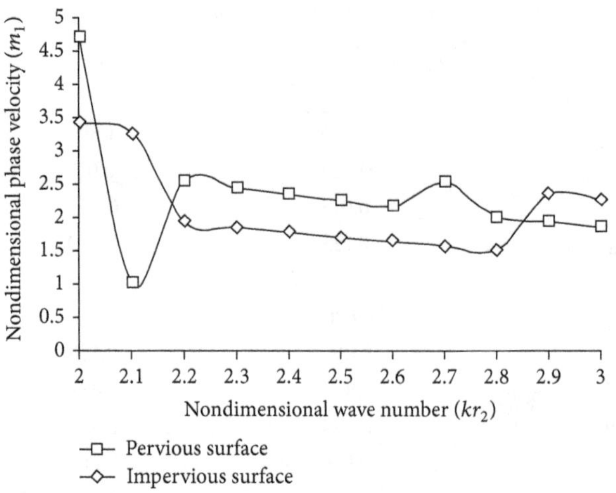

FIGURE 4: Variation of nondimensional phase velocity with wave number of spherical bone implanted with titanium for fixed $g = 2$.

1.003×10^5 inch/sec, and $V_3 = 0.842 \times 10^5$ inch/sec. Unlike the general case, here the second dilatational wave velocity is greater than shear wave velocity which is valid for the soft poroelastic solids [13]. Mass coupling parameter is taken to be zero [11]. The value of g is fixed and is taken to be 2. The nondimensional phase velocity (m_1) is computed against the nondimensional wave number (kr_2) and values are depicted in Figures 3–6. From Figure 3, it is clear that as wavenumber increases phase velocity decreases in the case of both pervious and impervious surfaces. Also, it is found that the phase velocity values of pervious surface are, in general, less than those of impervious surface.

From Figure 4, it is clear that the phase velocity values of the pervious surface are, in general, greater than those of impervious surface. There is a clear observation from Figures 3 and 4 that when the wave number lies between 0.1 and 1, the phase velocity values decrease, and when it exceeds 1, the nondimensional phase velocity values are periodic in the cases of both pervious and impervious surfaces.

Figures 5 and 6 depict plots of the nondimensional phase velocity (m_1) against the nondimensional wave number (kr_2) for $g = 2, 3$, and 4 in the case of both pervious and impervious surfaces. From the figures, it is clear that as g increases, the phase velocity increases for both pervious and impervious surfaces.

Figure 7 depicts the nondimensional frequency (Ω) against the ratio (g) for thick walled hollow spherical shells 1 and 2 in the case of ring mode. Spherical shell 1 is made up of sandstone saturated with water [9] and shell 2 is made up of sandstone saturated with kerosene [10]. From the figure, it is observed that as the ratio increases, frequency increases, and the frequency values of spherical shell 1 are, in general, greater than those of spherical shell 2. From the figures, it is clear that dispersive phenomena in the case of thick walled spherical shell and the case of its ring mode are different.

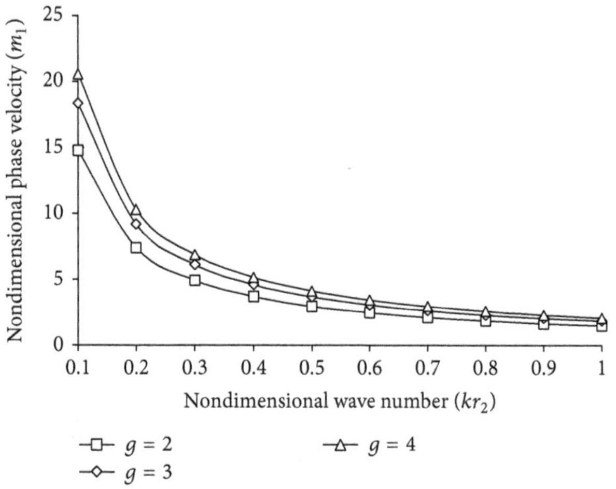

FIGURE 5: Variation of nondimensional phase velocity with wave number in spherical bone implanted with titanium for different values of g in the case of pervious surface.

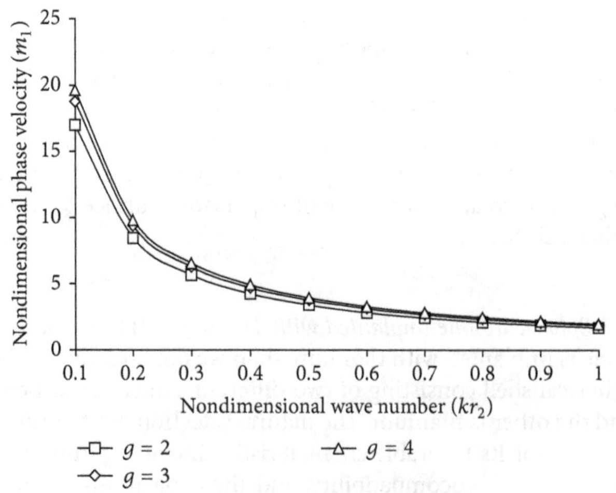

FIGURE 6: Variation of nondimensional phase velocity with wave number of spherical bone implanted with titanium for different values of g in the case of impervious surface.

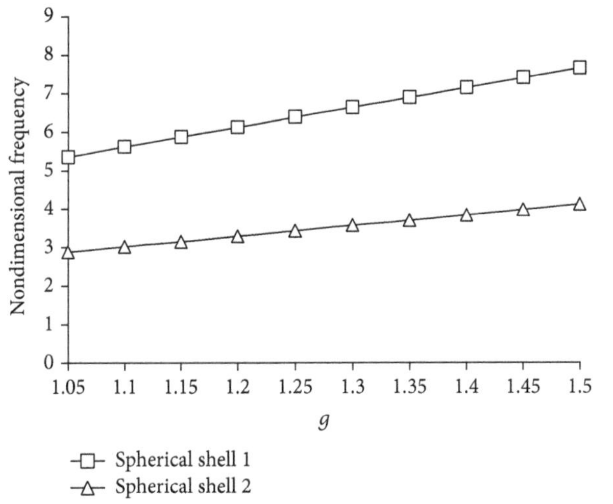

FIGURE 7: Variation of nondimensional frequency with ratio of outer and inner radii (g) in the case of ring mode.

6. Conclusion

In the framework of Biot's theory, axially symmetric vibrations of composite poroelastic spherical shell are investigated in the case of both pervious and impervious surfaces. Two parameters, frequency and phase velocity, are investigated. Limiting cases, namely, ring modes, are studied by using appropriate approximations. From the numerical results, it is clear that dispersive behavior in the case of shell and its ring modes is distinct though both are made of the same material. Similar analysis is made for any composite spherical shell made of two different poroelastic materials if their poroelastic constants are available. This kind of analysis is useful in obtaining the unknown data in indirect way of nondestructive evaluation (NDE).

Appendix

Consider the following:

$$
\begin{aligned}
M_{11}^{*}(r_1) = \frac{1}{r_1^3}\Bigg(& \left((P+Q)+(Q+R)\delta_1^2\right) \\
& \times\Big(\big(n(n+1)(n-2)+4 \\
& +(\xi_1 r_1)^2\left((\xi_1 r_1)^2-2n^2-n\right)\big) \\
& \times J_n(\xi_1 r_1)+\left(2-\frac{2n(n+4)}{(\xi_1 r_1)}\right) \\
& \times J_{n+1}(\xi_1 r_1)\Big) \\
& +\left((A+Q)+(Q+R)\delta_1^2\right) \\
& \times\Big(\big(2n^3-8n^2+16n+L \\
& \times\left(n^2-6n+6\right) \\
& -(2n+L-8)(\xi_1 r_1)^2\big) \\
& \times J_n(\xi_1 r_1)+\big(2(\xi_1 r_1)^2 \\
& +\left(6L-2n^2-12\right)\big) \\
& \times(\xi_1 r_1)J_{n+1}(\xi_1 r_1)\Big)\Bigg),
\end{aligned}
$$

$$
\begin{aligned}
M_{15}^{*}(r_1) = \frac{2NL_2}{r_1^3}\Big(& \left(n^3-7n^2+14n-8\right) \\
& +(5-n)(\xi_3 r_1)^2\Big)J_n(\xi_3 r_1) \\
& +\left((\xi_3 r_1)^3-\left(n^2+6\right)(\xi_3 r_1)\right)J_{n+1}(\xi_3 r_1),
\end{aligned}
$$

$$
\begin{aligned}
M_{21}^{*}(r_1) = \frac{2N}{r_1^3}\Big(& \left(-n^3+7n^2-2n-4\right)+(n-5) \\
& \times(\xi_1 r_1)^2\Big)J_n(\xi_1 r_1)+\left(2n-7-(\xi_1 r_1)^2\right) \\
& \times(\xi_1 r_1)J_{n+1}(\xi_1 r_1),
\end{aligned}
$$

$$
\begin{aligned}
M_{25}^{*}(r_1) = \frac{N}{r_1^3}\Big(& \Big(\left(L_1+n^2+n\right)\left(n^2-6n+8-(\xi_3 r_1)^2\right)\Big) \\
& -(\xi_3 r_1)^2\left(L_1+n^2-8n-2\right)\Big)J_n(\xi_3 r_1) \\
& -\Big(\left(2L_1+2n^3+6n^2+11n+4\right) \\
& -(\xi_3 r_1)(2n+1)\Big)(\xi_3 r_1)J_{n+1}(\xi_3 r_1),
\end{aligned}
$$

$$
\begin{aligned}
M_{31}^{*}(r_1) = & \frac{(Q+R)\delta_1^2}{r_1^3} \\
& \times\Bigg(\left(n^2+n+L\right) \\
& \left(n^2-2n-4-(\xi_1 r_1)^2-\frac{4n}{(\xi_1 r_1)}\right) \\
& -(\xi_1 r_1)^2\big(n^2-2n-2-(\xi_1 r_1)^2 \\
& +2(\xi_1 r_1)^3-16\big)J_n(\xi_1 r_1) \\
& +\Big(\left(n^2+n+L\right)((\xi_1 r_1)+4)+2(\xi_1 r_1)^3 \\
& \times((\xi_1 r_1)-1)+\left(n^2+3n+2\right) \\
& \times\Big(\frac{2}{(\xi_1 r_1)}-2(\xi_1 r_1)^2-(2n+6)(\xi_1 r_1) \\
& +\frac{4(6n+11)}{(\xi_1 r_1)}\Big)\Big)J_{n+1}(\xi_1 r_1)\Bigg).
\end{aligned}
\tag{A.1}
$$

$M_{12}^{*}(r_1)$, $M_{16}^{*}(r_1)$, $M_{22}^{*}(r_1)$, $M_{26}^{*}(r_1)$, and $M_{32}^{*}(r_1)$ are similar expressions as in $M_{11}^{*}(r_1)$, $M_{15}^{*}(r_1)$, $M_{21}^{*}(r_1)$, $M_{25}^{*}(r_1)$, and $M_{31}^{*}(r_1)$ with J_n, J_{n+1} replaced by

Y_n, Y_{n+1}, respectively; $M_{13}^*(r_1), M_{23}^*(r_1)$, and $M_{33}^*(r_1)$ are similar expressions to $M_{11}^*(r_1), M_{21}^*(r_1)$, and $M_{31}^*(r_1)$ with ξ_1, δ_1 replaced by ξ_2, δ_2, respectively; $M_{14}^*(r_1)$, $M_{24}^*(r_1)$, and $M_{34}^*(r_1)$ are similar expressions to $M_{11}^*(r_1)$, $M_{21}^*(r_1)$, and $M_{31}^*(r_1)$ with ξ_1, δ_1, J_n, and J_{n+1} replaced by ξ_2, δ_2, Y_n, and Y_{n+1}, respectively.

Conflict of Interests

The authors declare that there is no conflict of interests regarding the publication of this paper.

References

[1] R. Kumar, "Axially symmetric vibrations of a fluid- filled spherical shell," *Acustica*, vol. 21, no. 3, pp. 143–149, 1969.

[2] R. H. Rand, "Torsional vibrations of elastic prolate spheroids," *Journal of the Acoustical Society of America*, vol. 44, no. 3, pp. 749–751, 1968.

[3] S. Paul, "A note on the radial vibrations of a sphere of poroelastic material," *Indian Journal of Pure and Applied Mathematics*, vol. 7, no. 4, pp. 469–475, 1976.

[4] M. A. Biot, "The theory of propagation of elastic waves in fluid-saturated porous solid," *Journal of the Acoustical Society of America*, vol. 28, pp. 168–178, 1956.

[5] S. A. Shah and M. Tajuddin, "Torsional vibrations of poroelastic prolate spheroids," *International Journal of Applied Mechanics and Engineering*, vol. 16, pp. 521–529, 2011.

[6] S. A. Shah and M. Tajuddin, "On axially symmetric vibrations of fluid filled poroelastic spherical shells," *Open Journal of Acoustics*, vol. 1, pp. 15–26, 2011.

[7] B. Shanker, C. Nageswara Nath, S. Ahmed Shah, and J. Manoj Kumar, "Vibration analysis of a poroelastic composite hollow sphere," *Acta Mechanica*, vol. 224, no. 2, pp. 327–341, 2013.

[8] D. C. Gazis, "Exact analysis of plane-strain vibrations of thick-walled hollow cylinders," *Journal of the Acoustical Society of America*, vol. 30, pp. 786–794, 1957.

[9] C. H. Yew and P. N. Jogi, "Study of wave motions in fluid-saturated porous rocks," *Journal of the Acoustical Society of America*, vol. 60, no. 1, pp. 2–8, 1976.

[10] I. Fatt, "The Biot-Willis elastic coefficients for a sandstone," *Journal of Applied Mechanics*, vol. 26, pp. 296–297, 1957.

[11] J. L. Nowinski and C. F. Davis, "Propagation of longitudinal waves in circularly cylindrical bone elements," *Journal of Applied Mechanics, Transactions ASME*, vol. 38, no. 3, pp. 578–584, 1971.

[12] C. S. Jørgensen and T. Kundu, "Measurement of material elastic constants of trabecular bone: a micromechanical analytic study using a 1 GHz acoustic microscope," *Journal of Orthopaedic Research*, vol. 20, no. 1, pp. 151–158, 2002.

[13] C.-H. Lin, V. W. Lee, and M. D. Trifunac, "On the reflection of elastic waves in a poroelastic half-space saturated with non-viscous fluid," Tech. Rep. CE 01-04, Department of Civil Engineering, University of Southern California, Berkeley, Calif, USA, 2001.

13

An Application of Filtered Renewal Processes in Hydrology

Mario Lefebvre and Fatima Bensalma

Département de Mathématiques et de Génie Industriel, École Polytechnique, C.P. 6079, Succursale Centre-ville, Montréal, QC, Canada H3C 3A7

Correspondence should be addressed to Mario Lefebvre; mlefebvre@polymtl.ca

Academic Editor: George S. Dulikravich

Filtered renewal processes are used to forecast daily river flows. For these processes, contrary to filtered Poisson processes, the time between consecutive events is not necessarily exponentially distributed, which is more realistic. The model is applied to obtain one- and two-day-ahead forecasts of the flows of the Delaware and Hudson Rivers, both located in the United States. Better results are obtained than with filtered Poisson processes, which are often used to model river flows.

1. Introduction

Let $\{N(t), t \geq 0\}$ be a Poisson process with rate λ. A filtered Poisson (sometimes called *shot noise*) process is a continuous-time stochastic process $\{X(t), t \geq 0\}$ defined by

$$X(t) = \sum_{n=1}^{N(t)} w(t, \tau_n, Y_n) \quad (X(t) = 0 \text{ if } N(t) = 0), \quad (1)$$

in which the random variables $\tau_1, \tau_2, \ldots$ denote the arrival times of the events of the Poisson process and $Y_1, Y_2, \ldots$ are assumed to be independent and identically distributed random variables that are also independent of $\{N(t), t \geq 0\}$. The function $w(\cdot,\cdot,\cdot)$ is called the response function.

In many applications, the response function is chosen of the form

$$w(t, \tau_n, Y_n) = Y_n e^{-(t-\tau_n)/c}, \quad (2)$$

where c is a parameter that must be estimated. It then gives the value at time t of an event of magnitude Y_n of the Poisson process that occurred at time τ_n. Moreover, the random variables $Y_1, Y_2, \ldots$ are generally assumed to be exponentially distributed with parameter μ. With the above response function, the filtered Poisson process behaves as in Figure 1. Actually, this behavior depends on the form of the response function, but not on the distribution of the time between the events. Therefore, the same behavior would be observed in the case when $\{N(t), t \geq 0\}$ is a renewal process. Remember that a Poisson process is a particular renewal process.

To model the daily flows of rivers given their most recent observed values, conceptual and physical models based on the filtered Poisson process have been widely used successfully for many decades; see, for example, Weiss [1], Kelman [2], Koch [3], and Konecny [4]. Filtered Poisson processes are still being used to model various phenomena in civil engineering; see Yin et al. [5] and Miyamoto et al. [6].

Now, especially in hydrological applications, the form of the response function in (2) is taken for granted rather than being justified by the observations of the variable of interest. Actually, it is generally simply an assumption made to obtain a mathematically tractable model.

Similarly, the actual distribution of the Y_n's is not investigated. However, if one is only interested in forecasting the value of $X(t+\delta)$, based on the values of the process up to time t, this does not really cause a problem because the forecast is normally based on the mean of the Y_n's, and the distribution itself is not needed.

Finally, and even more importantly, the assumption that $\{N(t), t \geq 0\}$ is indeed a Poisson process is not tested. Again, this is a simplifying assumption, because one can then make use of the nice properties of the Poisson process, notably the fact that it has independent and stationary increments.

Next, notice that, with the response function in (2), the effect on the value of the process of an event that occurred at

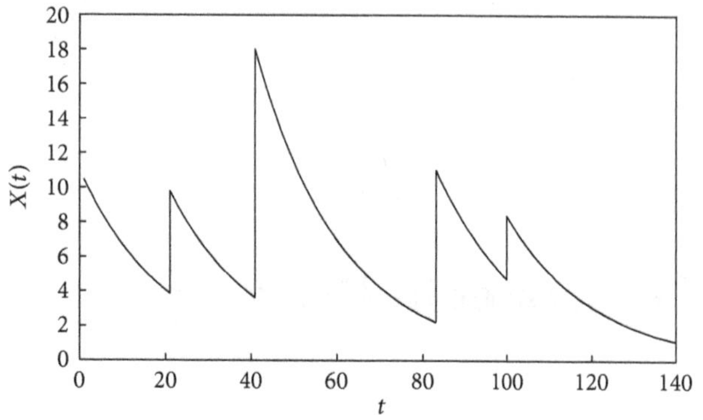

FIGURE 1: Example of a trajectory of a filtered Poisson process defined by (1) and (2).

time τ_n is maximum at that time instant and $\{X(t), t \geq 0\}$ has discontinuities at the arrival times of the events of the Poisson process. In the case of the application in hydrology that consists in modeling the flow of rivers, if one looks at real hydrographs, one sees that there is a more or less extended period of time during which the flow increases from a minimum to a maximum. Even if river flows data are generally available on a daily basis, one does not observe very sudden increases of the flow, followed by an exponential decrease. Rather, it takes on average two to three days for the flow to reach a maximum, and the decrease is more or less rapid.

To obtain a more realistic model for the variations of river flows, by taking into account the periods of flow increase, Lefebvre and Guilbault [7] used the following response function:

$$w(t, \tau_n, Y_n) = Y_n (t - \tau_n)^d e^{-(t-\tau_n)/c}, \quad (3)$$

where d is a positive parameter. With this response function, the river flow begins to increase immediately after time τ_n, but the maximum is reached after dc time units (i.e., generally days) and then it starts to decrease. Their work was improved by the authors, who provided a method to estimate the parameter d. They found that, for the applications that they considered, d was in the interval (0,1).

With a Poisson process being a continuous-time Markov chain, the time that the process spends in a given state is exponentially distributed, which is a strictly decreasing density function. Let $T_1, T_2, \ldots$ be the times between two successive events, that is, the interarrival times. In the case of river flows, the interarrival time is generally a random variable with a density function that is increasing at first. Hence, the assumption that $\{N(t), t \geq 0\}$ is a Poisson process is at least doubtful.

In Lefebvre [8], the author tested the hypothesis that the T_n's are exponentially distributed. He found that, for the Delaware River, it was not accurate. Instead, it was shown that both a gamma and a Rayleigh distribution fitted the data much better. The objective was to forecast the peak flows of the Delaware River, which is a difficult task due to the very weak correlation between the successive peaks. To try to forecast these peak flows, the author used a filtered renewal process as a model for the river flow. An advantage in using a Rayleigh distribution for the interarrival time is that the filtered renewal process can be transformed into a filtered Poisson process by working on a different time scale.

Poisson and renewal processes have been used in hydrology to model low flows (see, for instance, Loaiciga and Leipnik [9] and Yagouti et al. [10]) as well as various physical phenomena such as traffic noise (see Marcus [11]). Andersen [12, 13] gave new theoretical results on filtered renewal processes (see also the references therein).

The aim of the present paper is to show that we can obtain more accurate short-term forecasts of river flows by making use of filtered renewal processes rather than filtered Poisson processes. To do so, we will first need to develop a formula to forecast the flow at time $X(t + \delta)$, given the history of the process up to time t. Because we cannot appeal to the memoryless property of the exponential distribution, the task of finding an estimator for $X(t+\delta)$ is more difficult. The values of the river flow before time t must be taken into account, contrary to the case of filtered Poisson process, for which the forecasts depend only on $X(t)$.

Moreover, because we want to clearly see the improvement obtained by using a more realistic distribution for the interarrival time, we will take the response function defined in (2). Indeed, as mentioned above, filtered Poisson processes having this response function are commonly used in hydrology, and the forecasting formulas are then easy to derive and implement. Therefore, it will be easier to judge the quality of the model that we propose.

In the next section, filtered renewal processes will be formally defined and the formulas needed to forecast the river flows will be developed. Then, in Section 3, the results will be applied to forecast the flows of the Delaware and Hudson Rivers. Finally, a few concluding remarks will be made in Section 4.

2. River Flows Modeled as a Filtered Renewal Process

Let $X(t)$ denote the flow of a certain river at time t. We assume that

$$X(t) = \sum_{n=1}^{N(t)} Y_n e^{-(t-\tau_n)/c} \quad (X(t) = 0 \text{ if } N(t) = 0), \quad (4)$$

in which $\{N(t), t \geq 0\}$ is a renewal process. The τ_n's are the arrival times of the events of the process $\{N(t), t \geq 0\}$, and the times $T_n = \tau_n - \tau_{n-1}$ between the successive renewals are independent and identically distributed random variables. Finally, the (positive) random variables $Y_1, Y_2, \ldots$ are independent and identically distributed (and independent of $\{N(t), t \geq 0\}$).

Thus, the process is characterized by the sudden increases of the river flow caused by the events that occur at times $\tau_1, \tau_2, \ldots$. A renewal cycle is defined as the time T_n between two consecutive peaks. For the interarrival time, the following distributions (in addition to the exponential distribution) will be considered in Section 3: lognormal, gamma, Weibull, and Rayleigh.

Next, in order to be able to forecast the river flow at time $t + \delta$, we must first estimate the parameter c that appears in the response function. If there are no signals between t and $t + 1$, then $X(t + 1)$ is given by

$$X(t+1) = e^{-1/c} X(t). \tag{5}$$

Hence, to estimate the parameter c, we can consider all the days where the flow decreased in the data set and calculate the ratio $X(t+1)/X(t)$. The point estimate of c is then derived from the arithmetic mean $\overline{R}$ of this ratio:

$$\widehat{c} = -\frac{1}{\ln \overline{R}}. \tag{6}$$

In the case of the random variables Y_n, which constitute a random sample of a parent random variable Y, their exact distribution is not needed. Indeed, only the expected value of Y will be used to forecast the value of $X(t + \delta)$. Therefore, we simply have to compute the mean increase of the river flow during the calibration period of the model.

We will now derive formulas to estimate the river flow one and two days in advance. These formulas can be used with any distribution for the interarrival time T, which denotes the parent variable of the T_n's.

2.1. Forecasting the Flow at Time t + 1. We will first try to forecast the river flow at time $t + 1$, given all the information available up to t. To do so, we assume that there will be at most one event in the interval $(t, t + 1]$. At any rate, as mentioned above, in practice flow values are generally available on a daily basis. Therefore, it is not really possible to determine whether more than one event occurred in $(t, t + 1]$. When it does happen, these events can be considered as a single large event.

Apart from the case when the interarrival time T has an exponential distribution, we must take the history of the process into account to calculate the probability that there will indeed be an event in $(t, t + 1]$. Let

$$U_i = \tau_{N(t)+i} - \tau_{N(t)+i-1} \quad \text{for } i = 1, 2, \ldots. \tag{7}$$

That is, U_i is the time elapsed between the $(i-1)$st and the ith event after the most recent one that occurred before or at time t. The random variables T and U_1 are identically distributed. We must compute

$$p_k := P\left[U_1 \le k+1 \mid U_1 > k\right], \tag{8}$$

where k is the (integer) number of days since the most recent signal was observed. We have

$$p_k = \frac{\int_k^{k+1} f_{U_1}(u)\, du}{\int_k^{\infty} f_{U_1}(u)\, du}. \tag{9}$$

With the help of a mathematical software, we can obtain numerical values of this probability for any density function f_{U_1} and any value of k.

Next, if T is exponentially distributed, it is well known that, given that there is a signal (i.e., an event) between t and $t + 1$, the conditional density of T is uniform on the interval $(t, t + 1]$, so that the signal in question occurred on average at time $t + 1/2$. In the general case, we must compute the conditional density function of U_1 in the interval $(k, k + 1]$, given that $U_1 > k$.

If there is indeed a signal between t and $t + 1$, the forecast of the river flow at time $t + 1$ would be

$$Z_k := X(t)\, e^{-1/\widehat{c}} + E[Y] \frac{1}{P[U_1 > k]} \times \int_k^{k+1} e^{-(k+1-u)/\widehat{c}} f_{U_1}(u)\, du. \tag{10}$$

Therefore, the forecast of the river flow at time $t + 1$, given the entire history of the process up to t, is given by

$$\begin{aligned}\widehat{X}(t+1) &= X(t)\, e^{-1/\widehat{c}} (1 - p_k) + Z_k p_k \\ &= X(t)\, e^{-1/\widehat{c}} + p_k E[Y] \frac{1}{P[U_1 > k]} \\ &\quad \times e^{-(k+1)/\widehat{c}} \int_k^{k+1} e^{u/\widehat{c}} f_{U_1}(u)\, du.\end{aligned} \tag{11}$$

2.2. Forecasting the Flow at Time t + 2. Let $U_{1,k} = U_1 \mid \{U_1 > k\}$. To forecast the river flow at time $t + 2$, given all the information available up to time t, we need the distribution of the random variable $S_k := U_{1,k} + U_2$, where U_2 is defined in (7).

We now assume that the probability that there will be more than two signals in the interval $(t, t + 2]$ is negligible (irrespective of the value of k). The density function of S_k is given by the convolution of the density functions of $U_{1,k}$ and U_2:

$$f_{S_k}(s+k) = \int_k^{\infty} f_{U_{1,k}}(u) f_{U_2}(s+k-u)\, du. \tag{12}$$

Since U_2 is a nonnegative random variable, we can write that

$$f_{S_k}(s+k) = \int_k^{s+k} f_{U_{1,k}}(u) f_{U_2}(s+k-u)\, du \quad \text{for } s > 0. \tag{13}$$

In the case when T has an exponential distribution with parameter λ, we know that $(S_k - k)$ has a gamma distribution with parameters $\alpha = 2$ and λ. In the other cases, we must generally evaluate the above integral numerically.

The value of $X(t+2)$ is given by

$$X(t+2) = \begin{cases} e^{-2/c}X(t) & \text{if } U_{1,k} > k+2, \\ e^{-2/c}X(t) + Y_1^* e^{-(k+2-U_{1,k})/c} & \text{if } U_{1,k} \le k+2 \\ & \text{and } U_2 > 2-(U_{1,k}-k), \\ e^{-2/c}X(t) + Y_1^* e^{-(k+2-U_{1,k})/c} & \text{if } U_{1,k}+U_2 \le k+2 \\ \quad +Y_2^* e^{-(2-U_2)/c} & (\text{and } U_3 > 2-(U_{1,k}-k)-U_2), \end{cases} \tag{14}$$

where Y_i^* denotes the size of the ith signal in the interval $(t, t+2]$. As mentioned above, we assume that

$$P[U_3 > 2-(U_{1,k}-k)-U_2] \simeq 1. \tag{15}$$

We must first compute

$$p_{1,k} := P[U_{1,k} \in (k, k+2]]. \tag{16}$$

When U_1 has an exponential distribution with parameter λ,

$$p_{1,k} = P[U_1 \in (0, 2]] = 1 - e^{-2\lambda}. \tag{17}$$

In general, we have

$$p_{1,k} = \frac{\int_k^{k+2} f_{U_1}(u)\, du}{\int_k^{\infty} f_{U_1}(u)\, du}. \tag{18}$$

Next,

$$\begin{aligned} p_{2,k} &:= P[U_{1,k} + U_2 \in (k, k+2]] \equiv P[S_k \in (k, k+2]] \\ &= \int_0^2 f_{S_k}(s+k)\, ds \\ &= \int_0^2 \left[\int_k^{s+k} f_{U_{1,k}}(u) f_{U_2}(s+k-u)\, du \right] ds. \end{aligned} \tag{19}$$

Alternatively, using the fact that $U_{1,k}$ and U_2 (≥ 0) are independent random variables, we may write that

$$\begin{aligned} p_{2,k} &:= P[U_{1,k} + U_2 \in (k, k+2]] \\ &= \int_k^{k+2} P[U_{1,k} + U_2 \in (k, k+2] \mid U_{1,k} = u]\, f_{U_{1,k}}(u)\, du \\ &= \int_k^{k+2} P[U_2 \in (k-u, k+2-u]]\, f_{U_{1,k}}(u)\, du \\ &= \int_k^{k+2} \left[\int_0^{k+2-u} f_{U_2}(v)\, dv \right] f_{U_{1,k}}(u)\, du. \end{aligned} \tag{20}$$

Finally, by independence,

$$\begin{aligned} &P[\{U_{1,k} \le k+2\} \cap \{U_2 > 2-(U_{1,k}-k)\}] \\ &= \int_k^{k+2} P[U_2 \in (k+2-u, \infty)]\, f_{U_{1,k}}(u)\, du \\ &= \int_k^{k+2} \left[\int_{k+2-u}^{\infty} f_{U_2}(v)\, dv \right] f_{U_{1,k}}(u)\, du \\ &= \int_k^{k+2} \left[1 - \int_0^{k+2-u} f_{U_2}(v)\, dv \right] f_{U_{1,k}}(u)\, du \\ &= p_{1,k} - p_{2,k}. \end{aligned} \tag{21}$$

The forecast of the river flow at time $t+2$, given the entire history of the process up to time t, is thus given by

$$\begin{aligned} &\widehat{X}(t+2) \\ &= X(t)\, e^{-2/\widehat{c}} (1 - p_{1,k}) \\ &\quad + \left\{ X(t)\, e^{-2/\widehat{c}} + E[Y] \int_k^{k+2} e^{-(k+2-u)/\widehat{c}} f_{U_{1,k}}(u)\, du \right\} \\ &\quad \times (p_{1,k} - p_{2,k}) \\ &\quad + \left\{ X(t)\, e^{-2/\widehat{c}} + E[Y] \int_k^{k+2} e^{-(k+2-u)/\widehat{c}} f_{U_{1,k}}(u)\, du \right. \\ &\qquad + E[Y] \int_k^{k+2} \left[\int_0^{k+2-u} e^{-(2-v)/\widehat{c}} f_{U_2}(v)\, dv \right] \\ &\qquad \left. \times f_{U_{1,k}}(u)\, du \right\} p_{2,k} \\ &= X(t)\, e^{-2/\widehat{c}} + \left\{ E[Y] \int_k^{k+2} e^{-(k+2-u)/\widehat{c}} f_{U_{1,k}}(u)\, du \right\} p_{1,k} \\ &\quad + \left\{ E[Y] \int_k^{k+2} \left[\int_0^{k+2-u} e^{-(2-v)/\widehat{c}} f_{U_2}(v)\, dv \right] \right. \\ &\qquad \left. \times f_{U_{1,k}}(u)\, du \right\} p_{2,k} \end{aligned}$$

$$= X(t)\, e^{-2/\hat{c}} + \left\{ E[Y]\, e^{-(k+2)/\hat{c}} \int_k^{k+2} e^{u/\hat{c}} f_{U_{1,k}}(u)\, du \right\} p_{1,k} + \left\{ E[Y] \int_k^{k+2} \left[e^{-2/\hat{c}} \int_0^{k+2-u} e^{v/\hat{c}} f_{U_2}(v)\, dv \right] \times f_{U_{1,k}}(u)\, du \right\} p_{2,k}. \quad (22)$$

3. Forecasting the Flows of the Delaware and Hudson Rivers

To assess the quality of the forecasts obtained with the renewal filtered process, we need to apply the formulas developed in the previous section to real-life data. As in [7], we chose the Delaware and Hudson Rivers, located in the United States, at the Montague (01438500) and the North Creek (01315500) gage stations, respectively. The flow values are available on the Internet at the following address: http://nwis.waterdata.usgs.gov. To calibrate the model, we used the data from October 2008 to September 2009.

Remarks. (i) When one is dealing with real-life data, things are not as simple as the mathematical models suggest. Looking at the actual hydrograph of the Delaware River, we observe a number of weak peaks that occur while the flow is decreasing and for which the increasing period lasts only one day before the flow resumes its decline. Similarly, there are sometimes small increases of the flow value occurring just after a minimum was observed that last only one day. In building the data set, there is always a subjective part. We decided to neglect these weak peaks, thus considering only the peaks that appear quite clearly in the hydrograph.

(ii) Because the river flow is observed on a daily basis, we must discretize the set of possible values taken by the interarrival time T. We computed the number k of days elapsed between consecutive peaks in the data set. In the case of the first observed peak, k is the number of days elapsed since the beginning of the calibration period. Then, k is obtained by subtracting the arrival times of the consecutive peaks. Notice that k is greater than or equal to 1.

We first consider the Delaware River. For this river, we found that the average value $\overline{T}$ of the interarrival times in the data set is equal to 6.8889 days. Moreover, the standard deviation of the observations is given by $s_T = 3.8977$ days. Based on these values, one may at once conclude that T is very unlikely to be exponentially distributed. Indeed, remember that the mean and the standard deviation of an exponential distribution with parameter λ are both equal to $1/\lambda$.

Next, the value of the point estimate of the (unitless) parameter c (see (6)) is $\hat{c} = 9.2592$, and the average value of the magnitude Y of the signals is given by $\overline{Y} = 984.6121$ cubic feet per second.

For the distribution of the random variable T, we tested the following models:

(i) an exponential distribution with parameter λ,

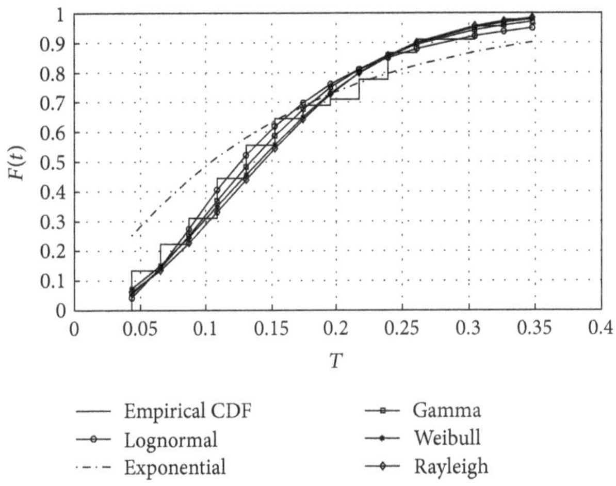

FIGURE 2: Empirical and fitted distribution functions for the random variable T (Delaware River).

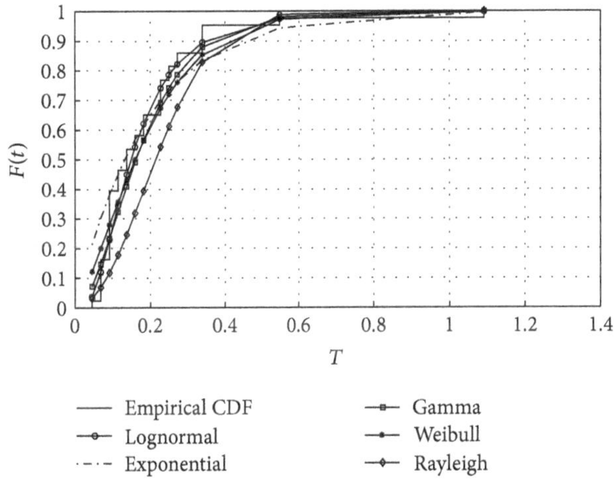

FIGURE 3: Empirical and fitted distribution functions for the random variable T (Hudson River).

(ii) a lognormal distribution with parameters μ and σ,

(iii) a gamma distribution with shape parameter α and scale parameter λ,

(iv) a Weibull distribution with shape parameter λ and scale parameter κ,

(v) a Rayleigh distribution with parameter σ.

Table 1 summarizes the results of the chi-square goodness-of-fit tests for each distribution and gives the respective parameter estimates. We find that, as expected, the exponential distribution is rejected, whereas the other distributions are all accepted, according to the P-values of the tests. The empirical cumulative distribution function as well as the distribution functions of the various models considered above is shown in Figure 2. We clearly see that the exponential distribution does not fit the data nearly as well as the other four models.

Now, in the case of the filtered Poisson process (1) with the response function defined in (2), to estimate the river flow at

Table 1: Goodness-of-fit tests for the distribution of the random variable T (Delaware River).

Distribution	Exponential	Lognormal	Gamma	Weibull	Rayleigh
Parameters	1/6.8889	1.7574, 0.6187	3.0544, 2.2554	7.7974, 1.8975	5.5817
P-value	0.0344	0.7887	0.7723	0.6927	0.7008

Table 2: Criteria for comparing the forecasting models (Delaware River).

		Filtered renewal process				Filtered Poisson process
		Lognormal	Gamma	Weibull	Rayleigh	Exponential
$X(t+1)$	Mape	12.58%	12.49%	12.55%	12.56%	12.76%
	Nash	0.8288	0.8290	0.8287	0.8287	0.8356
	Pc	0.1674	0.1664	0.1663	0.1669	0.2522
	Corr	0.9188	0.9186	0.9180	0.9188	0.9187
$X(t+2)$	Mape	20.36%	20.33%	20.34%	20.34%	21.83%
	Nash	0.5447	0.5445	0.5445	0.5445	0.5573
	Pc	0.2930	0.2893	0.2869	0.2863	0.3344
	Corr	0.7735	0.7732	0.7731	0.7730	0.7734

time $t+1$ we compute the expected value of $X(t+1)$ given all the past observations. Because of the memoryless property of the exponential distribution, we find that this conditional expectation only depends on the most recent value of the flow, that is, $X(t)$. We obtain (see [7]) that

$$E\left[X(t+1)\,X\right] = (t)\,e^{-1/c}X(t) + E\left[X(1)\right], \tag{23}$$

where

$$E\left[X(1)\right] = \frac{\lambda c}{\mu}\left(1 - e^{-1/c}\right). \tag{24}$$

Similarly, to estimate the river flow at time $t+2$, we only need to calculate

$$\begin{aligned} E\left[X(t+2) \mid X(t)\right] &= E\left[\sum_{n:0\le\tau_n\le t+2} Y_n e^{-(t+2-\tau_n)/c} \mid X(t)\right] \\ &= e^{-2/c}E\left[\sum_{n:0\le\tau_n\le t} Y_n e^{-(t-\tau_n)/c} \mid X(t)\right] \\ &\quad + E\left[\sum_{n:t<\tau_n\le t+2} Y_n e^{-(t+2-\tau_n)/c} \mid X(t)\right] \\ &= e^{-2/c}X(t) + E\left[X(2)\right], \end{aligned} \tag{25}$$

where

$$E\left[X(2)\right] = \frac{\lambda c}{\mu}\left(1 - e^{-2/c}\right). \tag{26}$$

Moreover, we have that

$$\lim_{t\to\infty} E\left[X(t)\right] = \frac{\lambda c}{\mu}, \tag{27}$$

$$\lim_{t\to\infty} \operatorname{Var}\left[X(t)\right] = \frac{\lambda c}{\mu^2}, \tag{28}$$

$$\lim_{t\to\infty} \rho_{X(t),X(t+\delta)} = e^{-\delta/c}, \tag{29}$$

where $\rho_{X(t),X(t+\delta)}$ denotes the correlation coefficient of $X(t)$ and $X(t+\delta)$.

By making use of these three equations, we can estimate the parameter λ of the Poisson process, the parameter μ of the assumed exponential distribution of the Y_n's, and the parameter c in the response function. Actually, we do not need point estimates of λ and μ to forecast the flow. From (27), we deduce that we can simply estimate the ratio $\lambda c/\mu$ that appears in both (24) and (26) by the mean value of the observed data when the process is in steady state. Finally, to estimate the parameter c, we first compute the empirical correlation coefficient r_1 between the flow values at times t and $t+1$ and then we set $\hat{c} = -1/\ln(r_1)$ (see (29)). With the value of r_1 being close to 1, the previous formula is well defined.

Remark. If one is only interested in forecasting the river flow at time $t+\delta$, one may estimate c by computing the empirical value of $\rho_{X(t),X(t+\delta)}$ and making use of (29).

We are now ready to compare the forecasts derived from both the filtered Poisson process and the filtered renewal process with the various distributions considered for the interarrival time T. We used the formulas given above to forecast the flow of the Delaware River for the 89-day period from February to April 2010.

To assess the accuracy of the forecasts, we consider four criteria commonly used in hydrology: the *mean absolute percentage error* (Mape), the *Nash criterion* (Nash), the *peak criterion* (Pc), and the *correlation coeffici t* (Corr) between the observed and forecasted values.

The Nash criterion is based on the mean square error and is used to assess the predictive power of hydrological models. It evaluates the quality of the forecasts from the differences

TABLE 3: Goodness-of-fit tests for the distribution of the random variable T (Hudson River).

Distribution	Exponential	Lognormal	Gamma	Weibull	Rayleigh
Parameters	1/8.3488	1.8754, 0.6625	9.2134, 1.3399	2.1785, 3.8324	7.9978
P-value	0.0013	0.6983	0.3203	0.2532	0.000139

TABLE 4: Criteria for comparing the forecasting models (Hudson River).

		Filtered renewal process			Filtered Poisson process
		Lognormal	Gamma	Weibull	Exponential
$X(t+1)$	Mape	12.33%	12.14%	12.33%	15.58%
	Nash	0.8240	0.8243	0.8240	0.8360
	Pc	0.2875	0.2895	0.2872	0.4818
	Corr	0.9146	0.9146	0.9146	0.9146
$X(t+2)$	Mape	20.84%	21.00%	21.15%	32.78%
	Nash	0.5841	0.5838	0.5835	0.6222
	Pc	0.4558	0.4533	0.4518	0.5575
	Corr	0.7908	0.7909	0.7910	0.7910

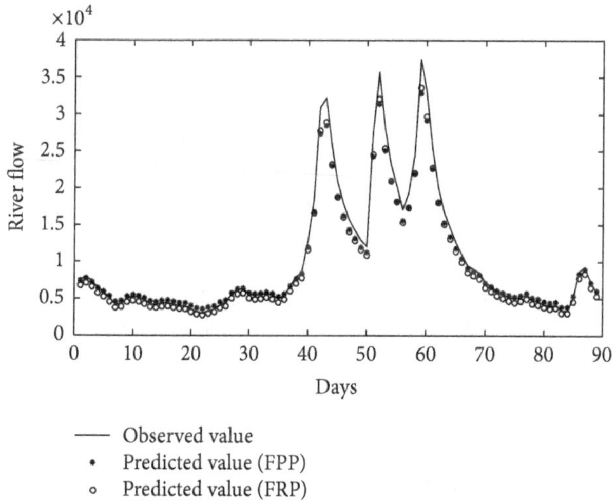

FIGURE 4: Observed and forecasted values of the flow of the Delaware River at time $t + 1$.

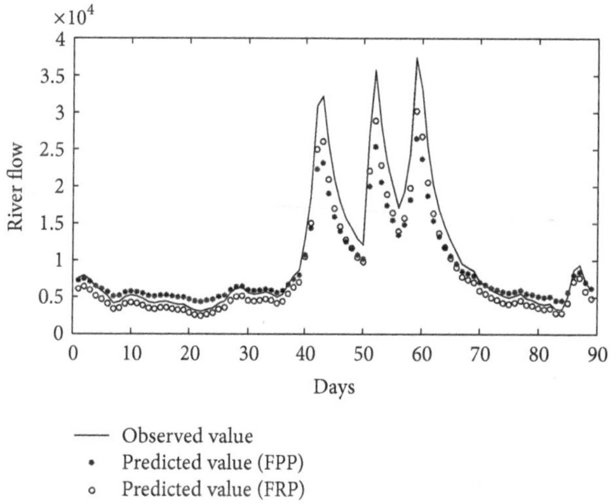

FIGURE 5: Observed and forecasted values of the flow of the Delaware River at time $t + 2$.

between the expected and observed daily values. It is equal to 1 in case of a perfect fit. For its part, the peak criterion is used to measure the quality of the forecasts during the critical peak period. The closer to 0 it is, the better the forecasts are.

Table 2 displays the results obtained with the two competing models. Looking at these results, we notice that the Mape, Nash, and Corr criteria yield practically the same values for the filtered renewal and filtered Poisson processes. However, the filtered renewal process did much better than the filtered Poisson process when we consider the peak criterion. This criterion is especially important because accurate forecasts are really needed during the peak period.

We now turn to the Hudson River. The value of the point estimate of the constant c for this river is $\hat{c} = 9.1172$ and the average magnitude of the signals is $\overline{Y} = 229.2533$, which is much smaller than in the case of the Delaware River.

Proceeding as above, we performed chi-square goodness-of-fit tests for the various distributions considered. The results are presented in Table 3 (see also Figure 3). This time, we see that both the exponential and Rayleigh distributions must be rejected, while the lognormal distribution clearly provides the best fit to the data.

The values of the four criteria used to compare the models are shown in Table 4. We discarded the Rayleigh distribution, because it yielded a very bad fit to the data. As in the case of the Delaware River, the values of the Nash and Corr criteria are quite similar, while the important peak criterion and the Mape criterion are much smaller for the filtered renewal process than for the filtered Poisson process.

Finally, we present the observed and forecasted values of the flows of the Delaware and Hudson Rivers at times $t + 1$ and $t+2$ in Figures 4, 5, 6, and 7, based on the filtered Poisson process (FPP) and the filtered renewal process (FRP) with a lognormal distribution for the interarrival time T. Notice that the filtered renewal process is generally better than the filtered

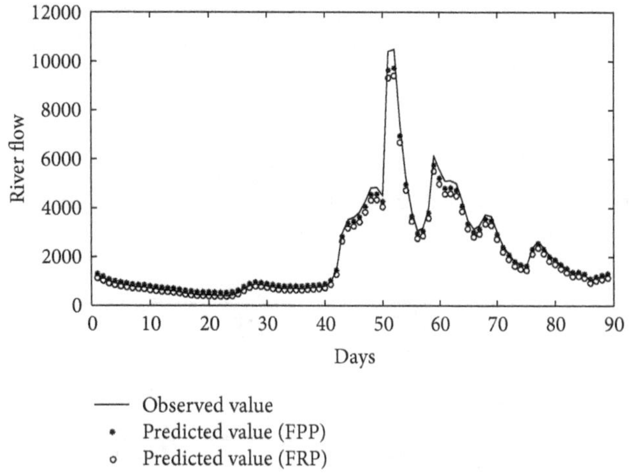

FIGURE 6: Observed and forecasted values of the flow of the Hudson River at time $t + 1$.

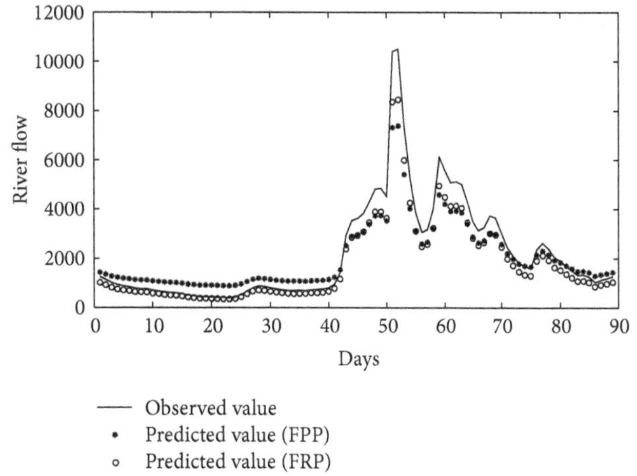

FIGURE 7: Observed and forecasted values of the flow of the Hudson River at time $t + 2$.

Poisson process at forecasting high flow values, which is our main concern.

4. Conclusion

In this paper, we were able to significantly improve the short-term hydrological forecasts produced by filtered Poisson processes by using a more realistic distribution for the inter-arrival time of the signals, thus working with filtered renewal processes. As we saw in the previous section, the filtered Poisson process, despite its lack of realism, is able to yield reasonable short-term forecasts. One possible explanation is the fact that the forecasting formulas derived from this model are very easy to implement. Also, contrary to the filtered renewal process, it does not require subjective decisions to estimate its various parameters. However, the filtered Poisson process was not able to produce forecasts as accurate as those derived from the filtered renewal process during the important peak period.

In addition to being used to forecast flow values, the models considered in this paper can also give us an estimate of the probability that the river flow will exceed a certain threshold in the next few days. This threshold may be a value that corresponds to a flow above which the risk of flooding is very high.

Finally, we could use the response function in (3) to further improve the forecasts. To do so, we would have to be able to estimate the parameter d in that response function and then obtain formulas that generalize the ones in (11) and (22), which is probably not an easy task.

Conflict of Interests

The authors declare that there is no conflict of interests regarding the publication of this paper.

Acknowledgment

The authors want to thank the anonymous reviewers for the constructive comments.

References

[1] G. Weiss, "Shot noise models for the generation of synthetic streamflow data," *Water Resources Research*, vol. 13, no. 1, pp. 101–108, 1977.

[2] J. Kelman, "A stochastic model for daily streamflow," *Journal of Hydrology*, vol. 47, no. 3-4, pp. 235–249, 1980.

[3] R. W. Koch, "A stochastic streamflow model based on physical principles," *Water Resources Research*, vol. 21, no. 4, pp. 545–553, 1985.

[4] F. Konecny, "On the shot-noise streamflow model and its applications," *Stochastic Hydrology and Hydraulics*, vol. 6, no. 4, pp. 289–303, 1992.

[5] Y.-J. Yin, Y. Li, and W. M. Bulleit, "Stochastic modeling of snow loads using a filtered Poisson process," *Journal of Cold Regions Engineering*, vol. 25, no. 1, pp. 16–36, 2010.

[6] H. Miyamoto, J. Morioka, K. Kanda et al., "A stochastic model for tree vegetation dynamics in river courses with interaction by discharge fluctuation impacts," *Journal of Japan Society of Civil Engineers B1: Hydraulic Engineering*, vol. 67, no. 4, pp. 1405–1410, 2012.

[7] M. Lefebvre and J.-L. Guilbault, "Using filtered Poisson processes to model a river flow," *Applied Mathematical Modelling*, vol. 32, no. 12, pp. 2792–2805, 2008.

[8] M. Lefebvre, "A filtered renewal process as a model for a river flow," *Mathematical Problems in Engineering*, vol. 2005, no. 1, pp. 49–59, 2005.

[9] H. A. Loaiciga and R. B. Leipnik, "Stochastic renewal model of low-flow streamflow sequences," *Stochastic Hydrology and Hydraulics*, vol. 10, no. 1, pp. 65–85, 1996.

[10] A. Yagouti, I. Abi-Zeid, T. B. M. J. Ouarda, and B. Bobée, "Revue de processus ponctuels et synthèse de tests statistiques pour le choix d'un type de processus," *Journal of Water Science*, vol. 14, no. 3, pp. 323–361, 2001.

[11] A. H. Marcus, "Traffic noise as a filtered Markov renewal process," *Journal of Applied Probability*, vol. 10, no. 2, pp. 377–386, 1973.

[12] P. K. Andersen, "Filtered renewal processes with a two-sided impact function," *Journal of Applied Probability*, vol. 16, no. 4, pp. 813–821, 1979.

[13] P. K. Andersen, "A note on filtered Markov renewal processes," *Journal of Applied Probability*, vol. 18, no. 3, pp. 752–756, 1981.

14

Developing Buoyancy Driven Flow of a Nanofluid in a Vertical Channel Subject to Heat Flux

Nirmal C. Sacheti,[1] Pallath Chandran,[1] Ashok K. Singh,[2] and Beer S. Bhadauria[2,3]

[1] *Department of Mathematics & Statistics, College of Science, Sultan Qaboos University, Al Khod, 123 Muscat, Oman*
[2] *Department of Mathematics, Banaras Hindu University, Varanasi 221005, India*
[3] *Department of Applied Mathematics, School of Physical Sciences, Babasaheb Bhimrao Ambedkar University, Lucknow 226025, India*

Correspondence should be addressed to Pallath Chandran; chandran@squ.edu.om

Academic Editor: Alberto Cardona

The developing natural convective flow of a nanofluid in an infinite vertical channel with impermeable bounding walls has been investigated. It is assumed that the nanofluid is dominated by two specific slip mechanisms and that the channel walls are subject to constant heat flux and isothermal temperature, respectively. The governing nonlinear partial differential equations coupling different transport processes have been solved numerically. The variations of velocity, temperature, and nanoparticles concentration have been discussed in relation to a number of physical parameters. It is seen that the approach to the steady-state profiles of velocity and temperature in the present work is different from the ones reported in a previous study corresponding to isothermal wall conditions.

1. Introduction

Analytical and numerical studies of natural convective processes involving viscous fluids have been the subject of comprehensive investigations during the last five decades or so, primarily stemming from their wide-ranging applications in several disciplines of engineering and technology. The scope and extent of such investigations have been enormous because of the sheer large number of aspects covered: steady or unsteady flows, permeable or impermeable boundaries, electrically conducting or nonconducting fluids, presence or absence of applied magnetic field, finite or infinite flow domain, adiabatic or isothermal boundary conditions, and two- or three-dimensional nature of flow, to name a few. In particular, a large body of literature exists on the transient free convective flows of viscous incompressible fluids through infinite vertical channels, subject to a variety of imposed conditions on the bounding walls (see, for example, [1–3]).

In recent years, an increasing number of theoretical as well as experimental investigations related to natural convection in nanofluids have drawn the attention of researchers because of the potentially large variety of real life applications in engineering and industry. As is known, nanofluids are engineered colloids of nanoparticles having diameters of the order 1–100 nanometers (nm) and a base fluid having very high thermal conductivity as well as high heat transfer coefficients [4–7]. They are mixtures of base fluids—for example, water or ethylene-glycol—pervaded by a very small amount of nanoparticles such as metallic or metallic oxide particles (Cu, CuO, Al_2O_3). Such fluids are thus considered suspensions. However, their particle size is so small that they approach the size of fluid particles. Because of this, nanofluids can be considered as single-phase fluids instead of two-phase. It has also been treated in the literature as a Newtonian fluid with a constant viscosity, in a number of applications. Also, nanofluids find a large number of industrial applications such as power manufacturing, transportation, cooling of high-energy electronic devices, nuclear systems, and nanodrugs, among others. Due to such wide-ranging applications, a vast body of literature related to nanofluids can be found in the recent literature, starting from mid-nineties [8–20]. Most of these works are devoted to the study of free, forced, or mixed convection in nanofluids in the presence or absence of porous media.

In the investigations of convective flows involving solid boundaries, the thermal conditions at the bounding surfaces are known to play key roles in the developing boundary layer flows. In a very recent work, Sacheti et al. [21] carried out a numerical study dealing with transient two-dimensional free convection in a nanofluid through a vertical channel with rigid walls. Assuming the channel walls to be isothermal—either heated or cooled—the authors discussed developing profiles of velocity, temperature, and nanoparticle concentration. The effects of a number of nondimensional variables were clearly brought out in this study, showcasing various transport mechanisms. The present work extends our earlier analysis [21] by considering the natural convection in a nanofluid dominated by two key slip mechanisms: Brownian diffusion and thermophoresis. It is assumed that one of the boundaries is subjected to heat flux. We investigate the transient nature of the flow when the heat is supplied at the constant rate at the left channel wall while the other wall is isothermal. The coupled nonlinear parabolic partial differential equations governing the flow subject to the relevant initial and boundary conditions have been solved by an implicit finite-difference scheme. The effects of the relevant transport phenomena characterized by Lewis number, buoyancy ratio parameter, Brownian motion parameter, and thermophoresis parameter have been analyzed. Furthermore, the temporal progression of the field variables—velocity, temperature, and nanoparticle volume fraction—from the unsteady state to the steady state has been examined. We have compared the results of our study involving heat flux wall condition at a boundary with the corresponding results for the isothermal wall conditions [21]. It has been observed that the approach to the steady-state values of fluid velocity and temperature follow different patterns in these two thermal cases.

2. Governing Equations

We consider a transient free convective flow between two vertical infinite walls situated at $y' = 0$ and $y' = L$. Each wall is assumed to be impermeable and perfectly thermally conducting. The coordinate system is chosen such that the x'-axis measures the distance along the left wall and the y'-axis measures the distance normal to it. Initially, the temperatures and volumetric fractions of nanoparticles as well as the walls are the same and are taken to be T'_c and ϕ'_c, respectively. At the time $t' > 0$, heat is supplied at a constant rate at the inner surface of the left wall while the inner surface of the right wall ($y' = L$) is maintained at a constant temperature T'_c which causes free convection currents in the channel. In addition, the volumetric fraction of nanoparticles is assumed to vary from ϕ'_h on the inner surface of the left wall to ϕ'_c on the inner surface of the right wall. Also, the Boussinesq approximation is assumed. The conservation equations of momentum, energy, and nanoparticles volume fraction are [8, 14, 15, 21]

$$\rho_f \frac{\partial u'}{\partial t'} = \mu \frac{\partial^2 u'}{\partial y'^2} + \left[(1 - \phi'_c) \rho_f \beta (T' - T'_c) - (\rho_p - \rho_f)(\phi' - \phi'_c) \right] g,$$

$$(\rho c)_f \frac{\partial T'}{\partial t'} = k \frac{\partial^2 T'}{\partial y'^2} + (\rho c)_p \left[D_B \frac{\partial \phi'}{\partial y'} \frac{\partial T'}{\partial y'} + \left(\frac{D_T}{T'_c} \right) \left(\frac{\partial T'}{\partial y'} \right)^2 \right],$$

$$\frac{\partial \phi'}{\partial t'} = D_B \frac{\partial^2 \phi'}{\partial y'^2} + \left(\frac{D_T}{T'_c} \right) \frac{\partial^2 T'}{\partial y'^2}. \tag{1}$$

Here u' is the velocity component of the fluid in the x'-direction, T' is the temperature of nanofluid, ϕ' is the nanoparticle volume fraction, g is the acceleration due to the gravity, β is the coefficient of thermal expansion, ρ_f is the density of the fluid, μ is the viscosity of the fluid assumed to be constant, ρ_p is the density of the particles, $(\rho c)_f$ is the heat capacity of the fluid, k is the thermal conductivity of the fluid, q is the constant heat flux on the left wall, and $(\rho c)_p$ is the effective heat capacity of the particles. Further, D_B and D_T are the coefficients of the Brownian diffusion and the thermophoresis diffusion, respectively.

Associated initial and boundary conditions on the velocity, temperature, and nanoparticle volume fraction are given, respectively, by

$$t' \le 0: \; u' = 0, \quad T' = T'_c, \quad \phi' = \phi'_c \quad \text{for } 0 \le y' \le L,$$

$$t' > 0: \; u' = 0, \quad q = -k \frac{\partial T'}{\partial y'}, \quad \phi' = \phi'_h \quad \text{at } y' = 0,$$

$$u' = 0, \quad T' = T'_c, \quad \phi' = \phi'_c \quad \text{at } y' = L. \tag{2}$$

We now introduce nondimensional variables:

$$y = \frac{y'}{L}, \quad t = \frac{t'\alpha}{L^2}, \quad u = \frac{u'\alpha}{\{\beta g L^2 (1 - \phi'_c)(qL/k)\}},$$

$$T = \frac{(T' - T'_c)}{(qL/k)}, \quad \phi = \frac{(\phi' - \phi'_c)}{(\phi'_h - \phi'_c)}. \tag{3}$$

In view of (3), the conservation equations (1) transform to

$$\frac{\partial u}{\partial t} = \Pr \frac{\partial^2 u}{\partial y^2} + T - N\phi,$$

$$\frac{\partial T}{\partial t} = \frac{\partial^2 T}{\partial y^2} + (n_b) \frac{\partial \phi}{\partial y} \frac{\partial T}{\partial y} + (n_t) \left(\frac{\partial T}{\partial y} \right)^2, \tag{4}$$

$$\frac{\partial \phi}{\partial t} = \frac{1}{\text{Le}} \frac{\partial^2 \phi}{\partial y^2} + \frac{(n_t)}{(\text{Le})(n_b)} \frac{\partial^2 T}{\partial y^2},$$

where $\alpha = k/(\rho c)_f$. The initial and boundary conditions transform to

$$t \le 0: \; u = 0, \quad T = 0, \quad \phi = 0 \quad \text{for } 0 \le y \le 1;$$

$$t > 0: \; u = 0, \quad \frac{\partial T}{\partial y} = -1, \quad \phi = 1 \quad \text{at } y = 0; \tag{5}$$

$$u = 0, \quad T = 0, \quad \phi = 0 \quad \text{at } y = 1.$$

The nondimensional physical parameters appearing in the above equations are the Prandtl number Pr, the Lewis number Le and the buoyancy ratio parameter N. The remaining parameters n_b and n_t are, respectively, the Brownian motion parameter and the thermophoresis parameter. The nondimensional parameters appearing in (4) are defined by

$$\text{Pr} = \frac{\mu}{\rho_f \alpha}, \qquad \text{Le} = \frac{\alpha}{D_B}, \qquad N = \frac{k(\rho_p - \rho_f)(\phi_h' - \phi_c')}{(1 - \phi_c')\, qL\beta\rho_f},$$
$$n_b = \frac{D_B(\rho c)_p}{\alpha(\rho c)_f}(\phi_h' - \phi_c'), \qquad n_t = \frac{D_T(\rho c)_p qL}{k\alpha(\rho c)_f T_c'}. \tag{6}$$

3. Results and Discussion

In order to solve the initial-boundary-value problem described by (4) and (5), we have used an implicit finite-difference method for the variables u, T, and ϕ, as functions of t and y. Using the central difference scheme for the space derivative and the backward difference scheme for the time derivative, the obtained finite difference equations have been reduced to a tridiagonal matrix which, in turn, has been solved by the Thomas algorithm. At the end of each time step, the sequence of function evaluations followed $\phi \rightarrow T \rightarrow u$ in that order. The unsteady values of the physical variables corresponding to a specific instant have been obtained by iterations on the nondimensional temporal variable. The computation is then advanced until a steady state is reached for each of the three physical variables. For the computation, we had chosen 21 grid points for the space variable while the time step was initially taken equal to 0.0004. In order to ensure the stability and convergence of the implicit finite difference method, the computations were repeated by taking smaller values of time step such as 0.0003 and 0.0002. In these cases, however, there were no significant changes in the results which confirms the stability and convergence of the implicit finite difference method used.

The results presented in this section correspond to the Prandtl number (Pr) equal to 7. In the following, we will focus our attention on discussing the effects of the physical parameters—Lewis number (Le), Brownian motion parameter (n_b), thermophoresis parameter (n_t), and buoyancy ratio parameter (N)—on the field variables. Before we begin our analyses of various features observed in Figures 1–10, we wish to draw the attention of the readers to the fact that these parameters—Le, n_b, n_t, and N—have specific physical relevance. They arise either from the nanofluid model in this study or the imposed wall conditions related to a nanofluid variable. For instance, the Lewis number incorporates the relative influence of thermal diffusion and the Brownian diffusion. The latter is one of the key slip mechanisms associated with the nanofluid. The parameter n_b, on the other hand, describes the combined influence of the Brownian diffusion and the imposed unequal wall conditions for the nanoparticle volume fraction. The parameter n_t is directly related to the thermophoresis which is yet another slip mechanism. The parameter N, the buoyancy parameter, arises due to differentials in the densities as well as imposed nanoparticle volume fractions.

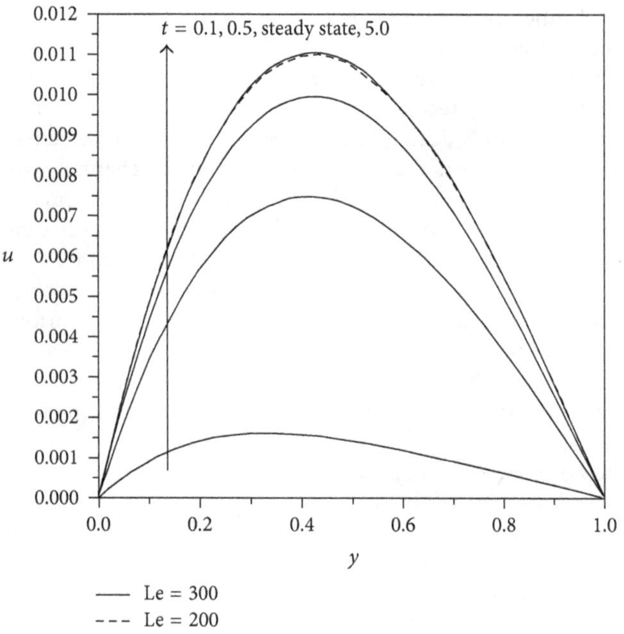

FIGURE 1: Variations of u with respect to y for $N = 0.1$, $n_b = 0.1$, and $n_t = 0.01$.

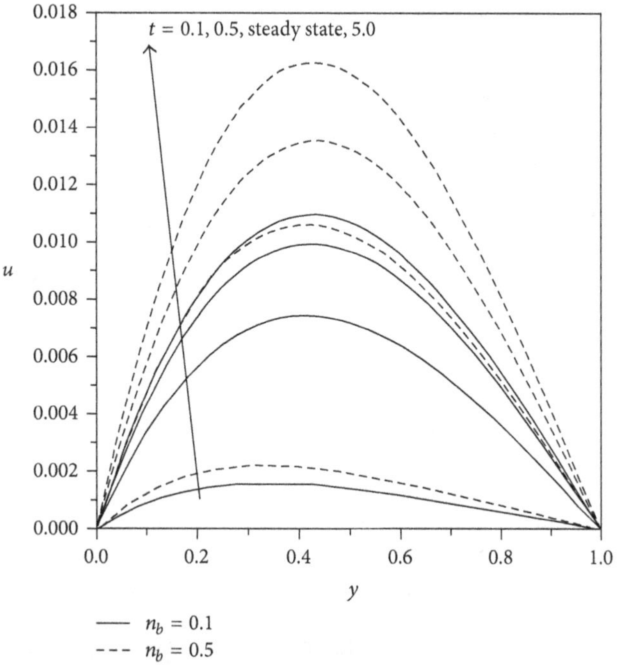

FIGURE 2: Variations of u with respect to y for $N = 0.1$, $n_t = 0.01$, and Le = 200.

In Figures 1–4, we have shown the effects of each of the above four parameters on the developing fluid velocity. The velocity profiles, as expected, tend to become parabolic with time. However, the profiles are not symmetric due to the interplay between the buoyancy forces and the nanofluid

TABLE 1: Skin friction, left-wall temperature, and Sherwood number (Pr = 7).

Le	N	n_b	n_t	t	τ	$T(0,t)$	Sh
200	0.1	0.1	0.01	0.1	0.0119	0.4267	29.213
				0.5	0.0399	0.9175	12.459
				5.0	0.0557	1.2008	3.5761
				147 (SS)	0.0514	1.1632	1.0086
200	0.1	0.1	0.1	0.1	0.0218	0.7742	29.155
				0.5	0.0764	1.7220	12.441
				5.0	0.1180	2.4249	3.7211
				147 (SS)	0.1128	2.3614	1.3778
200	0.1	0.5	0.01	0.1	0.0165	0.5786	29.218
				0.5	0.0572	1.2831	12.466
				5.0	0.0822	1.6817	3.5783
				147 (SS)	0.0680	1.4364	1.0091
200	0.5	0.1	0.01	0.1	0.0109	0.4267	29.213
				0.5	0.0369	0.9175	12.459
				5.0	0.0470	1.2008	3.5761
				147 (SS)	0.0324	1.1632	1.0086
300	0.1	0.1	0.01	0.1	0.0119	0.4270	32.639
				0.5	0.0400	0.9185	16.253
				5.0	0.0562	1.2043	4.3805
				210 (SS)	0.0514	1.1633	1.0123

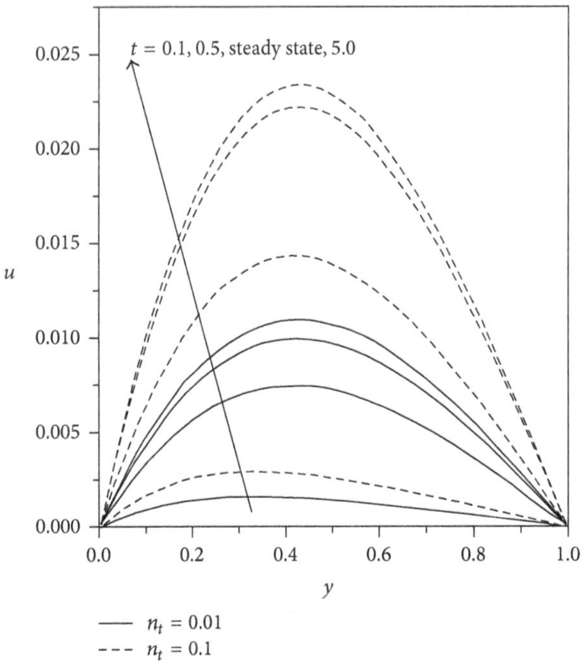

FIGURE 3: Variations of u with respect to y for $N = 0.1$, $n_b = 0.1$, and Le = 200.

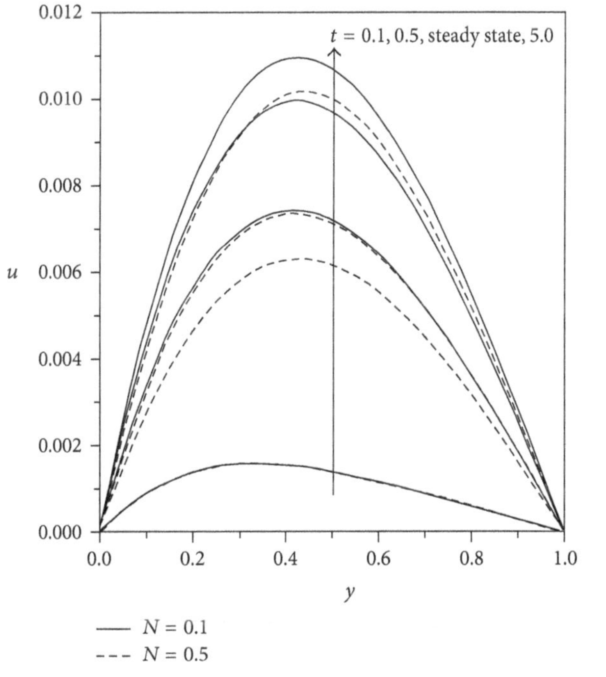

FIGURE 4: Variations of u with respect to y for $n_b = 0.1$, $n_t = 0.01$, and Le = 200.

characteristics. From Figure 1, we observe that the effect of the Lewis number on the velocity is almost negligible. As regards the effect of the Brownian motion parameter n_b—with the remaining three parameters fixed—we notice, from Figure 2, that the fluid velocity increases appreciably with increasing n_b. From these two figures, one may thus note that the velocity profiles are sensitive more with respect to the imposed particle volume fraction differentials than the Brownian diffusion. As regards the influence of the thermophoresis parameter n_t on the velocity (see Figure 3), we observe that the effect is similar to that of the parameter n_b. On the other hand, as can be seen from Figure 4, the nanofluid velocity in

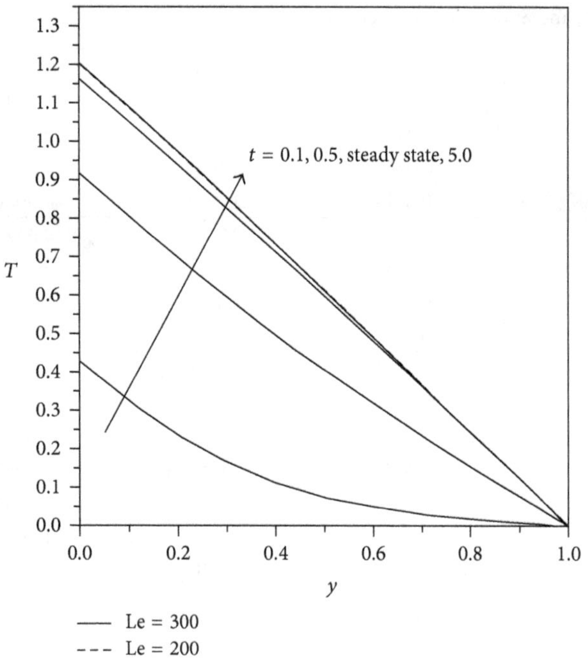

FIGURE 5: Variations of T with respect to y for $N = 0.1$, $n_b = 0.1$, and $n_t = 0.01$.

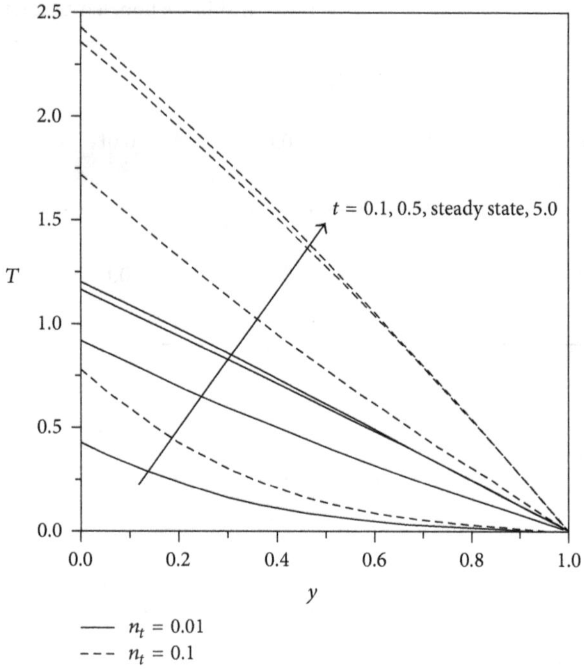

FIGURE 7: Variations of T with respect to y for $N = 0.1$, $n_b = 0.1$, and Le = 200.

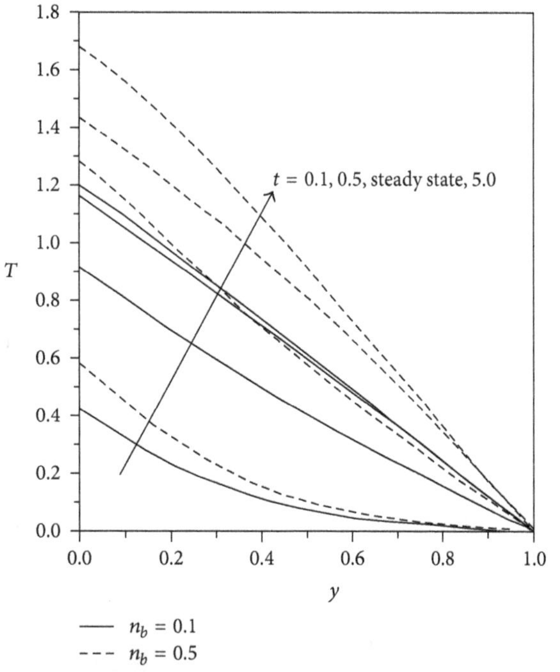

FIGURE 6: Variations of T with respect to y for $N = 0.1$, $n_t = 0.01$, and Le = 200.

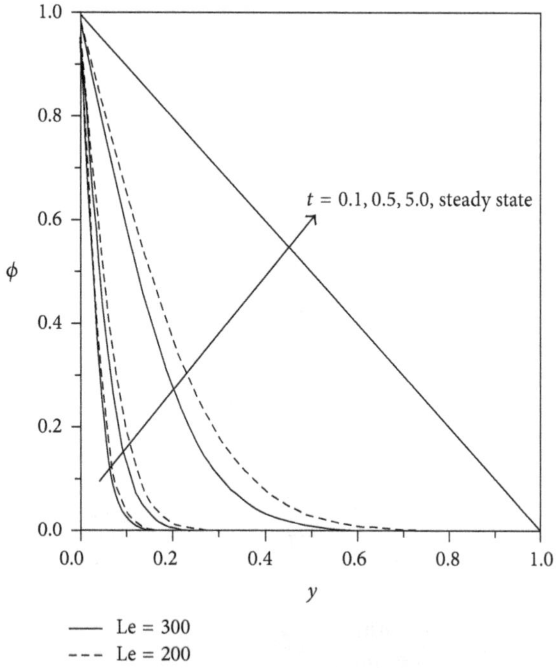

FIGURE 8: Variations of ϕ with respect to y for $N = 0.1$, $n_b = 0.1$, and $n_t = 0.01$.

the channel decreases with the enhancement of the buoyancy ratio parameter N. A common but interesting feature in all four figures is worth noting: while the transient velocity increases with time t at all points of the channel, the steady-state values show a bit of reversal from the peak values. This steady-state behavior is different to the case when the two walls of the channel are isothermal [21]. In the latter case, it was seen that the transition from the unsteady to the steady state followed a monotonically increasing pattern.

In Figures 5–7, we have shown the variation of nanofluid temperature T with respect to the parameters Le, n_b, and n_t, respectively. From Figure 5, we observe that the Lewis

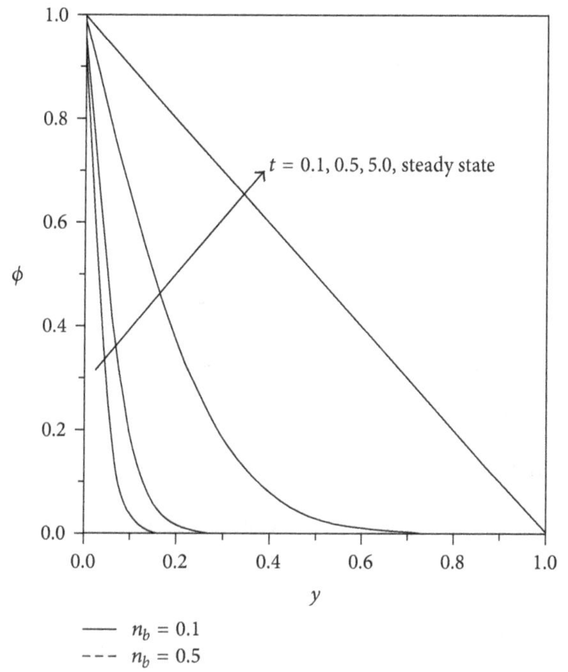

FIGURE 9: Variations of ϕ with respect to y for $N = 0.1$, $n_t = 0.01$, and Le = 200.

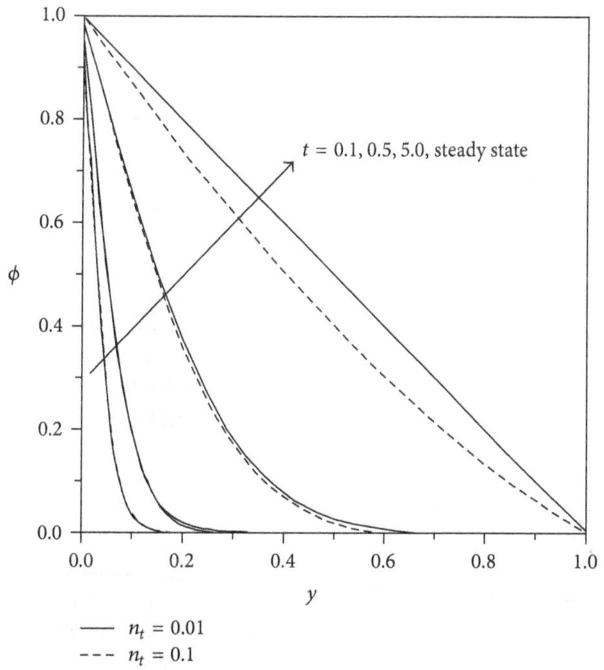

FIGURE 10: Variations of ϕ with respect to y for $N = 0.1$, $n_b = 0.1$, and Le = 200.

number has negligible effect on the temperature. However, Brownian motion and thermophoresis of the nanoparticles tend to increase the temperature in the channel (Figures 6 and 7). The temporal behavior of the temperature is similar to the one observed earlier for the velocity. Thus, qualitatively, the influence of these three parameters and time on temperature is similar to that on velocity.

As regards the influence of the parameters Le, n_b, and n_t on the nanoparticle volume fraction ϕ, one may note that the Brownian motion parameter has negligible effect on it (Figure 9), while the Lewis number and the thermophoresis parameter have both a decreasing effect (Figures 8 and 10). The profiles of ϕ with respect to time t are monotonic and attain steady-state with increasing time. This steady-state phenomenon is in contrast to what has been stated above for the nanofluid velocity and temperature.

3.1. Skin Friction, Surface Temperature, and Sherwood Number. In what follows, we will now introduce three specific quantities that are of practical interest for the convective flow considered here. These are the skin friction τ, surface temperature $T(0,t)$, and the Sherwood number Sh at the left channel wall. The skin friction and the Sherwood numbers are defined, respectively, by the absolute values of the first derivatives of u and ϕ with respect to the space variable y, evaluated at the wall $y = 0$. In order to assess the influence of the four governing parameters on these quantities, we have presented a set of computed values in Table 1. From the table, we observe the following features.

(a) The thermophoresis parameter tends to enhance the wall shear stress as well as the surface temperature, for all times. However, the Sherwood number exhibits nonmonotonic behavior with increasing thermophoresis parameter.

(b) The effect of Brownian motion parameter is to increase all three quantities τ, $T(0,t)$, and Sh. In terms of percentage increase, the shear stress is more sensitive to this parameter than the other two quantities. Within the transient time scales chosen, wall shear stress was seen to increase about 30 to 40 percent.

(c) The Lewis number tends to enhance all three quantities. However, the percentage change in this case is more pronounced for the Sherwood number. The percentage increase of this quantity was in the range of about 12 to 30.

(d) During the early part of the transient stage, the wall shear stress and the surface temperature increase with time while the Sherwood number decreases. The shear stress increases approximately 400 percent while the surface temperature increases nearly 200 percent. On the other hand, the Sherwood number decreases sharply by over 80 percent.

(e) Except the Lewis number, the other parameters do not influence the steady-state (SS) time of these quantities.

4. Conclusions

We have studied numerically using an implicit finite-difference scheme the transient free convection in a vertical channel saturated by a nanofluid. The model incorporates the effects of Brownian diffusion and thermophoretic diffusion of the nanoparticles. We have analyzed the effects of these processes and the application of different types of thermal conditions—uniform heat flux and temperature at opposing walls. It is seen that the nanofluid velocity and temperature increase with the Brownian motion and thermophoresis parameters, but the Lewis number has a negligible effect on them. Significant changes occur in the Sherwood number in a small time interval immediately after the start of the motion. However, such a phenomenon does not occur in the case of skin friction. The model analyzed enables one to assess the relative influence of different types of thermal conditions at the channel walls and the main nanoparticle diffusion processes on the flow and heat transfer. The enhanced thermal property of the nanofluid observed in the study is of great interest in the thermal management of systems in several applications in engineering and medicine.

Abbreviations

Nomenclature

D_B: Brownian diffusion
D_T: Thermophoresis diffusion
g: Acceleration due to gravity
k: Thermal conductivity of the nanofluid
L: Perpendicular distance between the walls
Le: Lewis number
N: Buoyancy ratio parameter
n_b: Brownian motion parameter
n_t: Thermophoresis parameter
Pr: Prandtl number
q: Constant heat flux on the left wall
Sh: Sherwood number
t: Nondimensional time
t': Time variable
T'_c: Constant temperature of the right wall
T': Temperature of the nanofluid
T: Nondimensional temperature of the nanofluid
u': Velocity of the nanofluid in the x'-direction
u: Nondimensional nanofluid velocity
x': Co-ordinate along the left wall
x: Nondimensional co-ordinate along the wall
y': Co-ordinate perpendicular to the left wall
y: Nondimensional co-ordinate perpendicular to the left wall.

Greek Symbols

α: $k/(\rho c)_f$
β: Coefficient of thermal expansion
ρ_f: Density of the nanofluid
ρ_p: Density of the nanoparticles
$(\rho c)_f$: Heat capacity of the fluid
$(\rho c)_p$: Effective heat capacity of the particles
ν: Kinematic viscosity of the nanofluid
ϕ: Nanoparticle volume fraction in nondimensional form
ϕ': Nanoparticle volume fraction
ϕ'_c: Volume fraction of nanoparticles at the right wall
ϕ'_h: Volume fraction of nanoparticles at the left wall
τ: Nondimensional skin friction at the left wall
μ: Viscosity of nanofluid.

Conflict of Interests

The authors declare that there is no conflict of interests regarding the publication of this paper.

References

[1] B. Gebhart, Y. Jaluria, R. L. Mahajan, and B. Sammakia, *Buoyancy-Induced Flows and Transport*, Hemisphere, Cambridge, UK, 1988.

[2] D. J. Nelson and B. D. Wood, "Combined heat and mass transfer natural convection between vertical parallel plates," *International Journal of Heat and Mass Transfer*, vol. 32, no. 9, pp. 1779–1787, 1989.

[3] A. K. Singh, H. R. Gholami, and V. M. Soundalgekar, "Transient free convection flow between two vertical parallel plates," *Heat and Mass Transfer*, vol. 31, no. 5, pp. 329–331, 1996.

[4] H. Masoda, A. Ebata, K. Teramae, and N. Hishinuma, "Alternation of thermal conductivity and viscosity of liquid by dispersing ultra-fine particles," *Netsu Bussei*, vol. 7, no. 4, pp. 227–223, 1993.

[5] S. Choi, "Enhancing thermal conductivity of fluids with nanoparticles," in *Developments and Applications of Non-Newtonian Flows: FED 231/MD* 66, D. A. Siginer and H. P. Wang, Eds., pp. 99–105, ASME, New York, NY, USA, 1995.

[6] J. A. Eastman, S. U. S. Choi, S. Li, W. Yu, and L. J. Thompson, "Anomalously increased effective thermal conductivities of ethylene glycol-based nanofluids containing copper nanoparticles," *Applied Physics Letters*, vol. 78, no. 6, pp. 718–720, 2001.

[7] S. K. Das, N. Putra, P. Thiesen, and W. Roetzel, "Temperature dependence of thermal conductivity enhancement for nanofluids," *Journal of Heat Transfer*, vol. 125, no. 4, pp. 567–574, 2003.

[8] J. Buongiorno, "Convective transport in nanofluids," *Journal of Heat Transfer*, vol. 128, no. 3, pp. 240–250, 2006.

[9] K. Khanafer, K. Vafai, and M. Lightstone, "Buoyancy-driven heat transfer enhancement in a two-dimensional enclosure utilizing nanofluids," *International Journal of Heat and Mass Transfer*, vol. 46, no. 19, pp. 3639–3653, 2003.

[10] N. Putra, W. Roetzel, and S. K. Das, "Natural convection of nano-fluids," *Heat and Mass Transfer*, vol. 39, no. 8-9, pp. 775–784, 2003.

[11] V. Trisaksri and S. Wongwises, "Critical review of heat transfer characteristics of nanofluids," *Renewable and Sustainable Energy Reviews*, vol. 11, no. 3, pp. 512–523, 2007.

[12] X.-Q. Wang and A. S. Mujumdar, "Heat transfer characteristics of nanofluids: a review," *International Journal of Thermal Sciences*, vol. 46, no. 1, pp. 1–19, 2007.

[13] D. Y. Tzou, "Instability of nanofluids in natural convection," *Journal of Heat Transfer*, vol. 130, no. 7, Article ID 072401, 9 pages, 2008.

[14] D. A. Nield and A. V. Kuznetsov, "The Cheng-Minkowycz problem for natural convective boundary-layer flow in a porous medium saturated by a nanofluid," *International Journal of Heat and Mass Transfer*, vol. 52, no. 25-26, pp. 5792–5795, 2009.

[15] D. A. Nield and A. V. Kuznetsov, "Thermal instability in a porous medium layer saturated by a nanofluid," *International Journal of Heat and Mass Transfer*, vol. 52, no. 25-26, pp. 5796–5801, 2009.

[16] E. Abu-Nada and A. J. Chamkha, "Effect of nanofluid variable properties on natural convection in enclosures filled with a CuO-EG—water nanofluid," *International Journal of Thermal Sciences*, vol. 49, no. 12, pp. 2339–2352, 2010.

[17] A. Chamkha, R. S. R. Gorla, and K. Ghodeswar, "Non-similar solution for natural convective boundary layer flow over a sphere embedded in a porous medium saturated with a nanofluid," *Transport in Porous Media*, vol. 86, no. 1, pp. 13–22, 2011.

[18] R. S. R. Gorla and A. Chamkha, "Natural convective boundary layer flow over a nonisothermal vertical plate embedded in a porous medium saturated with a nanofluid," *Nanoscale and Microscale Th rmophysical Engineering*, vol. 15, no. 2, pp. 81–94, 2011.

[19] A. J. Chamkha, S. Abbasbandy, A. M. Rashad, and K. Vajravelu, "Radiation effects on mixed convection over a wedge embedded in a porous medium filled with a nanofluid," *Transport in Porous Media*, vol. 91, no. 1, pp. 261–279, 2012.

[20] Z. Haddad, E. Abu-Nada, H. F. Oztop, and A. Mataoui, "Natural convection in nanofluids: are the thermophoresis and Brownian motion effects significant in nanofluid heat transfer enhancement?" *International Journal of Thermal Sciences*, vol. 57, pp. 152–162, 2012.

[21] N. C. Sacheti, P. Chandran, A. K. Singh, and B. S. Bhadauria, "Transient free convective flow of a nanofluid in a vertical channel," *International Journal of Energy & Technology*, vol. 4, pp. 1–7, 2012.

Inversion of Fourier Transforms by Means of Scale-Frequency Series

Nassar H. S. Haidar

Center for Research in Applied Mathematics & Statistics (CRAMS), AUL, Lebanon

Correspondence should be addressed to Nassar H. S. Haidar; nhaidar@suffolk.edu

Academic Editor: J. A. Tenreiro Machado

We report on inversion of the Fourier transform when the frequency variable can be scaled in a variety of different ways that improve the resolution of certain parts of the frequency domain. The corresponding inverse Fourier transform is shown to exist in the form of two dual scale-frequency series. Upon discretization of the continuous scale factor, this Fourier transform series inverse becomes a certain nonharmonic double series, a discretized scale-frequency (DSF) series. The DSF series is also demonstrated, theoretically and practically, to be rate-optimizable with respect to its two free parameters, when it satisfies, as an entropy maximizer, a pertaining recursive nonlinear programming problem incorporating the entropy-based uncertainty principle.

1. Introduction

We revisit the classical problem of recovering a Fourier transformable signal

$$f(t) = \frac{1}{2\pi}\int_R \hat{f}(\mu)\, e^{i\mu t} d\mu = \mathfrak{F}^{-1}\{\hat{f}(\mu)\} \qquad (1)$$

in $\mathcal{F}(R)$, $R = (-\infty, \infty)$, from its Fourier transform; see, for example, [1, 2], image $\hat{f}(\mu)$. The space $\mathcal{F}$ is the set of all real functions f for which the integral $\hat{f}(\mu) = \int_R f(t)e^{-i\mu t}dt$ exists.

Obviously, the function to be integrated in (1) is the product

$$\hat{f}(\mu)\, e^{i\mu t} = \hat{f}(\mu)\,[\cos\mu t + i \sin\mu t], \quad t \in R, \qquad (2)$$

and when the variable $|t| \rightarrow \infty$, the function $e^{i\mu t}$ will oscillate extremely rapidly. Then in order to follow the variations of the product $\hat{f}(\mu)e^{i\mu t}$ meaningfully in a μ-quadrature formula, see, for example, [3], for (1), there is a need for a large number of μ-points, even for slowly varying $\hat{f}(\mu)$. Consequently, the faster the $\hat{f}(\mu)$ decreases as $|\mu| \rightarrow \infty$, the more the tractable computationally the integral (1) will be, and the more accurate are its numerical evaluations.

Bandlimiting of signals has been a practical way for easing the previous problem. A fact that has so far been motivating the wide engineering interest in bandlimited signals, for which

$$\mathfrak{F}\{f(t)\} = \begin{cases} \hat{g}(\mu), & |\mu| \le 2\pi B, \\ 0, & \text{otherwise}, \end{cases} \\ = \hat{f}(\mu), \qquad (3)$$

where B is the band width of $f(t)$ with an image $\hat{f}(\mu)$ having the $[-2\pi B, 2\pi B]$ compact support of $\hat{g}(\mu)$. Inversion of bandlimited signals is currently performed by projections onto convex sets (POCS) algorithms [4], which encompass the Gerchberg-Papoulis algorithm [5].

In this work we report on a novel robust semianalytical method for integration of (1) to recover a signal $f(t) \in \mathcal{F}(R)$ that is not necessarily bandlimited. In this method, the frequency variable can be scaled in a variety of different ways that improve the resolution of certain parts of the frequency domain. The corresponding inverse Fourier transform is shown to exist in the form of two dual scale-frequency series. Upon discretization of the continuous scale factor, this Fourier transform series inverse becomes a certain nonharmonic [6] double series, a discretized scale-frequency (DSF) series. This DSF series is demonstrated, for the first time in this work, to be the proper framework for an entropy-maximizing inverse Fourier transformation. A fact has emerged after theoretically illustrating that the DSF series is rate-optimizable with respect to the scale factor step and the frequency gap size. The robustness of this optimization (which is illustrated by a simple example) reduces to the satisfaction, via entropy-maximization, of an emerging recursive nonlinear programming problem, incorporating the entropy-based uncertainty principle.

2. Theory of the Continuous Scale-Frequency Series Inverse

Let $\mathbb{Z}$ be the set of integers, $\mathbb{Q}$ the set of rational numbers, and $\mathbb{Q}^c$ the set of irrational numbers. $\lambda \in R \setminus \{0\}$ will be a scale factor, while the frequency γ can either be $m \in \mathbb{Z}$ or $\nu = \nu(m) = (m + \varsigma) \in \mathbb{Q}$ or $\mathbb{Q}^c$ (depending on whether $\varsigma \in \mathbb{Q}$ or $\mathbb{Q}^c$). The symbol $\square$ means either multiplication or division of γ by λ, and γ^c is the alternative (complementary) symbol to γ; that is, if $\gamma = m$, then $\gamma^c = \nu$ and vise versa. The signal $f(t) \in \mathcal{F}(R)$ is assumed to be expandable in the variably scaled trigonometric basis:

$$q_{\square\gamma\lambda}(t) = \left\{\lambda^{-1}\left[\cos\left(\gamma^c\square\lambda\right)t + \sin\left(\gamma\square\lambda\right)t\right]\right\}. \quad (4)$$

The ensemble of γ with $\square$ in (4) can be in any of their possible $C_2^4 = 6$ combinations of 4 quantities, taken two at a time. Accordingly throughout this work we will assume, as in [7], any of

$$\begin{aligned} q_{\succ m\lambda}(t) &= \lambda^{-1}\left[\cos\nu\lambda t + \sin m\lambda t\right]; \\ \square = \succ&: \gamma\square\lambda = m\lambda, \quad \gamma^c\square\lambda = \nu\lambda, \\ q_{\prec m\lambda}(t) &= \lambda^{-1}\left[\cos m\lambda t + \sin\nu\lambda t\right]; \\ \square = \prec&: \gamma\square\lambda = \nu\lambda, \quad \gamma^c\square\lambda = m\lambda \end{aligned} \quad (5)$$

for high frequency resolving scaling (as $\lambda \to 0$), any of

$$\begin{aligned} q_{\vdash m\lambda}(t) &= \lambda^{-1}\left[\cos\frac{\nu}{\lambda}t + \sin\frac{m}{\lambda}t\right]; \\ \square = \vdash&: \gamma\square\lambda = \frac{m}{\lambda}, \quad \gamma^c\square\lambda = \frac{\nu}{\lambda}, \\ q_{\dashv m\lambda}(t) &= \lambda^{-1}\left[\cos\frac{m}{\lambda}t + \sin\frac{\nu}{\lambda}t\right]; \\ \square = \dashv&: \gamma\square\lambda = \frac{\nu}{\lambda}, \quad \gamma^c\square\lambda = \frac{m}{\lambda} \end{aligned} \quad (6)$$

for low frequency resolving scaling, while any of

$$\begin{aligned} q_{\rhd m\lambda}(t) &= \lambda^{-1}\left[\cos\frac{m}{\lambda}t + \sin m\lambda t\right]; \\ \square = \rhd&: \gamma\square\lambda = m\lambda, \quad \gamma^c\square\lambda = \frac{m}{\lambda}, \\ q_{\lhd m\lambda}(t) &= \lambda^{-1}\left[\cos m\lambda t + \sin\frac{m}{\lambda}t\right]; \\ \square = \lhd&: \gamma\square\lambda = \frac{m}{\lambda}, \quad \gamma^c\square\lambda = m\lambda \end{aligned} \quad (7)$$

for mixed (high and low) frequency resolving scaling.

Next consider the sequence $(q_{\square\gamma\lambda})$ that supports the bases (4) when $R_o = (R\setminus\{0\})$, and let $\square$ be any one of the six symbols $\succ, \prec, \vdash, \dashv, \rhd, \lhd$, to invoke the following double parameterized Riemann-Lebesgue lemma (RLL).

Lemma 1 (see [7]). *For* $\Gamma_{\square\gamma\lambda}[\cdot] = \int_R[\cdot]q_{\square\gamma\lambda}(t)dt$, *when* $m \in \mathbb{Z}$,

$$\lim_{|m|\to\infty}\Gamma_{\square\gamma\lambda}[f] = 0, \quad (8)$$

for all $f \in \mathcal{F}(R)$, *for any scaling factor* $\lambda \in R_o$, *and for all six* $\square$ *symbols.*

This lemma justifies the existence and guarantees the convergence of the scale-frequency transform

$$F_\square(\lambda, m) := \int_R f(t)\,q_{\square\gamma\lambda}(t)\,dt = \left\langle f, q_{\square\gamma\lambda}(t)\right\rangle, \quad (9)$$

for all $\lambda \in R_o$, of $f \in \mathcal{F}(R)$ with respect to $q_{\square\gamma\lambda}$. Make then use of the notation

$$\begin{Bmatrix} F_c(\mu) \\ F_s(\mu) \end{Bmatrix} = \int_R f(t)\begin{Bmatrix} \cos\mu t \\ \sin\mu t \end{Bmatrix} dt \quad (10)$$

in

$$\hat{f}(\mu) = \int_R f(t)\,e^{-i\mu t}dt = \mathfrak{F}\{f(t)\} \quad (11)$$

to rewrite it as

$$\hat{f}(\mu) = F_c(\mu) - iF_s(\mu). \quad (12)$$

That is,

$$F_c(\mu) = \mathrm{Re}\left[\hat{f}(\mu)\right], \qquad F_s(\mu) = -\,\mathrm{Im}\left[\hat{f}(\mu)\right]. \quad (13)$$

Unlike the Fourier transformation $\hat{f}(\mu)$, which is a complex function of μ, $F_\square(\lambda, m)$ is a real function of the scaled frequencies

$$\rho_\square = \gamma\square\lambda, \qquad \mu_\square = \gamma^c\square\lambda. \quad (14)$$

Theorem 2. $F_\square^{-1}(\lambda, m) = f(t) \in \mathcal{F}(R)$ *satisfies*

$$f(t) = \frac{1}{2\pi}\int_R F_c(\mu_\square)\cos\mu_\square t\frac{d\mu_\square}{\lambda} + \frac{1}{2\pi}\int_R F_s(\rho_\square)\sin\rho_\square t\frac{d\rho_\square}{\lambda}. \quad (15)$$

Proof. The above result follows directly from the structural linearity of

$$F_\square(\lambda, m) = \frac{1}{\lambda}F_c(\mu_\square) + \frac{1}{\lambda}F_s(\rho_\square) \quad (16)$$

and the linearity of its inverse transformation.

The direct inversion formula (15) can be logarithmic in the scaling factor and is computationally of much larger complexity than (1). A simpler indirect inversion is, however, possible, and in a dual process, that is tied to the possible utilization of anyone of the components of $\hat{f}(\mu)$ in (13); namely,

$$F_{\square}(\lambda, m) \Longleftrightarrow \begin{Bmatrix} F_{\square c}^{-1}(\lambda, m) \\ F_{\square s}^{-1}(\lambda, m) \end{Bmatrix} = F_{\square}^{-1}(\lambda, m) = f(t), \tag{17}$$

$$\text{when } \begin{Bmatrix} F_c(\mu) \\ F_s(\mu) \end{Bmatrix} \text{ is utilized.}$$

Interestingly, this results in 12 distinct scale-frequency forms for the inverse Fourier transform $f(t)$.

Obviously, the symbols ⊢, ⊣ correspond to scaling sines and cosines in the same way, while ≻, ≺ or ⊳, ⊲ correspond to scaling cosines and sines in different ways. It should be noted here that although a Fourier transform is virtually unaffected by a shift of f in time (which mixes the roles of the cosines and sines), the corresponding scale-frequency transform could change drastically. This raises the question on the rather unclear merits of different-way scalings, to the favour of same-way (⊢, ⊣) scalings. It is believed, however, that different-way scalings can particularly reveal their merits when applied to discontinuous $f \in \mathcal{F}(R)$.

Theorem 3 (high frequency resolving scaling). *When* $\square = \succ$ *in* $F_{\square}(\lambda, m) \Leftrightarrow f(t)$, *the inverse is either*

$$\begin{aligned} f(t) &= F_{\succ c}^{-1}(\lambda, m) \\ &= \frac{1}{2\pi M} \sum_{m=0}^{M-1} \int_R m F_c(m\lambda) \cos(\lambda + 1) mt \, d\lambda \\ &\quad - \frac{1}{2\pi M} \sum_{m=0}^{M-1} \int_R m \{F_c(m\lambda) + F_s([m + \varsigma]\lambda) \\ &\qquad - F_c([m+\varsigma]\lambda)\} \times \sin(\lambda + 1) mt \, d\lambda, \end{aligned} \tag{18}$$

or the dual representation

$$\begin{aligned} f(t) &= F_{\succ s}^{-1}(\lambda, m) \\ &= \frac{1}{2\pi M} \\ &\quad \times \sum_{m=0}^{M-1} \int_R (m + \varsigma) \{F_c(m\lambda) + F_s([m + \varsigma]\lambda) - F_s(m\lambda)\} \\ &\quad \times \cos([m + \varsigma]\lambda + m) t \, d\lambda - \frac{1}{2\pi M} \\ &\quad \times \sum_{m=0}^{M-1} \int_R (m + \varsigma) F_s([m + \varsigma]\lambda) \sin([m + \varsigma]\lambda + m) t \, d\lambda. \end{aligned} \tag{19}$$

Proof. By incorporation of

$$F_{\succ}(\lambda, m) = \frac{1}{\lambda} F_c(\mu_{\succ}) + \frac{1}{\lambda} F_s(\rho_{\succ}) \tag{20}$$

in the details of the proofs of Theorem 1 and Corollary 1 of [7].

A similar theorem holds also for the case when $\square = \prec$.

Theorem 4 (low frequency resolving scaling). *For* $\square = \succ$ *in* $F_{\square}(\lambda, m) \Leftrightarrow f(t)$, *the inverse is either*

$$\begin{aligned} f(t) &= F_{\vdash c}^{-1}(\lambda, m) \\ &= -\frac{1}{2\pi M} \times \sum_{m=0}^{M-1} \int_R m \frac{1}{\lambda} F_c\left(\frac{m}{\lambda}\right) \cos\left(\frac{1}{\lambda} + 1\right) mt \, d(\ln \lambda) \\ &\quad + \frac{1}{2\pi M} \times \sum_{m=0}^{M-1} \int_R m \left\{ \frac{1}{\lambda} F_c\left(\frac{m}{\lambda}\right) + \frac{1}{\lambda} F_s\left(\frac{[m + \varsigma]}{\lambda}\right) \right. \\ &\qquad \left. - \frac{1}{\lambda} F_c\left(\frac{[m + \varsigma]}{\lambda}\right) \right\} \\ &\quad \times \sin\left(\frac{1}{\lambda} + 1\right) mt \, d(\ln \lambda), \end{aligned} \tag{21}$$

or the dual representation

$$\begin{aligned} f(t) &= F_{\vdash s}^{-1}(\lambda, m) \\ &= \frac{1}{2\pi M} \times \sum_{m=0}^{M-1} \int_R (m + \varsigma) \left\{ \frac{1}{\lambda} F_c\left(\frac{m}{\lambda}\right) + \frac{1}{\lambda} F_s\left(\frac{[m + \varsigma]}{\lambda}\right) \right. \\ &\qquad \left. - \frac{1}{\lambda} F_s\left(\frac{m}{\lambda}\right) \right\} \\ &\quad \times \cos\left(\frac{[m + \varsigma]}{\lambda} + m\right) t \, d(\ln \lambda) \\ &\quad + \frac{1}{2\pi M} \sum_{m=0}^{M-1} \int_R (m + \varsigma) \frac{1}{\lambda} F_s\left(\frac{[m + \varsigma]}{\lambda}\right) \\ &\quad \times \sin\left(\frac{[m + \varsigma]}{\lambda} + m\right) t \, d(\ln \lambda) \end{aligned} \tag{22}$$

when the logarithmic integrals are conceived in the Principal Value sense.

Proof. By incorporation of $F_{\vdash}(\lambda, m) = (1/\lambda)F_c(\mu_{\vdash}) + (1/\lambda)F_s(\rho_{\vdash})$ in the details of the proofs of Theorem 2 and Corollary 2 of [7].

A similar theorem holds also for the case when $\square = \dashv$.

3. Discretized Scale-Frequency Series Inverse

Here we discretize the λ axis uniformly over R_o according to $\lambda_k = k\delta \in [-\Lambda, \Lambda] \subset R$, where Λ is the scale factor band width, which is not apriori known. $k \in [-N, N] \setminus \{0\} = W_o \subset \mathbb{Z}$, with $\delta = \Lambda/N$ as the scale sample size, which can be arbitrarily small or large. In the special case when $\Lambda = N$, $\delta = 1$.

3.1. Discretized High Frequency Resolving Scaling. The above mentioned dicretization of the scale variable, with k as a scaling factor number, when $\square = \succ$, leads to

$$F_{\succ}(k\delta, m) = \frac{1}{k\delta} F_c(mk\delta) + \frac{1}{k\delta} F_s([m + \varsigma] k\delta), \tag{23}$$

the pertaining phase

$$\varphi_{k,m;\delta} = mk\delta + m, \tag{24}$$

and to the directly related discretized notation,

$$A_{k,m} = \frac{\delta}{2\pi M} m \operatorname{Re}[\hat{f}(mk\delta)], \tag{25}$$

$$B_{k,m} = \frac{\delta}{2\pi M} m [\operatorname{Im}[\hat{f}([m + \varsigma] k\delta)] + \operatorname{Re}[\hat{f}([m + \varsigma] k\delta)] - \operatorname{Re}[\hat{f}(mk\delta)]]. \tag{26}$$

Alternatively, for the dual representation (19), we employ

$$\overset{+}{\varphi}_{k,m;\delta} = (m + \varsigma) k\delta + m, \tag{27}$$

$$A^{+}_{k,m} = \frac{\delta}{2\pi M} (m + \varsigma) [\operatorname{Re}[\hat{f}(mk\delta)] - \operatorname{Im}[\hat{f}([m + \varsigma] k\delta)] + \operatorname{Im}[\hat{f}(mk\delta)]], \tag{28}$$

$$B^{+}_{k,m} = \frac{\delta}{2\pi M} (m + \varsigma) \operatorname{Im}[\hat{f}([m + \varsigma] k\delta)] \tag{29}$$

to state one of the main results of this work.

Theorem 5 (high frequency resolving discrete scaling). *When $\square = \succ$ in $F_{\square}(\lambda, m) \Leftrightarrow f(t)$, the DSF series inverse is either*

$$f(t) \approx \mathfrak{D}F^{-1}_{\succ c}(\lambda, m) = \sum_{k \in W_o} \sum_{m=0}^{M-1} A_{k,m} \cos \varphi_{k,m;\delta} t + B_{k,m} \sin \varphi_{k,m;\delta} t, \tag{30}$$

or the dual representation

$$f(t) \approx \mathfrak{D}F^{-1}_{\succ s}(\lambda, m) = \sum_{k \in W_o} \sum_{m=0}^{M-1} A^{+}_{k,m} \cos \overset{+}{\varphi}_{k,m;\delta} t + B^{+}_{k,m} \sin \overset{+}{\varphi}_{k,m;\delta} t. \tag{31}$$

Proof. By considering (13) when following the same arguments of the proofs of Theorem 3 and Corollary 3 of [7].

A similar result holds also for the case when $\square = \prec$.

3.2. Discretized Low Frequency Resolving Scaling. Here $\square = \vdash$ is associated, by discretizing the scale factor, with

$$\psi_{k,m;\delta} = \frac{m}{k\delta} + m, \tag{32}$$

$$\mathfrak{A}_{k,m} = -\frac{\delta}{2\pi M} m \left(\frac{1}{k\delta}\right)^2 \operatorname{Re}\left[\hat{f}\left(\frac{m}{k\delta}\right)\right], \tag{33}$$

$$\mathfrak{B}_{k,m} = \frac{\delta}{2\pi M} m \left(\frac{1}{k\delta}\right)^2 \times \left\{\operatorname{Re}\left[\hat{f}\left(\frac{m}{k\delta}\right)\right] - \operatorname{Im}\left[\hat{f}\left(\frac{[m + \varsigma]}{k\delta}\right)\right] - \operatorname{Re}\left[\hat{f}\left(\frac{[m + \varsigma]}{k\delta}\right)\right]\right\}. \tag{34}$$

Alternatively, for the dual representation (22), we employ

$$\overset{+}{\psi}_{k,m;\delta} = \frac{(m + \varsigma)}{k\delta} + m, \tag{35}$$

$$\mathfrak{A}^{+}_{k,m} = \frac{\delta}{2\pi M} (m + \varsigma) \left(\frac{1}{k\delta}\right)^2 \times \left\{\operatorname{Re}\left[\hat{f}\left(\frac{m}{k\delta}\right)\right] - \operatorname{Im}\left[\hat{f}\left(\frac{[m + \varsigma]}{k\delta}\right)\right] + \operatorname{Im}\left[\hat{f}\left(\frac{m}{k\delta}\right)\right]\right\}, \tag{36}$$

$$\mathfrak{B}^{+}_{k,m} = -\frac{\delta}{2\pi M} (m + \varsigma) \left(\frac{1}{k\delta}\right)^2 \operatorname{Im}\left[\hat{f}\left(\frac{[m + \varsigma]}{k\delta}\right)\right] \tag{37}$$

to state the following main result of this work.

Theorem 6 (low frequency resolving discrete scaling). *When $\square = \vdash$ in $F_{\square}(\lambda, m) \Leftrightarrow f(t)$, the DSF series inverse is either*

$$f(t) \approx \mathfrak{D}F^{-1}_{\vdash c}(\lambda, m) = \sum_{k \in W_o} \sum_{m=0}^{M-1} \mathfrak{A}_{k,m} \cos \psi_{k,m;\delta} t + \mathfrak{B}_{k,m} \sin \psi_{k,m;\delta} t \tag{38}$$

or the dual representation

$$f(t) \approx \mathfrak{D}F^{-1}_{\vdash s}(\lambda, m) = \sum_{k \in W_o} \sum_{m=0}^{M-1} \mathfrak{A}^{+}_{k,m} \cos \overset{+}{\psi}_{k,m;\delta} t + \mathfrak{B}^{+}_{k,m} \sin \overset{+}{\psi}_{k,m;\delta} t. \tag{39}$$

Proof. By considering (13) when following the same arguments of the proofs of Theorem 4 and Corollary 4 of [7].

A similar result holds also for the case when $\square = \dashv$.

4. Demonstrative Application

To illustrate the applicability of the previously reported results, we give here three illustrative examples of the high frequency resolving scaling in recovering three well-known simple Fourier preimages.

Example 7. The even real signal $f(t) = e^{-|t|}$ has a real image, $\hat{f}(\mu) = 2/(1+\mu^2)$; that is, $\mathrm{Im}[\hat{f}(\mu)] = 0$. According to (30) it admits the following family of nonharmonic series expansions:

$$\begin{aligned} e^{-|t|} &\approx \mathfrak{D}F^{-1}_{\succ c}(\lambda, m) \\ &= \frac{\delta}{\pi M} \sum_{\substack{k=-N \\ k\neq 0}}^{N} \sum_{m=0}^{M-1} \frac{m}{1+m^2k^2\delta^2} \cos(k\delta+1)mt \\ &+ \left\{ \frac{m}{1+[m+\varsigma]^2k^2\delta^2} - \frac{m}{1+m^2k^2\delta^2} \right\} \sin(k\delta+1)mt, \end{aligned} \tag{40}$$

for various values of ς, δ, M, and N. It should be noted here that δN defines the unknown scale factor band width Λ.

Obviously when $\varsigma = 0$, the expansions become purely cosinusoidal, and the dual DSF expansions $\mathfrak{D}F^{-1}_{\succ s}(\lambda, m)$ can similarly be constructed according to (31).

Example 8. For the odd signal $g(t) = te^{-|t|}$ we have a purely imaginary image, $\hat{g}(\mu) = (-4\mu/(1+\mu^2)^2)i$; that is, $\mathrm{Re}[\hat{g}(\mu)] = 0$. Consequently, according to (30) it admits the following family of nonharmonic sinusoidal series expansions:

$$\begin{aligned} te^{-|t|} &\approx \mathfrak{D}G^{-1}_{\succ c}(\lambda, m) \\ &= -\frac{2\delta}{\pi M} \\ &\times \sum_{\substack{k=-N \\ k\neq 0}}^{N} \sum_{m=0}^{M-1} \frac{m[m+\varsigma]k\delta}{\{1+[m+\varsigma]^2k^2\delta^2\}^2} \sin(k\delta+1)mt, \end{aligned} \tag{41}$$

in addition to the dual expansions $\mathfrak{D}G^{-1}_{\succ s}(\lambda, m)$.

Example 9. The nonsymmetric real signal $h(t) = (t+1)e^{-|t|}$ has the complex image $\hat{h}(\mu) = 2/(1+\mu^2) - (4\mu/(1+\mu^2)^2)i$. Then it admits, by (30), the family of nonharmonic series expansions:

$$\begin{aligned} (t+1)e^{-|t|} &\approx \mathfrak{D}H^{-1}_{\succ c}(\lambda, m) \\ &= \frac{\delta}{\pi M} \sum_{\substack{k=-N \\ k\neq 0}}^{N} \sum_{m=0}^{M-1} \frac{m}{1+m^2k^2\delta^2} \cos(k\delta+1)mt \\ &- \left\{ \frac{m}{1+m^2k^2\delta^2} + \frac{2m[m+\varsigma]k\delta}{\{1+[m+\varsigma]^2k^2\delta^2\}^2} \right. \\ &\left. - \frac{m}{1+[m+\varsigma]^2k^2\delta^2} \right\} \sin(k\delta+1)mt, \end{aligned} \tag{42}$$

in addition to the dual DSF expansions $\mathfrak{D}H^{-1}_{\succ s}(\lambda, m)$.

5. Rate-Optimization of the DSF Series Inverse

In DSF series inversion of $\hat{f}(\mu)$, the number M of frequency points is decided by the need for satisfactory followup of the variation of $\hat{f}(\mu)$ over its effective compact support. In contrast, the number of scaling points $2N = 2\Lambda/\delta \in \mathbb{Z}$ remains rather arbitrary.

Besides, these series can be made rate-optimal in the sense that the derivative of a certain norm of them will vanish at a high rate as M and N tend to infinity. Clearly when $N \to \infty$, $\delta \to 0$ for a given Λ. The two free parameters that can be optimized in this sense are (i) the scale factor step, δ, for a given N, and (ii) the frequency gap size, ς, for a given M.

Here we will study this optimization problem for the case of the DSF series inverse $\mathfrak{D}F^{-1}_{\succ c}(\lambda, m)$, (30), without loss of generality.

Definition 10. The DSF norm $\|f\|_\square$ of $f(t) \in \mathcal{F}(R)$ is the sum of the amplitudes of all scaled-harmonics appearing in the $\mathfrak{D}F^{-1}_{\succ c}(\lambda, m)$ expansion; that is,

$$\|f\|_\square = \sum_{\substack{k=-N \\ k\neq 0}}^{N} \sum_{m=0}^{M-1} \sqrt{A^2_{k,m;\square} + B^2_{k,m;\square}}, \tag{43}$$

where $A_{k,m;\square}$ and $B_{k,m;\square}$ are, respectively, defined by (25)-(26) or (28)-(29) and (33)-(34) or (36)-(37).

The concept of the similar (but different) trigonometric norm $\|f\|_*$ was first advanced by Lyaponov in 1954; see, for example, [8], for the standard orthonormal (over $[0, 2\pi]$) trigonometric system. Our DFS series $\mathfrak{D}F^{-1}_{\succ c}(\lambda, m)$ or $\mathfrak{D}F^{-1}_{\succ s}(\lambda, m)$ are, however, with respect to the nonharmonic trigonometric systems,

$$\left\{e^{i\varphi_{k,m;\delta}t}\right\}_{k,m} \text{ or } \left\{e^{i\varphi^{+}_{k,m;\delta}t}\right\}_{k,m;\varsigma}, \tag{44}$$

which, for certain δ and ς, may even not be closed in the space $L^2(R)$, and despite the fact that the coefficients $A_{k,m;\square}$ and $B_{k,m;\square}$ do satisfy the scale-frequency Riemann-Lebesgue lemma [7]. Nevertheless, $\|f\|_\square$ can easily be proved to satisfy the following defining properties of a norm:

(i) $\|cf\|_\square = |c|\|f\|_\square$, $c \in R$ (absolute homogeneity),

(ii) $\|f+g\|_\square \le \|f\|_\square + \|g\|_\square$ (subadditivity),

(iii) $\|f\|_\square = 0$, if and only if $f(t) = 0$ (separating points).

Moreover, as is well known, for piecewise continuous functions, the usual Fourier series (not necessarily (30)) may, in general, not converge absolutely and then the norm $\|f\|_*$ becomes unbounded. Besides, for sufficiently smooth functions, the pertaining Fourier coefficients satisfy the Riemann-Lebesgue lemma and this causes $\|f\|_*$ to approach the uniform norm $\|f\| = \sup_t|f(t)|$ as an upper bound, namely,

$$\|f\| \le \|f\|_*. \tag{45}$$

Relation (45) is called Lyaponov's norm inequality.

Conjecture 11. *Like* $\|f\|_*$*, the DSF norm* $\|f\|_\square$ *also satisfies*

$$\|f\| \le \|f\|_\square. \tag{46}$$

Proposition 12. *Stationarization of* $\|f\|_\square$ *with respect to* δ *and* ς *leads to stabilizing the DSF series inverse.*

Proof. Consider, without loss of generality, $\|f\|_\square$ for $\mathfrak{D}F_{\succ c}^{-1}(\lambda, m)$:

$$\|f\|_{\succ c} = \sum_{\substack{k=-N\\k\neq 0}}^{N} \sum_{m=0}^{M-1} \sqrt{A_{k,m}^2 + B_{k,m}^2}. \tag{47}$$

Substitution of (25) and (26) for $A_{k,m}^2 + B_{k,m}^2$ in (47), when adopting the simplifying notation

$$\widehat{f}_\delta = \widehat{f}\left(\frac{m}{k\delta}\right), \qquad \widehat{f}_{\delta,\varsigma} = \widehat{f}\left([m+\varsigma]\,k\delta\right), \tag{48}$$

leads to

$$\begin{aligned} A_{k,m}^2 + B_{k,m}^2 &= \frac{m^2\delta^2}{4\pi^2 M^2} \\ &\times \left\{\left\langle \operatorname{Re}\left[\widehat{f}_{\delta,\varsigma}\right] + \operatorname{Im}\left[\widehat{f}_{\delta,\varsigma}\right]\right\rangle\right. \\ &\quad \times \left.\left\langle \operatorname{Re}\left[\widehat{f}_{\delta,\varsigma}\right] + \operatorname{Im}\left[\widehat{f}_{\delta,\varsigma}\right] - 2\operatorname{Re}\left[\widehat{f}_\delta\right]\right\rangle + 2\operatorname{Re}^2\left[\widehat{f}_\delta\right]\right\}, \end{aligned} \tag{49}$$

then to

$$\begin{aligned} \|f\|_{\succ c} &= \frac{\delta}{2\pi M} \\ &\times \sum_{\substack{k=-N\\k\neq 0}}^{N} \sum_{m=0}^{M-1} m\left\{\left\langle \operatorname{Re}\left[\widehat{f}_{\delta,\varsigma}\right] + \operatorname{Im}\left[\widehat{f}_{\delta,\varsigma}\right]\right\rangle\right. \\ &\quad \times \left\langle \operatorname{Re}\left[\widehat{f}_{\delta,\varsigma}\right] + \operatorname{Im}\left[\widehat{f}_{\delta,\varsigma}\right] - 2\operatorname{Re}\left[\widehat{f}_\delta\right]\right\rangle \\ &\quad \left. + 2\operatorname{Re}^2\left[\widehat{f}_\delta\right]\right\}^{1/2}. \end{aligned} \tag{50}$$

Stationarization of $\|f\|_\square$ with respect to δ and ς leads, respectively, to

$$\begin{aligned} &\frac{\partial}{\partial\delta}\|f\|_{\succ c} \\ &= \frac{\delta^2}{4\pi^2 M^2} \\ &\times \sum_{\substack{k=-N\\k\neq 0}}^{N} \sum_{m=0}^{M-1} \frac{m^2}{\sqrt{A_{k,m}^2 + B_{k,m}^2}} \\ &\quad \times \left\{\left\langle \operatorname{Re}\left[\widehat{f}_{\delta,\varsigma}\right] + \operatorname{Im}\left[\widehat{f}_{\delta,\varsigma}\right] - \operatorname{Re}\left[\widehat{f}_\delta\right]\right\rangle\right. \\ &\quad \times \frac{\partial}{\partial\delta}\left\langle \operatorname{Re}\left[\widehat{f}_{\delta,\varsigma}\right] + \operatorname{Im}\left[\widehat{f}_{\delta,\varsigma}\right] - \operatorname{Re}\left[\widehat{f}_\delta\right]\right\rangle \\ &\quad \left. + \operatorname{Re}\left[\widehat{f}_\delta\right]\frac{\partial}{\partial\delta}\operatorname{Re}\left[\widehat{f}_\delta\right]\right\} = 0, \end{aligned} \tag{51}$$

$$\begin{aligned} &\frac{\partial}{\partial\varsigma}\|f\|_{\succ c} \\ &= \frac{\delta^2}{4\pi^2 M^2} \sum_{\substack{k=-N\\k\neq 0}}^{N} \sum_{m=0}^{M-1} \frac{m^2}{\sqrt{A_{k,m}^2 + B_{k,m}^2}} \\ &\quad \times \left\langle \operatorname{Re}\left[\widehat{f}_{\delta,\varsigma}\right] + \operatorname{Im}\left[\widehat{f}_{\delta,\varsigma}\right] - \operatorname{Re}\left[\widehat{f}_\delta\right]\right\rangle \\ &\quad \times \frac{\partial}{\partial\varsigma}\left\langle \operatorname{Re}\left[\widehat{f}_{\delta,\varsigma}\right] + \operatorname{Im}\left[\widehat{f}_{\delta,\varsigma}\right]\right\rangle = 0. \end{aligned} \tag{52}$$

Obviously, increasing M and N (or decreasing δ), which increases the smoothness of $\mathfrak{D}F_{\succ c}^{-1}(\lambda, m)$, accelerates the convergence of the series in (51) and (52) to zero. This is accompanied by accelerating the stationarization of $(\partial/\partial\delta)\|f\|_{\succ c}$ and $(\partial/\partial\varsigma)\|f\|_{\succ c}$, with a rate $\sim M^{-2}N^{-2}$, and to stabilizing $\mathfrak{D}F_{\succ c}^{-1}(\lambda, m)$ with respect to variations in δ and ς. This is a fact that illustrates the consistency of this proposition and its underlying Conjecture 11. The same arguments are also applicable to the rate optimization of every other $\mathfrak{D}F_\square^{-1}(\lambda, m)$.

Lemma 13. *A rate-optimal* (δ, ς) *pair for the DSF series inverse* $\mathfrak{D}F_{\succ c}^{-1}(\lambda, m)$ *exists when the system* (51)-(52) *is satisfied.*

Example 14. To illustrate the previous lemma, let us revisit Example 7 with $\widehat{f}(\mu) = 2/(1+\mu^2)$, to determine the rate-optimal values of δ and ς for it. In the notation of Proposition 12, we have

$$\begin{gathered} \operatorname{Re}\left[\widehat{f}_\delta\right] = \frac{2}{1 + m^2k^2\delta^2}, \\ \operatorname{Re}\left[\widehat{f}_{\delta,\varsigma}\right] = \frac{2}{1 + [m+\varsigma]^2 k^2\delta^2}, \qquad \operatorname{Im}\left[\widehat{f}_{\delta,\varsigma}\right] = 0. \end{gathered} \tag{53}$$

First, since

$$\begin{gathered} \frac{\partial}{\partial\delta}\operatorname{Re}\left[\widehat{f}_\delta\right] = -4\frac{m^2k^2\delta}{\left(1 + m^2k^2\delta^2\right)^2}, \\ \frac{\partial}{\partial\delta}\operatorname{Re}\left[\widehat{f}_{\delta,\varsigma}\right] = -4\frac{[m+\varsigma]^2 k^2\delta}{\left(1 + [m+\varsigma]^2 k^2\delta^2\right)^2}, \end{gathered} \tag{54}$$

then

$$\langle \mathrm{Re}[\hat{f}_{\delta,\varsigma}] - \mathrm{Re}[\hat{f}_\delta] \rangle = -2\frac{\varsigma[2m+\varsigma]}{(1+m^2k^2\delta^2)(1+[m+\varsigma]^2k^2\delta^2)},$$

$$\frac{\partial}{\partial\delta}\langle \mathrm{Re}[\hat{f}_{\delta,\varsigma}] - \mathrm{Re}[\hat{f}_\delta] \rangle = -4\left\langle \frac{[m+\varsigma]^2}{(1+[m+\varsigma]^2k^2\delta^2)^2} - \frac{m^2}{(1+m^2k^2\delta^2)^2} \right\rangle k^2\delta. \quad (55)$$

Consider further these relations in (51) to rewrite it as

$$\frac{\partial}{\partial\delta}\|f\|_{\succ c} = \frac{2\delta^3}{\pi^2M^2}\sum_{\substack{k=-N\\k\neq0}}^{N}\sum_{m=0}^{M-1}\frac{m^2k^2}{\sqrt{A_{k,m}^2+B_{k,m}^2}(1+m^2k^2\delta^2)} \times\left\{\frac{\varsigma[2m+\varsigma]}{(1+[m+\varsigma]^2k^2\delta^2)} \times\left\langle \frac{[m+\varsigma]^2}{(1+[m+\varsigma]^2k^2\delta^2)^2} - \frac{m^2}{(1+m^2k^2\delta^2)^2} \right\rangle\right\} = 0. \quad (56)$$

This nonlinear relation in δ can be solved only numerically, for each ς, to yield the rate-optimal δ (i.e., the optimal Λ) for any given N.

Second, we consider

$$\frac{\partial}{\partial\varsigma}\mathrm{Re}[\hat{f}_{\delta,\varsigma}] = -4\frac{k^2\delta^2[m+\varsigma]}{(1+[m+\varsigma]^2k^2\delta^2)^2}, \qquad \frac{\partial}{\partial\varsigma}\mathrm{Re}[\hat{f}_\delta] = 0, \quad (57)$$

in (52) to rewrite it as

$$\frac{\partial}{\partial\varsigma}\|f\|_{\succ c} = \frac{2\delta^4\varsigma}{\pi^2M^2} \times\sum_{\substack{k=-N\\k\neq0}}^{N}\sum_{m=0}^{M-1}\frac{k^2m^2}{\sqrt{A_{k,m}^2+B_{k,m}^2}(1+m^2k^2\delta^2)} \times\frac{[2m+\varsigma][m+\varsigma]}{(1+[m+\varsigma]^2k^2\delta^2)^3} = 0. \quad (58)$$

This can be easily resolved in ς and rearranged into the ratio

$$\varsigma = -\sum_{\substack{k=-N\\k\neq0}}^{N}\sum_{m=0}^{M-1}\frac{[2m+\varsigma]k^2m^3}{\sqrt{A_{k,m}^2+B_{k,m}^2}(1+m^2k^2\delta^2)(1+[m+\varsigma]^2k^2\delta^2)^3} \times\left(\sum_{\substack{k=-N\\k\neq0}}^{N}\sum_{m=0}^{M-1}\frac{[2m+\varsigma]k^2m^2}{\sqrt{A_{k,m}^2+B_{k,m}^2}(1+m^2k^2\delta^2)(1+[m+\varsigma]^2k^2\delta^2)^3}\right)^{-1} \quad (59)$$

for the rate-optimal ς.

Despite its complexity, the previous relation appears as a nonlinearly weighted negative mean of m, which may always be approximated by some fraction of M, like $-M/2$, say. Finally, the rate-optimal (δ,ς) pair for the $DF_{\succ c}^{-1}(\lambda,m)$ inverse emerges as the solution of the simultaneous system of nonlinear equations (56)-(59), a solution that may not necessarily be unique.

6. Recursive Maximum-Entropy Reconstruction of the DSF Series Inverse

The possible nonuniqueness of the rate-optimal (δ,ς) pair of Lemma 13 calls for imposing an additional weak (inequality) constraint to select a robust value for this pair, which may account for all approximations that are involved in the development of $\mathfrak{D}F_{\square}^{-1}(\lambda,m)$. The constraint must be the satisfaction of the Heisenberg uncertainty principle (HUP); see, for example, [9], when rate-optimizing δ and ς. The HUP essentially states that $f(t) = \mathfrak{D}F_{\square}^{-1}(\lambda,m)$ and $\hat{f}(\mu)$ cannot be both arbitrarily concentrated. In information theory, the definitive measure of concentration of a probability density function, $\phi(t)$, is the Boltzman entropy

$$H(\phi) = -\int_R \phi(t)\ln\phi(t)\,dt. \quad (60)$$

Hirschman Jr. [10, 11] observed in 1957 that Weyl's formulation of the HUP is a consequence of a stronger version of the inequality of this principle, namely, the entropy version

$$H(|\phi|^2)H(|\hat{\phi}|^2) \geq \ln\frac{e}{2}, \quad (61)$$

when $\|\phi\|_2 = 1$ and $\hat{\phi}(\mu) \geq 0$. Hirschman also conjectured that the minimizers of the above sharp inequality were Gaussians, as in the case of the HUP.

To state the main result of this section, we adopt the notation

$$Q = \int_R \mathfrak{D}F_{\succ c}^{-1}(\lambda, m)\, dt,$$
$$\|\hat{f}\|_2^2 = \int_R |\hat{f}|^2 dt, \tag{62}$$

the square of the $p = 2$ norm, which satisfies Parseval's identity: $\|\hat{f}\|_2 = \|f\|_2$.

Definition 15. The DSF Taxi-cab norm ${}_1\|f\|_\square$ of $f(t) \in \mathcal{F}(R)$ is the sum of the absolute values of all scaled-harmonics appearing in the $\mathfrak{D}F_{\succ c}^{-1}(\lambda, m)$ expansion; that is,

$${}_1\|\mathrm{f}\|_\square = \sum_{\substack{k=-N \\ k\neq 0}}^{N} \sum_{m=0}^{M-1} |A_{k,m;\square}| + |B_{k,m;\square}|, \tag{63}$$

where $A_{k,m;\square}$ and $B_{k,m;\square}$ are, respectively, defined by (25)-(26) or (27)-(28) and (33)-(34) or (36)-(37).

Lemma 16. *A rate-optimal pair (δ, ς) for $\mathfrak{D}F_{\succ c}^{-1}(\lambda, m)$ with a known Q should satisfy*

$${}_1\|\mathfrak{D}F_{\succ c}^{-1}(\lambda, m)\|_{\succ c}^2 \left[\ln\, {}_1\|\mathfrak{D}F_{\succ c}^{-1}(\lambda, m)\|_{\succ c}^2 - \ln Q^2\right] > \frac{Q^4 \ln(e/2)}{\left[H\left(Q^{-2}|\hat{f}|^2\right) - \|\hat{f}\|_2^2 \ln Q^{-2}\right]}. \tag{64}$$

Proof. Conceive $f(t)/Q$ as a probability density function $\phi(t)$. Moreover, if $f(t) \overset{FT}{\leftrightarrow} \hat{f}(\mu)$, then $f(t)/Q \overset{FT}{\leftrightarrow} \hat{f}(\mu)/Q$. Hence in the notation of our DSF series $\mathfrak{D}F_{\succ c}^{-1}(\lambda, m)$, even when it is not necessarily > 0, the previous entropy version for the uncertainty principle writes as

$$H\left(Q^{-2}\left|\mathfrak{D}F_{\succ c}^{-1}(\lambda, m)\right|^2\right) \geq \frac{\ln(e/2)}{H\left(Q^{-2}|\hat{f}|^2\right)}, \tag{65}$$

where

$$H\left(Q^{-2}\left|\mathfrak{D}F_{\succ c}^{-1}(\lambda, m)\right|^2\right) = 2\int_R \left|Q^{-2} \sum_{k\in W_o} \sum_{m=0}^{M-1} A_{k,m} \cos\varphi_{k,m;\delta} t + B_{k,m} \sin\varphi_{k,m;\delta} t\right|^2 \times \ln\left|Q^{-2} \sum_{k\in W_o} \sum_{m=0}^{M-1} A_{k,m} \cos\varphi_{k,m;\delta} t + B_{k,m} \sin\varphi_{k,m;\delta} t\right| dt, \tag{66}$$

$$H\left(Q^{-2}|\hat{f}|^2\right) = Q^{-2}\left[H\left(|\hat{f}|^2\right) - \|\hat{f}\|_2^2 \ln Q^2\right]. \tag{67}$$

Clearly

$$H\left(Q^{-2}\left|\mathfrak{D}F_{\succ c}^{-1}(\lambda, m)\right|^2\right) < \left|Q^{-2} \sum_{k\in W_o} \sum_{m=0}^{M-1} (A_{k,m} + B_{k,m})\right|^2 \times \ln\left|Q^{-2} \sum_{k\in W_o} (A_{k,m} + B_{k,m})\right|^2 T(k, m; \delta), \tag{68}$$

where

$$T(k, m; \delta) = \begin{cases} \dfrac{2\pi}{m(k\delta + 1)}, & m \neq 0 \\ 0, & m = 0. \end{cases} \tag{69}$$

Then replacement of $T(k, m; \delta)$ by 1 in the previous inequality reduces it to

$$H\left(Q^{-2}\left|\mathfrak{D}F_{\succ c}^{-1}(\lambda, m)\right|^2\right) < \left|Q^{-2} \sum_{k\in W_o} \sum_{m=0}^{M-1} (A_{k,m} + B_{k,m})\right|^2 \times \ln\left|Q^{-2} \sum_{k\in W_o} \sum_{m=0}^{M-1} (A_{k,m} + B_{k,m})\right|^2. \tag{70}$$

Now since

$$\left|Q^{-2} \sum_{k\in W_o} \sum_{m=0}^{M-1} A_{k,m} + B_{k,m}\right|^2 \leq {}_1\|Q^{-1}\mathfrak{D}F_{\succ c}^{-1}(\lambda, m)\|_{\succ c}^2 = \left\{\sum_{\substack{k=-N \\ k\neq 0}}^{N} \sum_{m=0}^{M-1} \left|\frac{A_{k,m}}{Q}\right| + \left|\frac{B_{k,m}}{Q}\right|\right\}^2, \tag{71}$$

then

$$H\left(Q^{-2}\left|\mathfrak{D}F_{\succ c}^{-1}(\lambda, m)\right|^2\right) < {}_1\|Q^{-1}\mathfrak{D}F_{\succ c}^{-1}(\lambda, m)\|_{\succ c}^2 \ln {}_1\|Q^{-1}\mathfrak{D}F_{\succ c}^{-1}(\lambda, m)\|_{\succ c}^2, \tag{72}$$

and relation (65) becomes

$${}_1\|Q^{-1}\mathfrak{D}F_{\succ c}^{-1}(\lambda, m)\|_{\succ c}^2 \ln {}_1\|Q^{-1}\mathfrak{D}F_{\succ c}^{-1}(\lambda, m)\|_{\succ c}^2 > \frac{\ln(e/2)}{H\left(Q^{-2}|\hat{f}|^2\right)}. \tag{73}$$

Furthermore, since

$${}_1\|Q^{-1}\mathfrak{D}F_{\succ c}^{-1}(\lambda, m)\|_{\succ c}^2 = Q^{-2}\, {}_1\|\mathfrak{D}F_{\succ c}^{-1}(\lambda, m)\|_{\succ c}^2, \tag{74}$$

then the left-hand side of (72) becomes

$$Q^{-2}\, {}_1\|\mathfrak{D}F_{\succ c}^{-1}(\lambda, m)\|_{\succ c}^2 \left[\ln\, {}_1\|\mathfrak{D}F_{\succ c}^{-1}(\lambda, m)\|_{\succ c}^2 - \ln Q^2\right]. \tag{75}$$

Finally, consideration of this relation together with (67) in (72) leads to the required result.

Since the entropy $H(Q^{-2}|\mathfrak{D}F_{\succ c}^{-1}(\lambda,m)|^2)$ is a function of the (δ, ς) free parameters, then variation of these parameters can always be forced towards maximization of this entropy. As is well known, see, for example, [12], a maximum-entropy $\mathfrak{D}F_{\succ c}^{-1}(\lambda,m)$ should be a most certain inverse of $\widehat{f}(\mu)$ and with best achievable smoothness.

Theorem 17. *A robust rate-optimal pair (δ, ς), for $\mathfrak{D}F_{\succ c}^{-1}(\lambda,m)$ with a known Q, is a solution to the nonlinear programming problem*

$$\begin{aligned} &Minimize \quad {}_1\left\|Q^{-1}\mathfrak{D}F_{\succ c}^{-1}(\lambda,m)\right\|_{\succ c}^2 \ln {}_1\left\|Q^{-1}\mathfrak{D}F_{\succ c}^{-1}(\lambda,m)\right\|_{\succ c}^2 \\ &Subject\ to: \quad (64), (51)\text{-}(52). \end{aligned} \tag{76}$$

Proof. By consideration of relation (72) of Lemma 16 when maximizing the entropy $H(Q^{-2}|\mathfrak{D}F_{\succ c}^{-1}(\lambda,m)|^2)$, subject to the inequality constraint (64) and the equality constraints (51) and (52). □

Q, however, is not apriori known. Hence, we may assume that δ_r, ς_r correspond to an rth iteration, Q_r, for Q and $\mathfrak{D}F_{\succ c}^{-1}(\lambda,m;\delta_r,\varsigma_r)$ is $\mathfrak{D}F_{\succ c}^{-1}(\lambda,m)$ that corresponds to Q_r. These concepts invoke a recursive determination process, indexed by $r = 0, 1, 2, 3, \ldots, L$, with L as the termination number, defined, namely, as

$$Q_{L+1} - Q_L \le \epsilon, \tag{77}$$

where $0 \le \epsilon \ll 1$ is an acceptable tolerance.

Proposition 18. *A robust rate-optimal pair $(\delta, \varsigma) = \lim_{r \to L}(\delta_r, \varsigma_r)$ for $\mathfrak{D}F_{\succ c}^{-1}(\lambda,m)$ exists when (δ_r, ς_r) is a solution to the recursive nonlinear programming problem*

$$\begin{aligned} &Minimize \quad {}_1\left\|Q_r^{-1}\mathfrak{D}F_{\succ c}^{-1}(\lambda,m;\delta_r,\varsigma_r)\right\|_{\succ c}^2 \\ &\qquad\qquad \times \ln {}_1\left\|Q_r^{-1}\mathfrak{D}F_{\succ c}^{-1}(\lambda,m;\delta_r,\varsigma_r)\right\|_{\succ c}^2 \\ &Subject\ to: \quad (64), (51)\text{-}(52), \\ &\qquad\qquad with\ \mathfrak{D}F_{\succ c}^{-1}(\lambda,m;\delta_r,\varsigma_r)\ replacing\ \mathfrak{D}F_{\succ c}^{-1}(\lambda,m), \\ &\qquad\qquad Q_{r+1} = \int_R \mathfrak{D}F_{\succ c}^{-1}(\lambda,m;\delta_r,\varsigma_r)\,dt, \end{aligned} \tag{78}$$

and arbitrary $Q_0 > 0$.

Due to convexity of the inequality constraint (64), the nonlinear programming problem of Theorem 17 should be solvable by the method of Kuhn-Tucker multipliers [13]. The rather large computational complexity of the above proposition is comparable to that of the alternative wavelet analysis. It should, however, not represent a difficulty in our present times, when powerful computing resources are available or accessible almost everywhere.

7. Conclusion

In this paper, we have demonstrated how evaluation of the inverse Fourier transform can be variably scaled. The reported DSF representation, for, say, high frequency resolving scaling, bypasses the need for quadratures in computing (1). Instead, a repeated evaluation of the real and imaginary parts of $\widehat{f}$, when evaluating the $A_{k,m}$ and $B_{k,m}$ (or the $A_{k,m}^{+}$ and $B_{k,m}^{+}$) coefficients, of (24)–(26), is needed at a large enough number of scale-frequency points (m, k), as illustrated in Sections 4 and 5. Low frequency resolving scaling also requires a similar computational effort with the real and imaginary parts of $\widehat{f}$ when evaluating the $\mathfrak{A}_{k,m}$ and $\mathfrak{B}_{k,m}$ (or the $\mathfrak{A}_{k,m}^{+}$ and $\mathfrak{B}_{k,m}^{+}$) coefficients of (32)–(34).

Determination of the rate-optimal scaling factor step δ and frequency gap number ς, in DSF series has been demonstrated to invoke solving a recursive nonlinear programming problem that maximizes entropy in the framework of the uncertainty principle. It should be emphasized that the scale-frequency approach to inversion of the Fourier transform is a new concept that is not necessariliy fast. In that sense it is not expected to pose a challenge to the FFT algorithm. Its error analysis is, however, of crucial importance and will be a subject for a future publication.

Conflict of Interests

The author declares that there is no conflict of interests regarding the publication of this paper.

Acknowledgment

The author is grateful to an anonymous referee for his critical reading of the original typescript.

References

[1] E. C. Titchmarsh, *Introduction to the Theory of Fourier Integrals*, Oxford University Press, Oxford, UK, 1948.

[2] L. Grafakos, *Classical and Modern Fourier Analysis*, Pearson Education, Cranbury, NJ, USA, 2004.

[3] V. I. Krylov and L. G. Kruglikova, *Handbook of Numerical Harmonic Analysis*, IPST Press, Jerusalem, Israel, 1969.

[4] H. Stark and Y. Yang, *Vector Space Methods: A Numerical Approach to Signal and Image Processing, Neural Nets, and Optics*, John Wiley & Sons, New York, NY, USA, 1998.

[5] A. Papoulis, "A new algorithm in spectral analysis and band-limited extrapolation," *IEEE Transactions on Circuits and Systems*, vol. 22, no. 9, pp. 735–742, 1975.

[6] R. Young, *An Introduction to Nonharmonic Fourier Series*, Academic Press, San Diego, Calif, USA, 2001.

[7] N. H. S. Haidar, "Applied scale-frequency analysis," *Advances and Applications in Mathematical Sciences*, vol. 7, no. 2, pp. 141–175, 2010.

[8] D. Šiljak, *Nonlinear Systems*, John Wiley & Sons, New York, NY, USA, 1969.

[9] N. H. S. Haidar, "A Venn diagram approach to Heisenberg inequalities," *Applied Physics Research*, vol. 4, no. 2, pp. 253–257, 2012.

[10] I. I. Hirschman Jr., "A note on entropy," *American Journal of Mathematics*, vol. 97, no. 1, pp. 152–156, 1957.

[11] D. L. Donoho and P. B. Stark, "Uncertainty principles and signal recovery," *SIAM Journal on Applied Mathematics*, vol. 49, no. 3, pp. 906–931, 1989.

[12] N. H. S. Haidar, "WKB theory for sampled unfolding," *Communications on Applied Nonlinear Analysis*, vol. 7, no. 3, pp. 31–46, 2000.

[13] G. Hadley, *Nonlinear and Dynamic Programming*, Adddison-Wesley, NY, New York, USA, 1964.

Several New Third-Order and Fourth-Order Iterative Methods for Solving Nonlinear Equations

Anuradha Singh and J. P. Jaiswal

Department of Mathematics, Maulana Azad National Institute of Technology, Bhopal 462051, India

Correspondence should be addressed to J. P. Jaiswal; asstprofjpmanit@gmail.com

Academic Editor: Viktor Popov

In order to find the zeros of nonlinear equations, in this paper, we propose a family of third-order and optimal fourth-order iterative methods. We have also obtained some particular cases of these methods. These methods are constructed through weight function concept. The multivariate case of these methods has also been discussed. The numerical results show that the proposed methods are more efficient than some existing third- and fourth-order methods.

1. Introduction

Newton's iterative method is one of the eminent methods for finding roots of a nonlinear equation:

$$f(x) = 0. \tag{1}$$

Recently, researchers have focused on improving the order of convergence by evaluating additional functions and first derivative of functions. In order to improve the order of convergence and efficiency index, many modified third-order methods have been obtained by using different approaches (see [1–3]). Kung and Traub [4] presented a hypothesis on the optimality of the iterative methods by giving 2^{n-1} as the optimal order. It means that the Newton iteration by two function evaluations per iteration is optimal with 1.414 as the efficiency index. By using the optimality concept, many researchers have tried to construct iterative methods of optimal higher order of convergence. The order of the methods discussed above is three with three function evaluations per full iteration. Clearly its efficiency index is $3^{1/3} \approx 1.442$, which is not optimal. Very recently, the concept of weight functions has been used to obtain different classes of third- and fourth-order methods; one can see [5–7] and the references therein.

This paper is organized as follows. In Section 2, we present a new class of third-order and fourth-order iterative methods by using the concept of weight functions, which includes some existing methods and also provides some new methods. We have extended some of these methods for multivariate case. Finally, we employ some numerical examples and compare the performance of our proposed methods with some existing third- and fourth-order methods.

2. Methods and Convergence Analysis

First we give some definitions which we will use later.

Definition 1. Let $f(x)$ be a real valued function with a simple root α and let x_n be a sequence of real numbers that converge towards α. The order of convergence m is given by

$$\lim_{n\to\infty} \frac{x_{n+1} - \alpha}{(x_n - \alpha)^m} = \zeta \neq 0, \tag{2}$$

where ζ is the asymptotic error constant and $m \in R^+$.

Definition 2. Let n be the number of function evaluations of the new method. The efficiency of the new method is measured by the concept of efficiency index [8, 9] and defined as

$$m^{1/n}, \tag{3}$$

where m is the order of convergence of the new method.

2.1. Third-Order Iterative Methods. To improve the order of convergence of Newton's method, some modified methods are given by Grau-Sánchez and Díaz-Barrero in [10], Weerakoon and Fernando in [1], Homeier in [2], Chun and Kim in [3], and so forth. Motivated by these papers, we consider the following two-step iterative method:

$$\begin{aligned} y_n &= x_n - a\frac{f(x_n)}{f'(x_n)}, \\ x_{n+1} &= x_n - A(t)\frac{f(x_n)}{f'(x_n)}, \end{aligned} \tag{4}$$

where $t = f'(y_n)/f'(x_n)$ and a is a real constant. Now we find under what conditions it is of order three.

Theorem 3. *Let α be a simple root of the function f and let f have suffici t number of continuous derivatives in a neighborhood of α. The method (4) has third-order convergence, when the weight function $A(t)$ satisfies the following conditions:*

$$A(1) = 1, \qquad A'(1) = -\frac{1}{2a}, \qquad |A''(1)| \le +\infty. \tag{5}$$

Proof. Suppose $e_n = x_n - \alpha$ is the error in the nth iteration and $c_h = f^{(h)}(\alpha)/h!f'(\alpha)$, $h \ge 1$. Expanding $f(x_n)$ and $f'(x_n)$ around the simple root α with Taylor series, then we have

$$\begin{aligned} f(x_n) &= f'(\alpha) \\ &\quad \times \left[e_n + c_2e_n^2 + c_3e_n^3 + c_4e_n^4 + c_5e_n^5 + O\left(e_n^6\right)\right], \\ f'(x_n) &= f'(\alpha) \\ &\quad \times \left[1 + 2c_2e_n + 3c_3e_n^2 + 4c_4e_n^3 + 5c_5e_n^4 + O\left(e_n^5\right)\right]. \end{aligned} \tag{6}$$

Now it can be easily found that

$$\frac{f(x_n)}{f'(x_n)} = e_n - c_2e_n^2 + \left(2c_2^2 - 2c_3\right)e_n^3 + O\left(e_n^4\right). \tag{7}$$

By using (7) in the first step of (4), we obtain

$$y_n = \alpha + (1-a)e_n + ac_2e_n^2 + 2a\left(c_3 - c_2^2\right)e_n^3 + O\left(e_n^4\right). \tag{8}$$

At this stage, we expand $f'(y_n)$ around the root by taking (8) into consideration. We have

$$\begin{aligned} &f'(y_n) \\ &= f'(\alpha)\left[1 + 2(1-a)c_2e_n + \left(2ac_2^2 + 3(1-a)^2c_3\right)e_n^2\right. \\ &\quad + \left(6(1-a)ac_2c_3 + 4ac_2\left(-c_2^2 + c_3\right) + 4(1-a)^3c_4\right) \\ &\quad \left.\times e_n^3 + O\left(e_n^4\right)\right]. \end{aligned} \tag{9}$$

Furthermore, we have

$$\begin{aligned} \frac{f'(y_n)}{f'(x_n)} &= 1 + \left\{-2c_2 + 2(1-a)c_2\right\}e_n \\ &\quad + \left\{4c_2^2 - 4(1-a)c_2^2 + 2ac_2^2 - 3c_3 + 3(1-a)^2c_3\right\} \\ &\quad \times e_n^2 + \cdots + O\left(e_n^4\right). \end{aligned} \tag{10}$$

By virtue of (10) and (4), we get

$$\begin{aligned} A(t)\times\frac{f(x_n)}{f'(x_n)} &= e_n - \left[\left(\frac{3a}{2} - 1\right)c_3 - 2c_2^2\left(-1 + a^2A''(1)\right)\right]e_n^3 \\ &\quad + O\left(e_n^4\right). \end{aligned} \tag{11}$$

Hence, from (11) and (4) we obtain the following general equation, which has third-order convergence:

$$\begin{aligned} e_{n+1} &= x_{n+1} - \alpha \\ &= x_n - A(t)\times\frac{f(x_n)}{f'(x_n)} - \alpha \\ &= \left[\left(\frac{3a}{2} - 1\right)c_3 - 2c_2^2\left(-1 + a^2A''(1)\right)\right]e_n^3 + O\left(e_n^4\right). \end{aligned} \tag{12}$$

This proves the theorem.

Particular Cases. To find different third-order methods we take $a = 2/3$ in (4).

Case 1. If we take $A(t) = (7 - 3t)/4$ in (4), then we get the formula:

$$\begin{aligned} y_n &= x_n - \frac{2}{3}\frac{f(x_n)}{f'(x_n)}, \\ x_{n+1} &= x_n - \left(\frac{7}{4} - \frac{3}{4}\frac{f'(y_n)}{f'(x_n)}\right)\frac{f(x_n)}{f'(x_n)}, \end{aligned} \tag{13}$$

and its error equation is given by

$$2c_2^2e_n^3 + \left(-9c_2^3 + 7c_2c_3 + \frac{c_4}{9}\right)e_n^4 + O\left(e_n^5\right). \tag{14}$$

Case 2. If we take $A(t) = 4t/(7t - 3)$ in (4), then we get the formula:

$$\begin{aligned} y_n &= x_n - \frac{2}{3}\frac{f(x_n)}{f'(x_n)}, \\ x_{n+1} &= x_n - \left(\frac{4f'(y_n)}{7f'(y_n) - 3f'(x_n)}\right)\frac{f(x_n)}{f'(x_n)}, \end{aligned} \tag{15}$$

and its error equation is given by

$$-\frac{1}{3}c_2^2e_n^3 + \frac{1}{9}\left(17c_2^3 - 21c_2c_3 + c_4\right)e_n^4 + O\left(e_n^5\right). \tag{16}$$

Case 3. If we take $A(t) = 4/(1+3t)$ in (4), then we get the formula:

$$y_n = x_n - \frac{2}{3}\frac{f(x_n)}{f'(x_n)}, \qquad x_{n+1} = x_n - \frac{4f(x_n)}{f'(x_n)+3f'(y_n)}, \tag{17}$$

and its error equation is given by

$$c_2^2 e_n^3 + \left(-3c_2^3 + 3c_2c_3 + \frac{c_4}{9}\right)e_n^4 + O\left(e_n^5\right). \tag{18}$$

Case 4. If we take $A(t) = (t+7)/(1+7t)$ in (4), then we get the formula:

$$y_n = x_n - \frac{2}{3}\frac{f(x_n)}{f'(x_n)}, \qquad x_{n+1} = x_n - \left(\frac{f'(y_n)+7f'(x_n)}{f'(x_n)+7f'(y_n)}\right)\frac{f(x_n)}{f'(x_n)}, \tag{19}$$

and its error equation is given by

$$\frac{5}{6}c_2e_n^3 + \frac{1}{36}\left(-79c_2^3 + 84c_2c_3 + 4c_4\right)e_n^4 + O\left(e_n^5\right). \tag{20}$$

Case 5. If we take $A(t) = (t+3)/4t$ in (4), then we get the formula:

$$y_n = x_n - \frac{2}{3}\frac{f(x_n)}{f'(x_n)}, \qquad x_{n+1} = x_n - \frac{f(x_n)}{4}\left(\frac{1}{f'(x_n)} + \frac{3}{f'(y_n)}\right), \tag{21}$$

which is Huen's formula [11].

Remark 4. By taking different values of a and weight function $A(t)$ in (4), one can get a number of third-order iterative methods.

2.2. Optimal Fourth-Order Iterative Methods. The order of convergence of the methods obtained in the previous subsection is three with three function evaluations (one function and two derivatives) per step. Hence its efficiency index is $3^{1/3} \approx 1.442$, which is not optimal. To get optimal fourth-order methods we consider

$$y_n = x_n - a\frac{f(x_n)}{f'(x_n)}, \qquad x_{n+1} = x_n - \{A(t)\times B(t)\}\frac{f(x_n)}{f'(x_n)}, \tag{22}$$

where $A(t)$ and $B(t)$ are two real-valued weight functions with $t = f'(y_n)/f'(x_n)$ and a is a real constant. The weight functions should be chosen in such a way that the order of convergence arrives at optimal level four without using additional function evaluations. The following theorem indicate the required conditions for the weight functions and constant a in (22) to get optimal fourth-order convergence.

Theorem 5. *Let α be a simple root of the function f and let f have suffici t number of continuous derivatives in a neighborhood of α. The method (22) has fourth-order convergence, when $a = 2/3$ and the weight functions $A(t)$ and $B(t)$ satisfy the following conditions:*

$$A(1) = 1, \quad A'(1) = -\frac{3}{4}, \quad \left|A^{(3)}(1)\right| \le +\infty,$$
$$B(1) = 1, \quad B'(1) = 0, \tag{23}$$
$$B''(1) = \frac{9}{4} - A''(1), \quad \left|B^{(3)}(1)\right| \le +\infty.$$

Proof. Using (6) and putting $a = 2/3$ in the first step of (22), we have

$$y_n = \alpha + \frac{e_n}{3} + \frac{2c_2e_n^2}{3} + \frac{4\left(c_3 - c_2^2\right)e_n^3}{3} + \cdots + O\left(e_n^5\right). \tag{24}$$

Now we expand $f'(y_n)$ around the root by taking (24) into consideration. Thus, we have

$$f'(y_n) = f'(\alpha)\left[1 + \frac{2c_2e_n}{3} + \frac{\left(4c_2^2 + c_3\right)e_n^2}{3} + \cdots + O\left(e_n^5\right)\right]. \tag{25}$$

Furthermore, we have

$$\frac{f'(y_n)}{f'(x_n)} = 1 - \frac{4c_2}{3}e_n + \left(4c_2^2 - \frac{8c_3}{3}\right)e_n^2 + \cdots + O\left(e_n^5\right). \tag{26}$$

By virtue of (26) and (22), we obtain

$$\begin{aligned}\{A(t)\times B(t)\}\frac{f(x_n)}{f'(x_n)} &= e_n - \frac{1}{81}\left[-81c_2c_3 + 9c_4 + \left(243 + 72A''(1) + 32A'''(1) + 32B'''(1)\right)c_2^3\right] \\ &\quad\times e_n^4 + O\left(e_n^5\right).\end{aligned} \tag{27}$$

Finally, from (27) and (22) we can have the following general equation, which reveals the fourth-order convergence:

$$\begin{aligned} e_{n+1} &= x_{n+1} - \alpha \\ &= x_n - \{A(t)\times B(t)\}\frac{f(x_n)}{f'(x_n)} - \alpha \\ &= \frac{1}{81}\left[-81c_2c_3 + 9c_4 + \left(243 + 72A''(1) + 32A'''(1) + 32B'''(1)\right)c_2^3\right]e_n^4 + O\left(e_n^5\right).\end{aligned} \tag{28}$$

It proves the theorem.

Particular Cases

Method 1. If we take $A(t) = (t+3)/4t$ and $B(t) = ((11/8) - (3/4)t + (3/8)t^2)$, where $t = f'(y)/f'(x)$, then the iterative method is given by

$$y_n = x_n - \frac{2}{3}\frac{f(x_n)}{f'(x_n)},$$
$$x_{n+1} = x_n - \left[\frac{11}{8} - \frac{3}{4}\frac{f'(y_n)}{f'(x_n)} + \frac{3}{8}\left(\frac{f'(y_n)}{f'(x_n)}\right)^2\right] \times \left(\frac{1}{f'(x_n)} + \frac{3}{f'(y_n)}\right)\frac{f(x_n)}{4}, \quad (29)$$

and its error equation is given by

$$e_{n+1} = \frac{1}{9}\left[23c_2^3 - 9c_2c_3 + c_4\right]e_n^4 + O\left(e_n^5\right). \quad (30)$$

Method 2. If we take $A(t) = (7-3t)/4$ and $B(t) = ((17/8) - (9/4)t + (9/8)t^2)$, where $t = f'(y)/f'(x)$, then the iterative method is given by

$$y_n = x_n - \frac{2}{3}\frac{f(x_n)}{f'(x_n)},$$
$$x_{n+1} = x_n - \left[\frac{17}{8} - \frac{9}{4}\frac{f'(y_n)}{f'(x_n)} + \frac{9}{8}\left(\frac{f'(y_n)}{f'(x_n)}\right)^2\right] \times \left(\frac{7}{4} - \frac{3}{4}\frac{f'(y_n)}{f'(x_n)}\right)\frac{f(x_n)}{f'(x_n)}, \quad (31)$$

and its error equation is given by

$$e_{n+1} = \left[3c_2^3 - c_2c_3 + \frac{c_4}{9}\right]e_n^4 + O\left(e_n^5\right). \quad (32)$$

Method 3. If we take $A(t) = 4t/(7t-3)$ and $B(t) = ((13/16) + (3/8)t - (3/16)t^2)$, where $t = f'(y)/f'(x)$, then the iterative method is given by

$$y_n = x_n - \frac{2}{3}\frac{f(x_n)}{f'(x_n)},$$
$$x_{n+1} = x_n - \left[\frac{13}{16} + \frac{3}{8}\frac{f'(y_n)}{f'(x_n)} - \frac{3}{16}\left(\frac{f'(y_n)}{f'(x_n)}\right)^2\right] \times \left(\frac{4f'(y_n)}{7f'(y_n) - 3f'(x_n)}\right)\frac{f(x_n)}{f'(x_n)}, \quad (33)$$

and its error equation is given by

$$e_{n+1} = \frac{1}{9}\left[-c_2^3 - 9c_2c_3 + c_4\right]e_n^4 + O\left(e_n^5\right). \quad (34)$$

Method 4. If we take $A(t) = 4/(1+3t)$ and $B(t) = ((25/8) - (9/8)t + (9/16)t^2)$, where $t = f'(y)/f'(x)$, then the iterative method is given by

$$y_n = x_n - \frac{2}{3}\frac{f(x_n)}{f'(x_n)},$$
$$x_{n+1} = x_n - \left[\frac{25}{16} - \frac{9}{8}\frac{f'(y_n)}{f'(x_n)} + \frac{9}{16}\left(\frac{f'(y_n)}{f'(x_n)}\right)^2\right] \times \left[\frac{4f(x_n)}{f'(x_n) + 3f'(y_n)}\right], \quad (35)$$

and its error equation is

$$e_{n+1} = \left[3c_2^3 - c_2c_3 + \frac{c_4}{9}\right]e_n^4 + O\left(e_n^5\right). \quad (36)$$

Method 5. If we take $A(t) = 4/(1+3t)$ and $B(t) = 1 + (9/16)(t-1)^2$, where $t = f'(y)/f'(x)$, then the iterative method is given by

$$y_n = x_n - \frac{2}{3}\frac{f(x_n)}{f'(x_n)},$$
$$x_{n+1} = x_n - \left[1 + \frac{9}{16}\left(\frac{f'(y_n)}{f'(x_n)} - 1\right)^2\right] \times \left[\frac{4f(x_n)}{f'(x_n) + 3f'(y_n)}\right], \quad (37)$$

which is same as the formula (11) of [12].

Method 6. If we take $A(t) = (t+7)/(1+7t)$ and $B(t) = ((47/32) - (15/16)t - (15/32)t^2)$, where $t = f'(y)/f'(x)$, then the iterative method is given by

$$y_n = x_n - \frac{2}{3}\frac{f(x_n)}{f'(x_n)},$$
$$x_{n+1} = x_n - \left[\frac{47}{32} - \frac{15}{16}\frac{f'(y_n)}{f'(x_n)} + \frac{15}{32}\left(\frac{f'(y_n)}{f'(x_n)}\right)^2\right] \times \left(\frac{7f'(x_n) + f'(y_n)}{f'(x_n) + 7f'(y_n)}\right)\frac{f(x_n)}{f'(x_n)}, \quad (38)$$

and its error equation is

$$e_{n+1} = \left[\frac{101c_2^3}{36} - c_2c_3 + \frac{c_4}{9}\right]e_n^4 + O\left(e_n^5\right). \quad (39)$$

Remark 6. By taking different values of $A(t)$ and $B(t)$ in (22), one can obtain a number of fourth-order iterative methods.

3. Further Extension to Multivariate Case

In this section, we extend some third- and fourth-order methods from our proposed methods to solve the nonlinear systems. Similarly we can extend other methods also. The multivariate case of our third-order method (15) is given by

$$Y^{(k)} = X^{(k)} - \frac{2}{3}\left[F'\left(X^{(k)}\right)\right]^{-1} F\left(X^{(k)}\right),$$
$$X^{(k+1)} = X^{(k)} - 4\left[7F'\left(Y^{(k)}\right) - 3F'\left(X^{(k)}\right)\right]^{-1} \times F'\left(Y^{(k)}\right)\left\{\left[F'\left(X^{(k)}\right)\right]^{-1} F\left(X^{(k)}\right)\right\}, \quad (40)$$

where $X^{(k)} = [x_1^{(k)}, x_2^{(k)}, \ldots, x_n^{(k)}]^T$, $(k = 0, 1, 2, \ldots)$; similarly $Y^{(k)}$; I is $n \times n$ identity matrix; $F(X^{(k)}) = [f_1(x_1^{(k)}, x_2^{(k)}, \ldots, x_n^{(k)}), f_2(x_1^{(k)}, x_2^{(k)}, \ldots, x_n^{(k)}), \ldots, f_n(x_1^{(k)}, x_2^{(k)}, \ldots, x_n^{(k)})]$; and $F'(X^{(k)})$ is the Jacobian matrix of F at $X^{(k)}$. Let $\xi + H \in \mathfrak{R}^n$ be any point of the neighborhood of exact solution $\xi \in \mathfrak{R}^n$ of the nonlinear system $F(X) = 0$. If Jacobian matrix $F'(\xi)$ is nonsingular, then Taylor's series expansion for multivariate case is given by

$$F(\xi + H) = F'(\xi)\left[H + C_2H^2 + C_3H^3 + \cdots + C_{p-1}H^{p-1}\right] + O\left(H^p\right), \quad (41)$$

where $C_i = [F'(\xi)]^{-1}(F^{(i)}(\xi)/i!)$, $i \geq 2$ and

$$F'(\xi + H) = F'(\xi)\left[I + 2C_2H + 3C_3H^2 + \cdots + (p-1)C_{p-1}H^{p-2}\right] + O\left(H^{p-1}\right), \quad (42)$$

where I is an identity matrix. From the previous equation we can find

$$\left[F'(\xi + H)\right]^{-1} = \left[F'(\xi)\right]^{-1}\left[I + L_1H + L_2H^2 + L_3H^3 + \cdots + L_{p-2}H^{p-2}\right] + O\left(H^{p-1}\right), \quad (43)$$

where $L_1 = -2C_2$, $L_2 = 4C_2^2 - 3C_3$, and $L_3 = -8C_2^3 + 6C_2C_3 + 6C_3C_2 - 4C_4$. Here we denote the error in kth iteration by $E^{(k)}$, that is, $E^{(k)} = X^{(k)} - \xi$. The order of convergence of method (40) can be proved by the following theorem.

Theorem 7. *Let $F : D \subseteq \mathfrak{R}^n \to \mathfrak{R}^n$ be suffici tly Frechet diff rentiable in a convex set D, containing a root ξ of $F(X) = 0$. Let one suppose that $F'(X)$ is continuous and nonsingular in D and $X^{(0)}$ is close to ξ. Then the sequence $\{X^{(k)}\}_{k\geq 0}$ obtained by the iterative expression (40) converges to ξ with order three.*

Proof. For the convenience of calculation, we replace 2/3 by β in the first step of (40). From (41), (42), and (43), we have

$$F\left(X^{(k)}\right) = F'(\xi) \times \left[E^{(k)} + C_2E^{(k)^2} + C_3E^{(k)^3} + C_4E^{(k)^4} + C_5E^{(k)^5}\right] + O\left(E^{(k)^6}\right), \quad (44)$$

$$F'\left(X^{(k)}\right) = F'(\xi) \times \left[I + 2C_2E^{(k)} + 3C_3E^{(k)^2} + 4C_4E^{(k)^3} + 5C_5E^{(k)^4}\right] + O\left(E^{(k)^5}\right), \quad (45)$$

$$\left[F'\left(X^{(k)}\right)\right]^{-1} = \left[F'(\xi)\right]^{-1}\left\{I - 2C_2E^{(k)} + \left(4C_2^2 - 3C_3\right)E^{(k)^2} + \left(-8C_2^3 + 6C_2C_3 + 6C_3C_2 - 4C_4\right) \times E^{(k)^3}\right\} + O\left(E^{(k)^4}\right), \quad (46)$$

where $C_i = [F'(\xi)]^{-1}(F^{(i)}(\xi)/i!)$, $i \geq 2$. Now from (46) and (44), we can obtain

$$S = \left[F'\left(X^{(k)}\right)\right]^{-1} F\left(X^{(k)}\right) = G_1E^{(k)} + G_2E^{(k)^2} + G_3E^{(k)^3} + G_4E^{(k)^4} + O\left(E^{(k)^5}\right), \quad (47)$$

where

$$G_1 = I, \quad G_2 = -C_2, \quad G_3 = -2C_3 + 2C_2^2, \quad G_4 = -3C_4 - 4C_2C_3 + 3C_3C_2 - 4C_2^3. \quad (48)$$

By virtue of (47) the first step of the method (40) becomes

$$Y^{(k)} = (1 - \beta)E^{(k)} + \beta C_2E^{(k)^2} + \beta\left(-2C_2^2 + 2C_3\right)E^{(k)^3} + \beta\left(4C_2^3 - 4C_2C_3 - 3C_3C_2 + 3C_4\right)E^{(k)^4} + O\left(E^{(k)^5}\right). \quad (49)$$

Taylor's series expansion for Jacobian matrix $F'(Y^{(k)})$ can be given as

$$\begin{aligned} F'\left(Y^{(k)}\right) &= F'(\xi)\left[I + 2C_2(1-\beta)E^{(k)} \right. \\ &\quad + \left(2\beta C_2^2 + 3C_3(1-\beta)^2\right)E^{(k)^2} \\ &\quad + \left(-4\beta C_2^3 + 4\beta C_2C_3 + 6\beta(1-\beta)C_3C_2 \right. \\ &\quad \left. + 4C_4(1-\beta)^3\right)E^{(k)^3} \\ &\quad + \left(8\beta C_2^4 - 8\beta C_2^2C_3 - 6\beta C_2C_3C_4 \right. \\ &\quad + 6\beta C_2C_4 - 3\beta^2C_3C_2 \\ &\quad - 12\beta(1-\beta)C_3C_2^2 \\ &\quad + 12\beta(1-\beta)C_3^2 \\ &\quad \left. + 12\beta(1-\beta)^2C_4C_2\right) \\ &\quad \left. + 5C_5(1-\beta)^4\right)E^{(k)^4}\Big] + O\left(E^{(k)^5}\right). \end{aligned} \quad (50)$$

Now

$$\begin{aligned} &\left[7F'\left(Y^{(k)}\right) - 3F'\left(X^{(k)}\right)\right] \\ &\quad = 4\left[F'(\xi)\right] \\ &\quad \times \left[I + \frac{1}{4}\left[A_1E^{(k)} + A_2E^{(k)^2} + A_3E^{(k)^3}\right]\right] \\ &\quad + O\left(E^{(k)^4}\right), \end{aligned} \quad (51)$$

where

$$\begin{aligned} A_1 &= C_2(8 - 14\beta), \\ A_2 &= 14\beta C_2^2 + 21C_3(1-\beta)^2 - 9C_3, \\ A_3 &= -28\beta C_2^3 + 28\beta C_2C_3 + 42\beta(1-\beta)C_3C_2 \\ &\quad + 28C_4(1-\beta)^3 - 12C_4. \end{aligned} \quad (52)$$

Taking inverse of both sides of (51), we get

$$\begin{aligned} &4\left[7F'(Y^{(k)}) - 3F'\left(X^{(k)}\right)\right]^{-1} \\ &\quad = \left[F'(\xi)\right]^{-1}\left[I + B_1E^{(k)} + B_2E^{(k)^2} + B_3E^{(k)^3}\right] \\ &\quad + O\left(E^{(k)^4}\right), \end{aligned} \quad (53)$$

where

$$\begin{aligned} B_1 &= -\frac{A_1}{4}, \\ B_2 &= \left(-\frac{A_2}{4} + \frac{A_1^2}{16}\right), \\ B_3 &= \left(-\frac{A_3}{4} - \frac{A_1^3}{64} + \frac{A_1A_2}{16} + \frac{A_2A_1}{16}\right). \end{aligned} \quad (54)$$

By multiplying (53) and (50), we get

$$\begin{aligned} &4\left[7F'\left(Y^{(k)}\right) - 3F'\left(X^{(k)}\right)\right]^{-1}\left[F'\left(Y^{(k)}\right)\right] \\ &\quad = \left(I + E_1E^{(k)} + E_2E^{(k)^2} + E_3E^{(k)^3}\right) + O\left(E^{(k)^4}\right), \end{aligned} \quad (55)$$

where

$$\begin{aligned} E_1 &= B_1 + D_1, \\ E_2 &= B_1D_1 + B_2 + D_2, \\ E_3 &= B_2D_1 + B_1D_2 + B_3 + D_3, \end{aligned} \quad (56)$$

and the values of D_1, D_2, and D_3 are mentioned below:

$$\begin{aligned} D_1 &= 2C_2(1-\beta), \\ D_2 &= 2\beta C_2^2 + 3C_3(1-\beta)^2, \\ D_3 &= -4\beta C_2^3 + 4\beta C_2C_3 + 6\beta(1-\beta)C_3C_2 + 4C_4(1-\beta)^3. \end{aligned} \quad (57)$$

From multiplication of (47) and (55), we achieve

$$\begin{aligned} &4\left[7F'\left(Y^{(k)}\right) - 3F'\left(X^{(k)}\right)\right]^{-1}\left[F'\left(Y^{(k)}\right)\right]S \\ &\quad = \left[G_1E^{(k)} + \{G_2 + E_1G_1\}E^{(k)^2} + \{E_1G_2 + E_2G_1 + G_3\}E^{(k)^3}\right] \\ &\quad + O\left(E^{(k)^4}\right). \end{aligned} \quad (58)$$

After replacing the value of the above equation in second part of (40), we get

$$\begin{aligned} &E^{(k+1)} - \{I - G_1\}E^{(k)} - \{G_2 + E_1G_1\}E^{(k)^2} \\ &\quad - \{E_1G_2 + E_2G_1 + G_3\}E^{(k)^3} + O\left(E^{(k)^4}\right). \end{aligned} \quad (59)$$

The final error equation of method (40) is given by

$$E^{(k+1)} = -\left(\frac{C_2^2}{3}\right)E^{(k)^3} + O\left(E^{(k)^4}\right). \quad (60)$$

Thus, we end the proof of Theorem 7.

TABLE 1: Functions and their roots.

$f(x)$	α
$f_1(x) = [\sin(x)]^2 + x$	$\alpha_1 = 0$
$f_2(x) = [\sin(x)]^2 - x^2 + 1$	$\alpha_2 \approx 1.4044916482153412260350868177786$
$f_3(x) = e^{-x} + \sin(x) - 1$	$\alpha_3 \approx 2.0768312745331126130700442444750$
$f_4(x) = x^2 + \sin(x) + x$	$\alpha_4 = 0$
$f_5(x) = \sin[2\cos(x)] - 1 - x^2 + e^{\sin(x^3)}$	$\alpha_5 \approx 1.3061752018468278250148429 09066$
$f_6(x) = x^6 - 10x^3 + x^2 - x + 3$	$\alpha_6 \approx 0.6586048471181404367638600147 10$
$f_7(x) = x^4 - x^3 + 11x - 7$	$\alpha_7 \approx 0.8035111991107776889781376602 93$
$f_8(x) = x^3 - \cos(x) + 2$	$\alpha_8 \approx -1.1725779647539700126733327148 68$
$f_9(x) = \sqrt{x} - \cos(x)$	$\alpha_9 \approx 0.6417143708728826583985653003 16$
$f_{10}(x) = \log(x) - x^3 + 2\sin(x)$	$\alpha_{10} \approx 1.2979977432803718471644792382 86$

The multivariate case of (33) is given by

$$Y^{(k)} = X^{(k)} - \frac{2}{3}\left[\left[F'\left(X^{(k)}\right)\right]^{-1} F\left(X^{(k)}\right)\right],$$
$$X^{(k+1)} = X^{(k)} - \left[\frac{13}{16}I + \frac{3}{8}\left(\left[F'\left(X^{(k)}\right)\right]^{-1} F'\left(Y^{(k)}\right)\right) - \frac{3}{16}\left(\left[F'\left(X^{(k)}\right)\right]^{-1} F'\left(Y^{(k)}\right)\right)^2\right] \cdot 4\left[7F'\left(Y^{(k)}\right) - 3F'\left(X^{(k)}\right)\right]^{-1} F'\left(Y^{(k)}\right) \cdot \left[\left[F'\left(X^{(k)}\right)\right]^{-1} F\left(X^{(k)}\right)\right]. \quad (61)$$

The following theorem shows that this method has fourth-order convergence.

Theorem 8. *Let $F : D \subseteq \mathfrak{R}^n \rightarrow \mathfrak{R}^n$ be suffici tly Frechet diff rentiable in a convex set D, containing a root ξ of $F(X) = 0$. Let one suppose that $F'(X)$ is continuous and nonsingular in D and $X^{(0)}$ is close to ξ. Then the sequence $\{X^{(k)}\}_{k\geq 0}$ obtained by the iterative expression (61) converges to ξ with order four.*

Proof. For the convenience of calculation we replace $2/3$ by β and put $a_1 = 13/16$, $a_2 = 3/8$, and $a_3 = -3/16$ in (61). From (46) and (50), we have

$$t = \left[F'\left(X^{(k)}\right)\right]^{-1} F'\left(Y^{(k)}\right) = I - 2\beta C_2 E^{(k)} + \left\{6\beta C_2^2 + 3C_3\left(\beta^2 - 2\beta\right)\right\} E^{(k)^2} + \left\{-16\beta C_2^3 + \left(-6\beta^2 + 16\beta\right) C_2 C_3 + 6\beta\left(2 - \beta\right) C_3 C_2 + \left(4(1-\beta)^3 - 4\right) C_4\right\} E^{(k)^3} + O\left(E^{(k)^4}\right). \quad (62)$$

From the above equation we have

$$t^2 = \left(\left[F'\left(X^{(k)}\right)\right]^{-1} F'\left(Y^{(k)}\right)\right)^2 = I - 4\beta C_2 E^{(k)} + \left\{\left(12\beta + 4\beta^2\right) C_2^2 + 6\left(\beta^2 - 2\beta\right) C_3\right\} E^{(k)^2} + \left\{\left(-32\beta - 24\beta^2\right) C_2^3 + \left(-6\beta^3 + 32\beta\right) C_2 C_3 + \left(-6\beta^3 + 24\beta\right) C_3 C_2 + 2\left(4(1-\beta)^3 - 4\right)\right\} E^{(k)^3} + O\left(E^{(k)^4}\right). \quad (63)$$

With the help of (62) and (63), we can obtain

$$a_1 I + a_2 t + a_3 t^2 = \left(a_1 + a_2 + a_3\right) I + \left(-2\beta a_2 - 4\beta a_3\right) C_2 E^{(k)} + \left\{\left(3\left(\beta^2 - 2\beta\right) a_2 + 6\left(\beta^2 - 2\beta\right) a_3\right) C_3 + \left(6\beta a_2 + \left(12\beta + 4\beta^2\right) a_3\right) C_2^2\right\} E^{(k)^2} + \left\{\left(-16\beta a_2 + \left(-32\beta - 24\beta^2\right) a_3\right) C_2^3 + \left(\left(-6\beta^2 + 16\beta\right) a_2 + \left(-6\beta^3 + 32\beta\right) a_3\right) C_2 C_3 + \left(6\beta\left(2 - \beta\right) a_2 + \left(-6\beta^3 + 24\beta\right) a_3\right) C_3 C_2 + \left(\left(4(1-\beta)^3 - 4\right) a_2 + 2\left(4(1-\beta)^3 - 4\right) a_3\right) C_4\right\} \times E^{(k)^3} + O\left(E^{(k)^4}\right). \quad (64)$$

By multiplying (64) to (58), we have

$$\left(a_1 I + a_2 t + a_3 t^2\right) 4\left[7F'\left(Y^{(k)}\right) - 3F'\left(X^{(k)}\right)\right]^{-1}\left[F'\left(Y^{(k)}\right)\right] S = \left(T_1 E^{(k)} + T_2 E^{(k)^2} + T_3 E^{(k)^3} + T_4 E^{(k)^4} + O\left(E^{(k)^5}\right)\right), \quad (65)$$

TABLE 2: Comparison of absolute value of the functions by different methods after fourth iteration (TNFE-12).

$\lvert f\rvert$	Guess	HN3	M3	SL4	JM4	M4
$\lvert f_1\rvert$	0.3	$0.1e-57$	$0.3e-93$	$0.2e-172$	$0.4e-162$	$0.5e-199$
	0.2	$0.5e-69$	$0.1e-91$	$0.1e-186$	$0.2e-198$	$0.1e-245$
	0.1	$0.1e-90$	$0.6e-107$	$0.5e-241$	$0.2e-266$	$0.6e-339$
	−0.1	$0.2e-85$	$0.1e-93$	$0.1e-198$	$0.1e-247$	$0.7e-278$
	−0.2	$0.3e-59$	$0.9e-64$	$0.1e-99$	$0.8e-161$	$0.5e-165$
$\lvert f_2\rvert$	1.3	$0.1e-92$	$0.3e-102$	$0.1e-244$	$0.4e-278$	$0.6e-297$
	1.2	$0.1e-67$	$0.7e-75$	$0.2e-152$	$0.2e-197$	$0.7e-200$
	1.1	$0.8e-53$	$0.1e-57$	$0.4e-94$	$0.5e-147$	$0.2e-132$
	1.4	$0.6e-205$	$0.7e-217$	$0.1e-613$	$0.5e-634$	$0.1e-672$
	1.5	$0.3e-99$	$0.9e-114$	$0.1e-296$	$0.2e-300$	$0.7e-374$
$\lvert f_3\rvert$	2.0	$0.1e-112$	$0.2e-122$	$0.8e-362$	$0.1e-325$	$0.1e-418$
	2.3	$0.1e-81$	$0.1e-102$	$0.1e-215$	$0.1e-229$	$0.5e-275$
	2.1	$0.7e-157$	$0.1e-169$	$0.6e-493$	$0.1e-466$	$0.3e-543$
	2.2	$0.3e-100$	$0.1e-116$	$0.4e-288$	$0.2e-288$	$0.1e-343$
	1.9	$0.1e-81$	$0.5e-88$	$0.1e-223$	$0.1e-224$	$0.3e-272$
$\lvert f_4\rvert$	0.3	$0.4e-78$	$0.9e-101$	$0.4e-157$	$0.3e-219$	$0.2e-257$
	0.2	$0.1e-90$	$0.3e-109$	$0.5e-201$	$0.4e-258$	$0.2e-301$
	0.1	$0.2e-113$	$0.2e-128$	$0.2e-279$	$0.1e-328$	$0.1e-379$
	−0.2	$0.9e-84$	$0.4e-90$	$0.7e-223$	$0.2e-229$	$0.1e-286$
	−0.1	$0.8e-110$	$0.5e-119$	$0.1e-285$	$0.3e-314$	$0.2e-483$
$\lvert f_5\rvert$	1.35	$0.1e-101$	$0.6e-112$	$0.3e-252$	$0.1e-312$	$0.2e-320$
	1.31	$0.1e-77$	$0.1e-86$	$0.3e-170$	$0.1e-236$	$0.1e-233$
	1.29	$0.1e-69$	$0.4e-78$	$0.1e-141$	$0.5e-211$	$0.5e-203$
	1.15	$0.8e-39$	$0.2e-42$	$0.7e-28$	$0.8e-107$	$0.1e-510$
	1.20	$0.1e-46$	$0.4e-52$	$0.2e-54$	$0.3e-135$	$0.1e-101$
$\lvert f_6\rvert$	0.7	$0.7e-109$	$0.1e-122$	$0.1e-288$	$0.2e-334$	$0.1e-380$
	0.6	$0.4e-94$	$0.7e-104$	$0.9e-229$	$0.3e-286$	$0.8e-300$
	0.5	$0.3e-57$	$0.2e-63$	$0.3e-95$	$0.4e-166$	$0.2e-154$
	0.8	$0.1e-68$	$0.2e-87$	$0.2e-171$	$0.3e-207$	$0.2e-282$
	1.2	$0.6e-36$	$0.6e-52$	$0.1e-151$	$0.2e-97$	$0.1e-112$
$\lvert f_7\rvert$	0.65	$0.2e-294$	$0.1e-306$	$0.2e-588$	$0.2e-807$	$0.8e-810$
	0.75	$0.2e-177$	$0.8e-187$	$0.5e-250$	$0.4e-462$	$0.1e-457$
	0.95	$0.3e-129$	$0.5e-130$	$0.1e-134$	$0.2e-295$	$0.2e-290$
	0.90	$0.1e-137$	$0.3e-140$	$0.2e-153$	$0.3e-322$	$0.9e-318$
	0.80	$0.2e-160$	$0.3e-167$	$0.3e-207$	$0.9e-399$	$0.6e-395$
$\lvert f_8\rvert$	−1.0	$0.2e-64$	$0.4e-72$	$0.1e-112$	$0.1e-193$	$0.5e-182$
	−1.1	$0.3e-96$	$0.2e-106$	$0.6e-224$	$0.8e-297$	$0.8e-301$
	−1.2	$0.6e-132$	$0.1e-144$	$0.3e-345$	$0.8e-412$	$0.6e-431$
	−1.5	$0.5e-50$	$0.6e-71$	$0.5e-100$	$0.2e-155$	$0.5e-275$
	−0.9	$0.1e-47$	$0.2e-52$	$0.5e-46$	$0.1e-135$	$0.6e-105$
$\lvert f_9\rvert$	0.9	$0.1e-127$	$0.6e-133$	$0.1e-152$	$0.5e-315$	$0.2e-307$
	0.7	$0.2e-178$	$0.1e-186$	$0.1e-315$	$0.5e-455$	$0.4e-451$
	0.6	$0.2e-189$	$0.1e-206$	$0.1e-351$	$0.6e-482$	$0.2e-479$
	0.8	$0.6e-144$	$0.1e-149$	$0.5e-206$	$0.9e-356$	$0.3e-350$
	1.0	$0.2e-117$	$0.2e-123$	$0.3e-117$	$0.7e-295$	$0.1e-284$
$\lvert f_{10}\rvert$	1.2	$0.4e-74$	$0.2e-81$	$0.3e-154$	$0.1e-213$	$0.8e-229$
	2.0	$0.7e-26$	$0.1e-52$	$0.2e-75$	$0.5e-76$	$0.1e-107$
	1.5	$0.2e-57$	$0.1e-79$	$0.3e-139$	$0.1e-170$	$0.2e-229$
	1.3	$0.9e-214$	$0.7e-226$	$0.1e-612$	$0.4e-660$	$0.7e-708$
	1.8	$0.3e-33$	$0.2e-76$	$0.1e-83$	$0.1e-97$	$0.7e-134$

TABLE 3: Comparison of CPU time (in seconds) between some existing methods and our proposed methods.

Function	CPU time					
	Guess	HN3	M3	SL4	JM4	M4
f_1	0.3	0.2867	0.2644	0.3060	0.2449	0.2449
f_2	1.5	0.2943	0.2510	0.3049	0.2682	0.3043
f_3	2.3	0.3019	0.3658	0.3457	0.3562	0.3483
f_4	0.3	0.3091	0.2850	0.2832	0.2399	0.2428
f_5	1.35	0.3399	0.3694	0.3938	0.4149	0.3940
f_6	0.7	0.2896	0.2708	0.2388	0.2613	0.2550
f_7	0.65	0.2517	0.2356	0.2938	0.2644	0.2880
f_8	−1.00	0.2697	0.2279	0.2739	0.2934	0.2900

where

$$\begin{aligned}
T_1 &= G_1\left(a_1 + a_2 + a_3\right), \\
T_2 &= G_1\left(-2\beta a_2 - 4\beta a_3\right) C_2 + \left(a_1 + a_2 + a_3\right)\left(G_2 + E_1 G_1\right), \\
T_3 &= \left(a_1 + a_2 + a_3\right)\left(E_1 G_2 + E_2 G_1 + G_3\right) \\
&\quad + \left(-2\beta a_2 - 4\beta a_3\right) C_2 \left(G_2 + E_1 G_1\right) \\
&\quad + G_1\left[\left(3\left(\beta^2 - 2\beta\right) a_2 + 6\left(\beta^2 - 2\beta\right) a_3\right) C_3 \right. \\
&\qquad \left. + \left(6\beta a_2 + \left(12\beta + 4\beta^2\right) a_3\right) C_2^2\right], \\
T_4 &= \left(G_2 + E_1 G_1\right)\left(\left[\left(3\left(\beta^2 - 2\beta\right) a_2 + 6\left(\beta^2 - 2\beta\right) a_3\right) C_3\right] \right. \\
&\qquad \left. + \left[\left(6\beta a_2 + \left(12\beta + 4\beta^2\right) a_3\right) C_2^2\right]\right) \\
&\quad + \left(E_1 G_2 + E_2 G_1 + G_3\right)\left[\left(-2\beta a_2 - 4\beta a_3\right) C_2\right] \\
&\quad + \left(a_1 + a_2 + a_3\right)\left(E_2 G_2 + E_3 G_1 + E_1 G_3 + G_4\right) \\
&\quad + G_1\left(\left\{-16\beta a_2 + \left(-32\beta - 24\beta^2\right) a_3\right\} C_2^3 \right. \\
&\qquad + \left\{\left(-6\beta^2 + 16\beta\right) a_2 + \left(-6\beta^3 + 32\beta\right) a_3\right\} C_2 C_3 \\
&\qquad + \left\{6\beta\left(2 - \beta\right) a_2 + \left(-6\beta^3 + 24\beta\right) a_3\right\} C_3 C_2 \\
&\qquad \left. + \left\{\left(4(1-\beta)^3 - 4\right) a_2 + 2\left(4(1-\beta)^3 - 4\right) a_3\right\} C_4\right).
\end{aligned} \tag{66}$$

The final error equation of method (61) is given by

$$E^{(k+1)} = \left(-\frac{1}{9} C_2^3 + 8 C_2 C_3 - C_3 C_2 + \frac{1}{9} C_4\right) E^{(k)^4} + O\left(E^{(k)^5}\right), \tag{67}$$

which confirms the theorem.

4. Numerical Testing

4.1. Single Variate Case. In this section, ten different test functions have been considered in Table 1 for single variate case to illustrate the accuracy of the proposed iterative methods. The root of each nonlinear test function is also listed. All computations presented here have been performed in *MATHEMATICA 8*. Many streams of science and engineering require very high precision degree of scientific computations. We consider 1000 digits floating point arithmetic using "*SetAccuracy* []" command. Here we compare the performance of our proposed methods with some well-established third-order and fourth-order iterative methods. In Table 2, we have represented Huen's method by HN3, our proposed third-order method (15) by M3, fourth-order method (17) of [5] by SL4, fourth-order Jarratt's method by JM4, and proposed fourth-order method by M4. The results are listed in Table 2.

An effective way to compare the efficiency of methods is CPU time utilized in the execution of the programme. In present work, the CPU time has been computed using the command "*TimeUsed* []" in *MATHEMATICA*. It is well known that the CPU time is not unique and it depends on the specification of the computer. The computer characteristic is Microsoft Windows 8 Intel(R) Core(TM) i5-3210M CPU@ 2.50 GHz with 4.00 GB of RAM, 64-bit operating system throughout this paper. The mean CPU time is calculated by taking the mean of 10 performances of the programme. The mean CPU time (in seconds) for different methods is given in Table 3.

4.2. Multivariate Case. Further, six nonlinear systems (Examples 9–14) are considered for numerical testing of system of nonlinear equations. Here we compare our proposed third-order method (40) (MM3) with Algorithm (2.2) (NR1) and Algorithm (2.3) (NR2) of [13] and fourth-order method (61) (MM4) with (22) (SH4) of [14] and method (3.4) (BB4) of [15]. The comparison of norm of the function for different iterations is given in Table 4.

Example 9. Consider

$$\begin{aligned}
x_1^2 - x_2 - 19 &= 0, \\
-x_1^2 + \frac{x_2^3}{6} + x_2 - 17 &= 0,
\end{aligned} \tag{68}$$

with initial guess $X^{(0)} = (5.1, 6.1)^T$, and one of its solutions is $\alpha = (5, 6)^T$.

Example 10. Consider

$$\begin{aligned}
-\operatorname{Sin}\left(x_1\right) + \operatorname{Cos}\left(x_2\right) &= 0, \\
-\frac{1}{x_2} + \left(x_3\right)^{x_1} &= 0, \\
e^{x_1} - \left(x_3\right)^2 &= 0,
\end{aligned} \tag{69}$$

with initial guess $X^{(0)} = (1, 0.5, 1.5)^T$, and one of its solutions is $\alpha = (0.9095 \cdots 0.6612 \cdots 1.5758 \cdots)^T$.

TABLE 4: Norm of the functions by different methods after first, second, third, and fourth iteration.

Example	Guess	Method	$\|\|F(x^{(1)})\|\|$	$\|\|F(x^{(2)})\|\|$	$\|\|F(x^{(3)})\|\|$	$\|\|F(x^{(4)})\|\|$
Example 9	(5.1, 6.1)	NR1	$3.8774e-4$	$9.2700e-15$	$7.7652e-47$	$4.0858e-143$
		NR2	$3.8774e-4$	$9.2700e-15$	$7.7652e-47$	$4.0858e-143$
		MM3	$1.2657e-4$	$1.0705e-16$	$3.9789e-53$	$1.8320e-162$
		BB4	$2.1416e-5$	$1.1267e-24$	$4.6477e-102$	$1.2561e-411$
		SH4	$1.2923e-5$	$9.2420e-26$	$1.2710e-106$	$4.2184e-430$
		MM4	$3.0768e-6$	$3.9419e-29$	$8.7758e-121$	$2.1039e-487$
Example 10	(1, 0.5, 1.5)	NR1	$3.0006e-2$	$1.3681e-4$	$1.3174e-11$	$1.1754e-32$
		NR2	$2.9765e-2$	$1.3230e-4$	$1.1848e-11$	$8.5484e-33$
		MM3	$9.9051e-3$	$3.7473e-6$	$9.1835e-17$	$1.3035e-48$
		BB4	$2.1133e-2$	$6.9602e-6$	$7.1401e-20$	$7.7987e-76$
		SH4	$1.5676e-2$	$1.1309e-6$	$2.4814e-23$	$5.5195e-90$
		MM4	$6.1451e-3$	$8.1169e-8$	$2.3264e-28$	$6.9642e-110$
Example 11	(0.5, 0.5, 0.5, −0.2)	NR1	$2.3097e-3$	$7.5761e-10$	$5.6516e-30$	$2.8662e-91$
		NR2	$2.3097e-3$	$7.5761e-10$	$5.6516e-30$	$2.8662e-91$
		MM3	$9.4336e-4$	$1.6380e-11$	$1.7401e-35$	$2.5268e-108$
		BB4	$9.1400e-4$	$1.8627e-14$	$8.7599e-59$	$6.9713e-238$
		SH4	$5.3618e-4$	$1.4537e-15$	$2.1746e-63$	$1.7744e-256$
		MM4	$7.7084e-5$	$2.1932e-20$	$3.4979e-84$	$3.6487e-341$
Example 12	(1.0, 2.0)	NR1	$2.1427e-3$	$1.7987e-10$	$1.0504e-31$	$2.0958e-95$
		NR2	$2.1498e-3$	$1.8262e-10$	$1.1001e-31$	$2.4077e-95$
		MM3	$7.6174e-4$	$2.3592e-12$	$7.9632e-38$	$3.0435e-114$
		BB4	$5.3124e-4$	$8.2104e-16$	$3.8411e-63$	$2.8216e-252$
		SH4	$2.9895e-4$	$6.5567e-17$	$1.8332e-67$	$1.5913e-269$
		MM4	$1.0131e-4$	$2.8562e-19$	$3.4019e-77$	$2.4842e-308$
Example 13	(−0.8, 1.1, 1.1)	NR1	$2.9692e-4$	$5.5149e-14$	$3.8063e-40$	$1.2456e-118$
		NR2	$2.9718e-5$	$5.5137e-14$	$3.8044e-40$	$1.2438e-118$
		MM3	$9.8775e-6$	$6.7719e-16$	$2.3491e-46$	$9.7596e-138$
		BB4	$4.0723e-6$	$1.6873e-21$	$9.1974e-83$	$7.8287e-328$
		SH4	$2.1907e-6$	$8.6294e-23$	$3.6734e-87$	$1.1711e-349$
		MM4	$1.0838e-6$	$8.4404e-25$	$1.2327e-96$	$4.3437e-384$
Example 14	(0.5, 1.5)	NR1	$1.8661e-1$	$7.1492e-4$	$4.5647e-11$	$1.1889e-32$
		NR2	$1.7417e-1$	$5.7596e-4$	$2.3870e-11$	$1.7000e-33$
		MM3	$1.2770e-1$	$4.6794e-5$	$4.1708e-15$	$3.0222e-45$
		BB4	$9.8299e-2$	$9.2624e-6$	$8.3046e-22$	$5.3716e-86$
		SH4	$1.0359e-1$	$5.4166e-6$	$4.6302e-23$	$2.4821e-91$
		MM4	$1.4490e-1$	$1.0558e-5$	$6.7122e-22$	$1.0964e-86$

Example 11. Consider

$$\begin{aligned} x_2x_3 + x_4(x_2+x_3) &= 0,\\ x_1x_3 + x_4(x_1+x_3) &= 0,\\ x_1x_2 + x_4(x_1+x_2) &= 0,\\ x_1x_2 + x_1x_3 + x_2x_3 &= 1, \end{aligned} \tag{70}$$

with initial guess $X^{(0)} = (0.5, 0.5, 0.5, -0.2)^T$, and one of its solutions is $\alpha \approx (0.577350, 0.577350, 0.577350, -0.288675)^T$.

Example 12. Consider

$$\begin{aligned} -e^{x_1} + \tan^{-1}(x_2) + 2 &= 0,\\ \tan^{-1}\left(x_1^2 + x_2^2 - 5\right) &= 0, \end{aligned} \tag{71}$$

with initial guess $X^{(0)} = (1.0, 2.0)^T$, and one of its solutions is $\alpha = (1.12906503\cdots 1.930080863\cdots)^T$.

Example 13. Consider

$$\begin{aligned} -e^{-x_1} + x_2 + x_3 &= 0,\\ -e^{-x_2} + x_1 + x_3 &= 0,\\ -e^{-x_3} + x_1 + x_2 &= 0, \end{aligned} \tag{72}$$

with initial guess $X^{(0)} = (-0.8, 1.1, 1.1)^T$, and one of its solutions is $\alpha = (-0.8320\cdots 1.1489\cdots 1.1489\cdots)^T$.

Example 14. Consider

$$\begin{aligned} \log(x_2) - x_1^2 + x_1x_2 &= 0, \\ \log(x_1) - x_2^2 + x_1x_2 &= 0, \end{aligned} \tag{73}$$

with initial guess $X^{(0)} = (0.5, 1.5)^T$, and one of its solutions is $\alpha = (1, 1)^T$.

5. Conclusion

In the present work, we have provided a family of third- and optimal fourth-order iterative methods which yield some existing as well as many new third-order and fourth-order iterative methods. The multivariate case of these methods has also been considered. The efficiency of our methods is supported by Table 2 and Table 4.

Conflict of Interests

The authors declare that there is no conflict of interests regarding the publication of this paper.

Acknowledgments

The authors would like to express their sincerest thanks to the editor and reviewer for their constructive suggestions, which significantly improved the quality of this paper. The authors would also like to record their sincere thanks to Dr. F. Soleymani for providing his efficient cooperation.

References

[1] S. Weerakoon and T. G. I. Fernando, "A variant of Newton's method with accelerated third-order convergence," *Applied Mathematics Letters*, vol. 13, no. 8, pp. 87–93, 2000.

[2] H. H. H. Homeier, "On Newton-type methods with cubic convergence," *Journal of Computational and Applied Mathematics*, vol. 176, no. 2, pp. 425–432, 2005.

[3] C. Chun and Y. Kim, "Several new third-order iterative methods for solving nonlinear equations," *Acta Applicandae Mathematicae*, vol. 109, no. 3, pp. 1053–1063, 2010.

[4] H. T. Kung and J. F. Traub, "Optimal order of one-point and multipoint iteration," *Journal of Computational and Applied Mathematics*, vol. 21, no. 4, pp. 643–651, 1974.

[5] F. Soleymani, "Two new classes of optimal Jarratt-type fourth-order methods," *Applied Mathematics Letters*, vol. 25, no. 5, pp. 847–853, 2012.

[6] M. Sharifi, D. K. R. Babajee, and F. Soleymani, "Finding the solution of nonlinear equations by a class of optimal methods," *Computers and Mathematics with Applications*, vol. 63, no. 4, pp. 764–774, 2012.

[7] S. K. Khattri and S. Abbasbandy, "Optimal fourth order family of iterative methods," *Matematicki Vesnik*, vol. 63, no. 1, pp. 67–72, 2011.

[8] W. Gautschi, *Numerical Analysis: An Introduction*, Birkhauser, Boston, Mass, USA, 1997.

[9] J. F. Traub, *Iterative Methods for Solution of Equations*, Chelsea Publishing, New York, NY, USA, 1997.

[10] M. Grau-Sánchez and J. L. Díaz-Barrero, "Zero-finder methods derived using Runge-Kutta techniques," *Applied Mathematics and Computation*, vol. 217, no. 12, pp. 5366–5376, 2011.

[11] K. Huen, "Neue methode zur approximativen integration der differentialge-ichungen einer unabhngigen variablen," *Zeitschrift für angewandte Mathematik und Physik*, vol. 45, pp. 23–38, 1900.

[12] F. Soleymani and D. K. R. Babajee, "Computing multiple zeros using a class of quartically convergent methods," *Alexandria Engineering Journal*, vol. 52, pp. 531–541, 2013.

[13] M. A. Noor and M. Waseem, "Some iterative methods for solving a system of nonlinear equations," *Computers and Mathematics with Applications*, vol. 57, no. 1, pp. 101–106, 2009.

[14] J. R. Sharma, R. K. Guha, and R. Sharma, "An efficient fourth-order weighted-Newton method for systems of nonlinear equations," *Numerical Algorithms*, vol. 62, pp. 307–323, 2013.

[15] D. K. R. Babajee, A. Cordero, F. Soleymani, and J. R. Torregrosa, "On a novel fourth-order algorithm for solving systems of nonlinear equations," *Journal of Applied Mathematics*, vol. 2012, Article ID 165452, 12 pages, 2012.

H_∞ Controller Design for an Observer-Based Modified Repetitive-Control System

Lan Zhou,[1] Jinhua She,[2] Shaowu Zhou,[1] and Qiwei Chen[1]

[1] *School of Information and Electrical Engineering, Hunan University of Science and Technology, Xiangtan 411201, China*
[2] *School of Computer Science, Tokyo University of Technology, Tokyo 192-0982, Japan*

Correspondence should be addressed to Lan Zhou; zhoulan75@163.com

Academic Editor: Elio Usai

This paper presents a method of designing a state-observer based modified repetitive-control system that provides a given H_∞ level of disturbance attenuation for a class of strictly proper linear plants. Since the time delay in a repetitive controller can be treated as a kind of disturbance, we convert the system design problem into a standard state-feedback H_∞ control problem for a linear time-invariant system. The Lyapunov functional and the singular-value decomposition of the output matrix are used to derive a linear-matrix-inequality (LMI) based design algorithm for the parameters of the feedback controller and the state-observer. A numerical example demonstrates the validity of the method.

1. Introduction

In control engineering practice, many systems exhibit repetitive behavior, such as a robot manipulator, a hard disk drive, and many other servo systems. Repetitive control [1], or RC for short, has proven to be a useful control strategy for a system with a periodic reference input and/or disturbance signal [2–4]. The distinguishing feature of RC is that it contains a pure-delay positive-feedback loop, which is the internal model of a periodic signal. For a given periodic reference input, a repetitive controller gradually reduces the tracking error through repeated learning actions [5], which involves adding the control input of the previous period to that of the present period to regulate the present control input. This theoretically guarantees gradual improvement and finally eliminates any tracking error and provides very precise control, which is a chief characteristic of the human learning process.

From the standpoint of system theory, an RC system (RCS) is a neutral-type delay system. Asymptotic tracking and stabilization of the control system are possible only when the relative degree of the compensated plant is zero [5]. To use RC on a strictly proper plant, that is, the case that most control engineering applications deal with, the repetitive controller has to be modified by the insertion of a low-pass filter into the time-delay feedback line. The resulting system is called a modified RCS (MRCS). Since a modified repetitive controller is just an approximate model of a periodic signal, there exists a steady-state tracking error; that is, in an MRCS, the low-pass filter relaxes the stabilization condition but degrades the tracking precision [6].

RC is similar to iterative learning control (ILC), which is another well-known method that makes use of previous control trials. However, as pointed out by [7–9] and others, there are significant differences between them. First of all, the initial state of a period is different. In an RCS, the state at the beginning of a period is the same as the final state in the previous period. However, in an ILC system (ILCS), the reference trajectory is defined over a finite time interval, and the state of an ILCS is usually reset after a trial. In the literatures of ILC, the initial or boundary conditions, that is, the initial state on each trial and the initial trial profile, are commonly taken to be zero. The difference in initial-condition resetting leads to the different analysis techniques and results.

One problem with an RCS is that the improved disturbance rejection at the periodic frequency and its harmonics is achieved at the expense of degraded system sensitivity at intermediate frequencies. In other words, an RCS cannot reject, and may even amplify, an aperiodic disturbance.

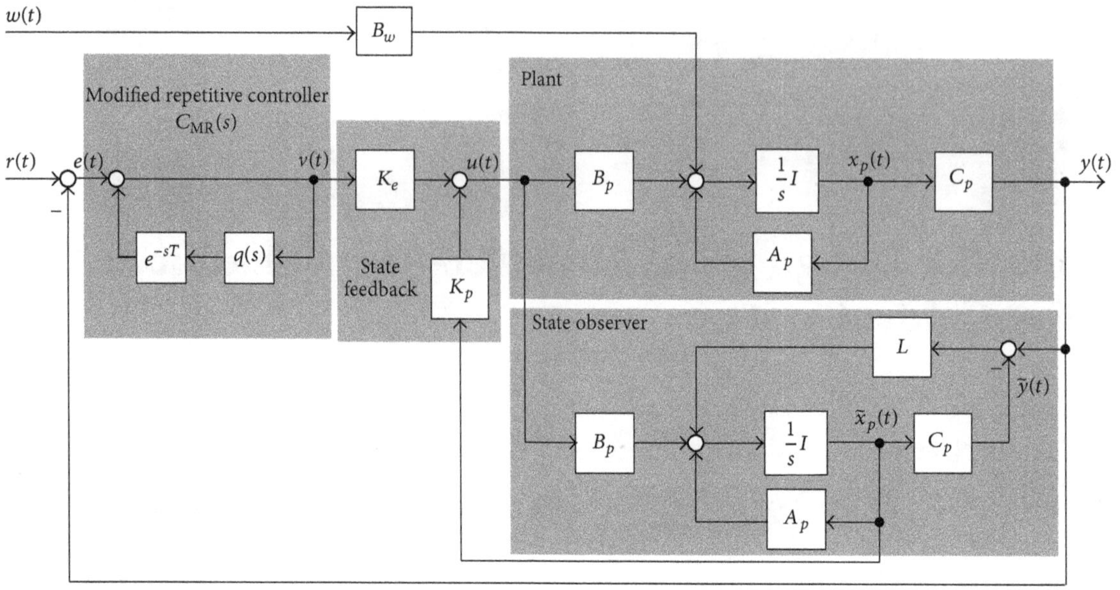

FIGURE 1: Configuration of the state-observer based MRCS.

Various solutions have been presented in the literatures addressing this problem. Time-varying and adaptive RC was proposed in [10]. Kim et al. presented a design method of a two-parameter robust RCS that used discrete-time m-synthesis and H_∞ control to reject both periodic and aperiodic disturbances [11]. However, the high order of the controller makes it difficult to implement. She et al. devised an H_∞ control method to design the parameters of the stabilization controller for a robust MRCS with time-varying uncertainties [12]. But it only considers the case that there is no disturbance input to the controlled plant. In addition, the full states of a plant are needed to design the state-feedback controller. In many practical applications, the complete state of a plant is not available because of running cost and difficulty of installation.

Zhou et al. presented a method based on two-dimensional system theory of designing a robust observer based MRCS [13]. But it only considered the robust stability of the system. To enable RC method to handle a larger class of systems, this paper presents a method of designing an MRCS with a prescribed bound on disturbance attenuation for a class of strictly proper linear systems. First, we present a configuration of an observer based MRCS and formulate the design problem as an equivalent H_∞ state-feedback control problem. Then, combining the Lyapunov stability theory and the singular-value decomposition (SVD) of the output matrix, we derive a linear-matrix-inequality (LMI) based sufficient stability condition. The condition can be used directly to design the parameters of the feedback controller and the state-observer. Compared to the one in [13] and other RC methods, the advantage of this method is that it takes into consideration the transient performance of the MRCS. Finally, a numerical example demonstrates the validity of the method.

Throughout this paper, $L_2[0, +\infty)$ is the linear space of square integrable functions from $[0, +\infty)$ to C^p; $\|G(s)\|_\infty := \sup_{0 \le w < \infty} \|G(j\omega)\|$ is the H_∞ norm of a transfer function $G(s)$; and $*$ denotes the transpose of a block entry in a matrix.

2. Problem Description

Consider the MRCS in Figure 1. The compensated single-input, single-output (SISO) plant is

$$\begin{aligned} \dot{x}_p(t) &= A_p x_p(t) + B_p u(t) + B_w w(t), \\ y(t) &= C_p x_p(t), \end{aligned} \tag{1}$$

where $x_p(t) \in \mathbb{R}^n$ is the state of the plant; $u(t), y(t) \in \mathbb{R}$ are the control input and output, respectively; and $w(t) \in L_2[0, +\infty)$ is the disturbance input. Setting $B_\omega \neq 0$ adds the disturbance to the system, and setting $B_\omega = 0$ removes it. A_p, B_p, and C_p are real constant matrices. Assume that (A_p, B_p) is controllable and (C_p, A_p) is observable, which are standard for a servo system.

In Figure 1, $r(t)$ is a periodic reference signal with a period of T, and

$$e(t) = r(t) - y(t) \tag{2}$$

is the tracking error. The modified repetitive controller is

$$C_{MR}(s) = \frac{1}{1 - q(s) e^{-sT}}. \tag{3}$$

$q(s)$ is a first-order low-pass filter that ensures the stability of the control system. Without loss of generality, we assume that $q(s)$ is a first-order filter; that is

$$q(s) = \frac{\omega_c}{s + \omega_c}, \quad (4)$$

where ω_c is the cutoff angular frequency of the filter. The inverse Laplace-transform of (3) yields

$$\dot{v}(t) = -\omega_c v(t) + \omega_c v(t - T) + \omega_c e(t) + \dot{e}(t), \quad (5)$$

where the output, $v(t)$, of $C_{MR}(s)$ is chosen to be the state of the modified repetitive controller.

The following state-observer is used to reproduce the state of the plant

$$\begin{aligned} \dot{\tilde{x}}_p(t) &= A_p\tilde{x}_p(t) + B_p u(t) + L[y(t) - \tilde{y}(t)] \\ \tilde{y}(t) &= C_p\tilde{x}_p(t), \end{aligned} \quad (6)$$

where L is the observer gain.

The error between the states of the actual plant and the observer is

$$x_e(t) = \tilde{x}_p(t) - x_p(t). \quad (7)$$

Thus, from (1) and (6), we have

$$\dot{x}_e(t) = (A_p - LC_p)x_e(t) - B_w w(t). \quad (8)$$

A linear control law based on the states of the observer and the repetitive controller is

$$u(t) = K_e v(t) + K_p\tilde{x}_p(t), \quad K_e \in \mathbb{R}, \; K_p \in \mathbb{R}^{1\times n}. \quad (9)$$

where K_e is the feedback gain of the repetitive controller and K_p is the reconstructed state-feedback gain.

Since the stability of the system does not depend on an exogenous signal, we set $r(t) = 0$. Choose $x_p(t)$, $x_e(t)$, and $v(t)$ to be the state variables of the MRCS in Figure 1. Then, from the dynamic equations (1), (5), (8), and (9) yields the following state equation:

$$\dot{x}(t) = Ax(t) + A_1x(t - T) + Bu(t) + \tilde{B}w(t), \quad (10)$$

where

$$x^T(t) = \begin{bmatrix} x_p^T(t) & x_e^T(t) & v^T(t) \end{bmatrix},$$

$$A = \begin{bmatrix} A_p & 0 & 0 \\ 0 & A_p - LC_p & 0 \\ -\omega_c C_p - C_p A_p & 0 & -\omega_c \end{bmatrix},$$

$$A_1 = \begin{bmatrix} 0 & 0 & 0 \\ 0 & 0 & 0 \\ 0 & 0 & \omega_c \end{bmatrix}, \quad B = \begin{bmatrix} B_p \\ 0 \\ -C_p B_p \end{bmatrix}, \quad (11)$$

$$\tilde{B} = \begin{bmatrix} B_w \\ -B_w \\ -C_p B_w \end{bmatrix}.$$

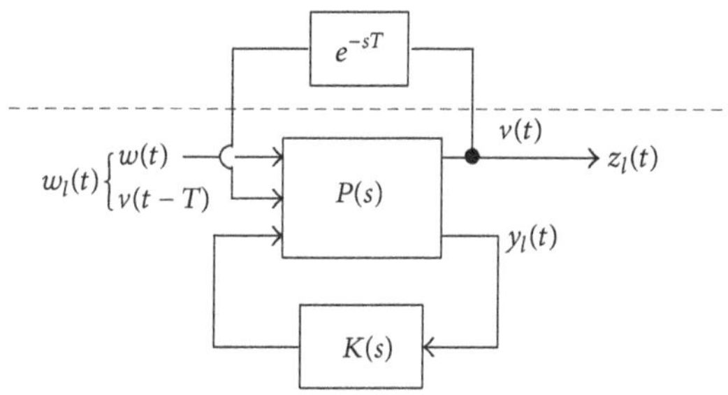

FIGURE 2: Argument system.

To enhance the system convergence, applying the method in [12], we treat the delay item $v(t - T)$ as a disturbance. We construct a generalized plant, $P(s)$, and redraw Figure 1 as Figure 2. Considering that the H_∞ norm of the time delay is one, we formulate the problem of the controller design as an H_∞ state-feedback one (Figure 3). The state-space representation of the generalized plant, $P(s)$, is

$$\begin{aligned} \dot{x}(t) &= Ax(t) + B_1 w_l(t) + B_2 u(t), \\ z_l(t) &= C_1 x(t), \\ y_l(t) &= C_2 x(t), \end{aligned} \quad (12)$$

where

$$w_l(t) = \begin{bmatrix} w^T(t) & v^T(t - T) \end{bmatrix}^T, \qquad y_l(t) = \begin{bmatrix} x_p^T(t) & 0 & v^T(t) \end{bmatrix}^T,$$

$$B_1 = \begin{bmatrix} B_w & 0 \\ -B_w & 0 \\ -C_p B_w & \omega_c \end{bmatrix}, \quad B_2 = \begin{bmatrix} B_p \\ 0 \\ -C_p B_p \end{bmatrix},$$

$$C_1 = \begin{bmatrix} 0 & 0 & 0 \\ 0 & 0 & 1 \end{bmatrix}, \quad C_2 = \begin{bmatrix} 1 & 0 & 0 \\ 0 & 0 & 0 \\ 0 & 0 & 1 \end{bmatrix}. \quad (13)$$

Substituting the control input (9) into (12) yields a representation of the closed-loop system in Figure 1:

$$\begin{aligned} \dot{x}(t) &= (A + B_2 K)x(t) + B_1 w_l(t), \\ z(t) &= C_1 x(t), \end{aligned} \quad (14)$$

where

$$K = \begin{bmatrix} K_p & K_p & K_e \end{bmatrix}. \quad (15)$$

This paper considers the following H_∞ disturbance attenuation problem.

Design suitable control gains, K_e and K_p, and the state-observer gain, L, such that

(1) the MRCS in Figure 1 is internally stable; and

(2) the H_∞ norm from w_l to $z(t)$ is less than γ; that is,

$$\left\| G_{w_l z}(s) \right\|_\infty = \left\| C_1 [sI - (A + B_2 K)]^{-1} B_1 \right\|_\infty < \gamma, \quad (16)$$

where γ is a positive number.

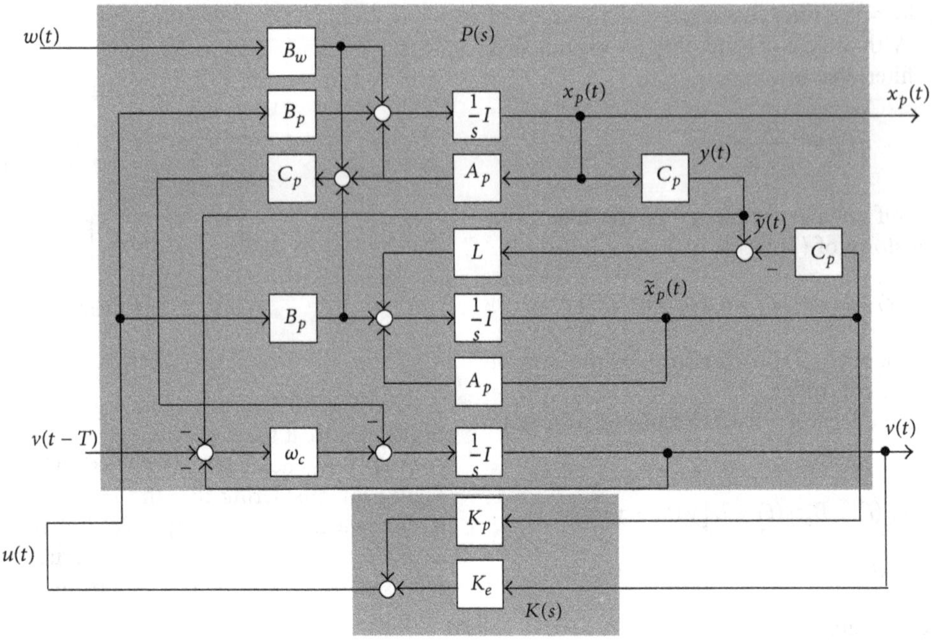

FIGURE 3: Formulation of design problem for H_∞ state-feedback control.

3. Design of the H_∞ Controller

In this section we construct an LMI-based design algorithm for the H_∞ state-feedback controller by employing a Lyapunov functional and the SVD of the output matrix.

Definition 1 (see [14]). The matrix Π has full row rank ($\text{rank}(\Pi) = p$). The SVD of Π is

$$\Pi = U\left[S \;\; 0\right]V^T, \tag{17}$$

where $S \in \mathbb{R}^{p\times p}$ is a diagonal matrix with positive, diagonal elements in decreasing order; $0 \in \mathbb{R}^{p\times(n-p)}$ is a zero matrix; and $U \in \mathbb{R}^{p\times p}$ and $V \in \mathbb{R}^{n\times n}$ are unitary matrices.

The following lemma presents an equivalent condition for matrix equation

$$\Pi X = \overline{X}\Pi. \tag{18}$$

Lemma 2 (see [15]). *For a given $\Pi \in \mathbb{R}^{p\times n}$ with $\text{rank}(\Pi) = p$, if $X \in \mathbb{R}^{n\times n}$ is a symmetric matrix, then there exists a matrix, $\overline{X} \in \mathbb{R}^{p\times p}$, such that $\Pi X = \overline{X}\Pi$ holds if and only if*

$$X = V\,\text{diag}\left\{X_{11}, X_{22}\right\}V^T, \tag{19}$$

where $X_{11} \in \mathbb{R}^{p\times p}$ and $X_{22} \in \mathbb{R}^{(n-p)\times(n-p)}$.

The following two lemmas are also employed in the derivation of the existence condition for the H_∞ controller of the system (14).

Lemma 3 (Schur complement [16]). *For any real matrix $\Sigma = \Sigma^T$, the following assertions are equivalent:*

(1) $\Sigma = \left[\begin{smallmatrix} S_{11} & S_{12} \\ * & S_{22}\end{smallmatrix}\right] < 0$;

(2) $S_{11} < 0$ *and* $S_{22} - S_{12}^T S_{11}^{-1} S_{12} < 0$;

(3) $S_{22} < 0$ *and* $S_{11} - S_{12} S_{22}^{-1} S_{12}^T < 0$.

Lemma 4 (see [17]). *For the system*

$$\begin{aligned} \dot{x}(t) &= Ax(t) + Bw(t), \\ z(t) &= Cx(t) + Dw(t), \end{aligned} \tag{20}$$

the following assertions are equivalent.

(1) *A is stable; and the H_∞ norm of the transfer function $G_{wz}(s)$, from $w(t)$ to $z(t)$ satisfies $\|G_{wz}(s)\|_\infty < \gamma$.*

(2) *There exists a symmetric matrix $P > 0$ such that*

$$\begin{bmatrix} PA + A^TP & PB & C^T \\ * & -I & D^T \\ * & * & -\gamma^2 I \end{bmatrix} < 0, \tag{21}$$

holds.

So, we have the following theorem.

Theorem 5. *For a given positive scalar γ, if there exist symmetric, positive-definite matrices X_1, X_{11}, X_{22}, and X_3, and arbitrary matrices W_1, W_2, W_3, and W_4 such that the LMI*

$$\begin{bmatrix} \Theta_{11} & B_pW_2 & \Theta_{13} & B_w & 0 & 0 & 0 \\ * & \Theta_{22} & -W_2^TB_p^TC_p^T & B_w & 0 & 0 & 0 \\ * & * & \Theta_{33} & -C_pB_w & \omega_c & 0 & X_3 \\ * & * & * & -I & 0 & 0 & 0 \\ * & * & * & * & -I & 0 & 0 \\ * & * & * & * & * & -\gamma^2 I & 0 \\ * & * & * & * & * & * & -\gamma^2 I \end{bmatrix} < 0, \tag{22}$$

holds, where the SVD of the output matrix C_p is

$$C_p = U[S \quad 0]V^T, \qquad X_2 = V\,\mathrm{diag}\{X_{11}, X_{22}\}V^T, \tag{23}$$

$$\begin{aligned}
\Theta_{11} &= A_pX_1 + X_1A_p^T + B_pW_1 + W_1^TB_p^T,\\
\Theta_{13} &= B_pW_3 - W_1^TB_p^TC_p^T - \omega_cX_1C_p^T - X_1A_p^TC_p^T,\\
\Theta_{22} &= A_pX_2 + X_2A_p^T - W_4C_p - C_p^TW_4^T,\\
\Theta_{33} &= -C_pB_pW_3 - W_3^TB_p^TC_p^T - 2\omega_cX_3,
\end{aligned} \tag{24}$$

then the closed-loop repetitive-control system (14) is asymptotically stable and provides a prescribed H_∞ disturbance attenuation level, γ. Furthermore, the parameters in control law (15) are

$$K_p = W_1X_1^{-1}, \qquad K_e = W_3X_3^{-1}, \tag{25}$$

and the state-observer gain in (6) is

$$L = W_4USX_{11}^{-1}S^{-1}U^T. \tag{26}$$

Proof. Applying Lemma 2 to (23), there exists

$$\overline{X}_2 = USX_{11}S^{-1}U^T, \tag{27}$$

such that

$$C_pX_2 = \overline{X}_2C_p, \qquad \overline{X}_2^{-1} = USX_{11}^{-1}S^{-1}U^T. \tag{28}$$

Thus, the observer gain L in (26) is equal to

$$L = W_4\overline{X}_2^{-1}. \tag{29}$$

Let

$$\begin{aligned}
P_1 &= X_1^{-1}, \quad P_2 = X_2^{-1}, \quad P_3 = X_3^{-1}\\
W_1 &= K_pX_1, \quad W_2 = K_pX_2,\\
W_3 &= K_eX_3, \quad W_4 = L\overline{X}^2.
\end{aligned} \tag{30}$$

Pre- and postmultiplying the matrix on the left-side of LMI (22) by

$$P = \mathrm{diag}\{P_1, P_2, P_3, I, I, I, I\}, \tag{31}$$

yields the following matrix inequality that is equivalent to LMI (22)

$$\begin{bmatrix}
\Lambda_{11} & P_1B_pK_p & \Lambda_{13} & P_1B_w & 0 & 0 & 0\\
* & \Lambda_{22} & -K_p^TB_p^TC_p^TP_3 & P_2B_w & 0 & 0 & 0\\
* & * & \Lambda_{33} & -P_3C_pB_w & P_3\omega_c & 0 & 1\\
* & * & * & -I & 0 & 0 & 0\\
* & * & * & * & -I & 0 & 0\\
* & * & * & * & * & -\gamma^2I & 0\\
* & * & * & * & * & * & -\gamma^2I
\end{bmatrix} < 0, \tag{32}$$

where

$$\begin{aligned}
\Lambda_{11} &= P_1A_p + A_p^TP_1 + P_1B_pK_p + K_p^TB_p^TP_1,\\
\Lambda_{13} &= P_1B_pK_e - K_p^TB_p^TC_p^TP_3 - \omega_cC_p^TP_3 - A_p^TC_p^TP_3,\\
\Lambda_{22} &= P_2A_p + A_p^TP_2 - P_2LC_p - C_p^TL^TP_2,\\
\Lambda_{33} &= -P_3C_pB_pK_e - K_e^TB_p^TC_p^TP_3 - 2\omega_cP_3.
\end{aligned} \tag{33}$$

Two steps bring us to the conclusion.

First, we consider the stability of the closed-loop MRCS (14) without a disturbance ($B_w = 0$).

Substituting the control input (9) into (10) yields the closed-loop system

$$\dot{x}(t) = (A + BK)x(t) + A_1x(t - T), \tag{34}$$

where A and A_1 are defined in (10).

Choose a Lyapunov functional candidate to be

$$V(t) = x^T(t)Px(t), \tag{35}$$

where

$$P = \mathrm{diag}\{P_1, P_2, P_3\}. \tag{36}$$

Along the time trajectory of (34)

$$\frac{dV(t)}{dt} = 2x^T(t)P\dot{x}(t) = \xi^T(t)\widetilde{\Lambda}\xi(t), \tag{37}$$

where

$$\xi(t) = \left[x^T(t) \;\; x^T(t-T)\right]^T,$$

$$\widetilde{\Lambda} = \begin{bmatrix} \Lambda_{11} & P_1B_pK_p & \Lambda_{13} & 0 & 0 & 0 \\ * & \Lambda_{22} & -K_p^TB_p^TC_p^TP_3 & 0 & 0 & 0 \\ * & * & \Lambda_{33} & 0 & 0 & P_3\omega_c \\ * & * & * & 0 & 0 & 0 \\ * & * & * & * & 0 & 0 \\ * & * & * & * & * & 0 \end{bmatrix}, \quad (38)$$

and $\Lambda_{11}, \Lambda_{13}, \Lambda_{22}, \Lambda_{33}$ are defined in (33).

From (37), $\widetilde{\Lambda} < 0$ implies that, for any $\xi(t) \neq 0$, $dV(t)/dt < 0$. Also, $\widetilde{\Lambda} < 0$ is equivalent to the matrix inequality

$$\begin{bmatrix} \Lambda_{11} & P_1B_pK_p & \Lambda_{13} & 0 \\ * & \Lambda_{22} & -K_p^TB_p^TC_p^TP_3 & 0 \\ * & * & \Lambda_{33} & P_3\omega_c \\ * & * & * & 0 \end{bmatrix} < 0. \quad (39)$$

Furthermore, since $B_w = 0$, applying Sylvester Criterion [18] and Schur-complement Lemma 3 to (32) yields $\widetilde{\Lambda} < 0$. So, if LMI (22) holds, then MRCS (34) is asymptotically stable; that is, the closed-loop system (14) is asymptotically stable when $B_w = 0$.

Next, we consider the case $B_w \neq 0$. Note that closed-loop MRCS (14) is asymptotically stable.

Applying Lemma 4 to (14), it follows that if the matrix inequality

$$\Omega = \begin{bmatrix} P(A+B_2K) + (A+B_2K)^TP & PB_1 & C_1^T \\ * & -I & 0 \\ * & * & -\gamma^2 I \end{bmatrix} < 0 \quad (40)$$

holds, then the system (14) satisfies the disturbance attenuation performance $\|G_{wlz}(s)\|_\infty < \gamma$.

Obviously, from the coefficient matrices and controller parameters in (14), the matrix inequality (40) is equal to (32). Also, LMI (32) is equivalent to (22). So, if LMI (22) holds, then $\Omega < 0$. □

Remark 6. Theorem 5 provides an LMI-based sufficient condition for the closed-loop repetitive-control system (14) with a prescribed H_∞ disturbance-attenuation level γ. The condition can be used directly to design the parameters of the controller and the state-observer in Figure 1.

In addition, from Theorem 5, we obtain a sufficient condition for an H_∞ disturbance-attenuation level for the system (1) under the state-feedback-based control law:

$$u(k) = K_e v(t) + K_p x_p(k). \quad (41)$$

The representation of the corresponding closed-loop system in Figure 1 is

$$\dot{\overline{x}}(t) = \left(\overline{A} + \overline{B}_2\overline{K}\right)x(t) + \overline{B}_1 w_l(t),$$
$$\overline{z}(t) = \overline{C}_1\overline{x}(t), \quad (42)$$

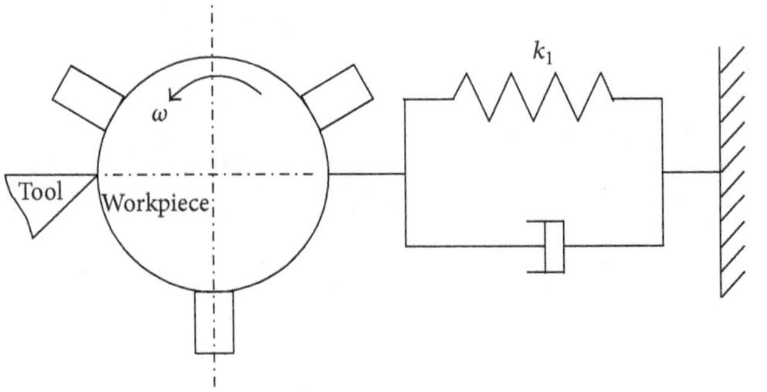

FIGURE 4: Chuck-workpiece system.

where

$$\overline{x}(t) = \left[x_p^T(t) \;\; v^T(t)\right]^T,$$

$$\overline{A} = \begin{bmatrix} A_p & 0 \\ -\omega_cC_p - C_pA_p & -\omega_c \end{bmatrix}, \quad \overline{B}_1 = \begin{bmatrix} B_w & 0 \\ -C_pB_p & \omega_c \end{bmatrix},$$

$$\overline{B}_2 = \begin{bmatrix} B_p \\ -C_pB_p \end{bmatrix}, \quad \overline{C}_1 = \begin{bmatrix} 0 & 0 \\ 0 & 1 \end{bmatrix}, \quad \overline{K} = [K_p \;\; K_e]. \quad (43)$$

Corollary 7. *For given positive scalar γ, if there exist symmetric, positive-definite matrices X_1 and X_2 and arbitrary matrices W_1 and W_2 such that the LMI*

$$\begin{bmatrix} \Theta_{11} & \Theta_{12} & B_w & 0 & 0 & 0 \\ * & \Theta_{22} & -C_pB_w & \omega_c & 0 & X_3 \\ * & * & -I & 0 & 0 & 0 \\ * & * & * & -I & 0 & 0 \\ * & * & * & * & -\gamma^2 I & 0 \\ * & * & * & * & * & -\gamma^2 I \end{bmatrix} < 0, \quad (44)$$

holds, where

$$\Theta_{11} = A_pX_1 + X_1A_p^T + B_pW_1 + W_1^TB_p^T,$$
$$\Theta_{12} = B_pW_2 - W_1^TB_p^TC_p^T - \omega_cX_1C_p^T - X_1A_p^TC_p^T, \quad (45)$$
$$\Theta_{22} = -C_pB_pW_2 - W_2^TB_p^TC_p^T - 2\omega_cX_2,$$

then the closed-loop MRCS (42) is asymptotically stable and the H_∞ disturbance-attenuation performance $\|G_{w_lz}(s)\| < \gamma$ is guaranteed. Furthermore, the feedback gains in control law (41) are

$$K_p = W_1X_1^{-1}, \quad K_e = W_2X_2^{-1}. \quad (46)$$

4. Numerical Example

In this section, we apply our method to the position control of a chuck-workpiece system with a three-jak chuck (Figure 4, [19]).

Assume that the parameters of plant (1) are

$$A_p = \begin{bmatrix} 0 & 1 \\ -1 & -1 \end{bmatrix}, \quad B_p = B_w = \begin{bmatrix} 0.5 \\ 0 \end{bmatrix},$$
$$C_p = [1 \;\; 0]. \quad (47)$$

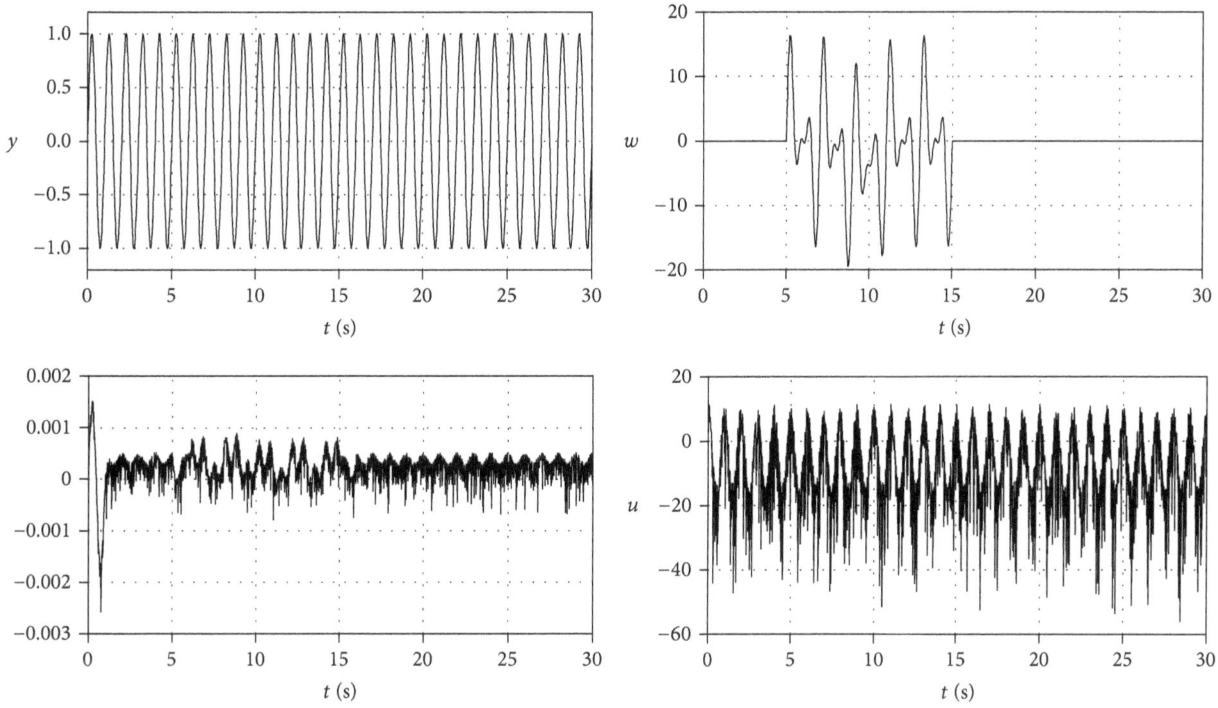

FIGURE 5: Simulation results for observer based H_∞ repetitive-control system.

We consider the problem of tracking the reference input

$$r(t) = \sin(2\pi t), \tag{48}$$

and rejecting the aperiodic disturbance

$$w = \begin{cases} 0, & t < 5 \\ \sin(2\pi t) + 5\sin(\pi t) + 5\sin(3\pi t) & \\ \quad +5(\tan(t-10) - \tan(t-9)), & 5 \le t \le 15 \\ 0, & t > 15, \end{cases} \tag{49}$$

below the level $\gamma = 0.2$.

Thus, the repetition period is $T = 1$.

Carrying out the design procedure in Theorem 5, yields the corresponding control gains

$$K_e = 4.0882 \times 10^4, \qquad K_p = [-63.5735 \ \ -1.9221], \tag{50}$$

and the state-observer gain

$$L = [0.5003 \ \ -0.3630]^T. \tag{51}$$

Simulation results in Figure 5 show that the system is asymptotically stable. After it enters the steady state in the third period for the reference input, the disturbance is added to the system. And the largest steady-state peak-to-peak relative tracking error is as small as 0.0893%.

To better assess the disturbance attenuation performance of the observer based MRCS, we design an H_∞ output-feedback controller [20] (Figure 6) for the plant (47), where the PI controller in Figure 6 is

$$K_{pi} = 650 + \frac{1400}{s}. \tag{52}$$

Let

$$w_l(t) = \left[v^T(t-T) \ \ w^T(t)\right]^T, \qquad z(t) = e(t). \tag{53}$$

The H_∞ disturbance rejection problem was formulated as follows.

Find an H_∞ output controller $K_\infty(s)$ such that the MRCS in Figure 6 is internally stable and provides the disturbance attenuation level γ; that is, $\|G_{wlz}(s)\|_\infty < \gamma$.

The problem of designing $K_\infty(s)$ is shown in Figure 7, and the resulting controller is

$$K_\infty(s) = \frac{541667.3035(s+1)(s+2.154)}{(s+8.157\times 10^4)(s+2.154)(s+1)}. \tag{54}$$

Choose the index

$$J_e = \sup_{t \ge t_s} |e(t)| \tag{55}$$

to evaluate the steady-state tracking performance, where t_s is the setting time of the control system.

Figure 8 shows the tracking error of the system in Figure 6 for the reference input (48) and the aperiodic disturbance (49). We find that $J_e = 0.0098$ for the observer based H_∞ in Figure 1, but $J_{e_{\text{output}}} = 0.022$ for the H_∞ MRCS in Figure 6. Moreover, the former has better transient performance.

5. Conclusion

In a conventional repetitive-control system, the tracking performance may be degraded by aperiodic disturbance even if it contains an internal model of a periodic signal,

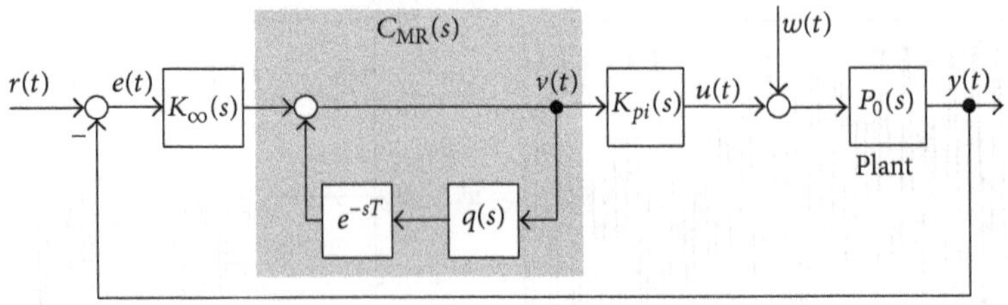

FIGURE 6: Block diagram of H_∞ MRCS.

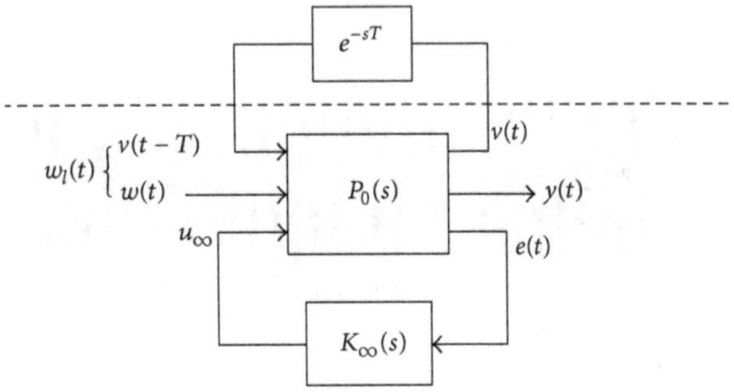

FIGURE 7: Block diagram for designing $K_\infty(s)$ in Figure 6.

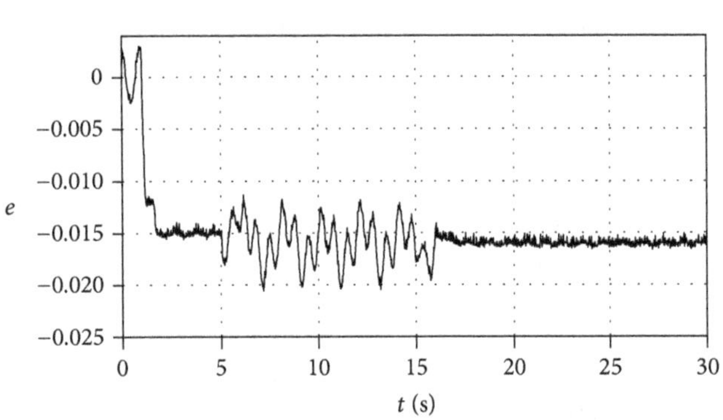

FIGURE 8: Tracking error of H_∞ MRCS in Figure 6 for reference input (48) and input disturbance (49).

which theoretically guarantees asymptotic tracking for the periodic reference input. This paper describes a method of designing an observer based repetitive-control system with a prescribed H_∞ disturbance attenuation level for a class of strictly proper linear plants. An equivalent system is established to convert the design problem to a static state-feedback H_∞ control problem. The Lyapunov stability theory and the SVD of the output matrix are used to derived an LMI-based design algorithm for the parameters of the feedback controller and the state-observer. The validity of the method was demonstrated using the position control of a noncircular cutting workpiece system. The simulation results show that the designed MRCS is asymptotically stable and exhibits satisfactory disturbance-attenuation performance.

Conflict of Interests

The authors declare that there is no conflict of interests regarding the publication of this paper.

Acknowledgments

This work was supported in part by the National Natural Science Foundation of China under Grant nos. 61203010, 61210011, and 51374107, Hunan Natural Science Foundation under Grant no. 11JJ4059, and Scientific Research Fund of Hunan Provincial Education Department under Grant no. 12B044.

References

[1] T. Inoue, M. Nakano, and S. Iwai, "High accuracy control of servomechanism for repeated contouring," in *Proceedings of the 10th Annual Symposium on Incremental Motion Control Systems and Devices*, pp. 285–292, Urbana-Champaign, 1981.

[2] P. Roncero-Sanchez, E. Acha, J. E. Ortega-Calderon, V. Feliu, and A. Garcia-Cerrada, "A versatile control scheme for a dynamic voltage restorer for power-quality improvement," *IEEE Transactions on Power Delivery*, vol. 24, no. 1, pp. 277–284, 2009.

[3] T.-Y. Doh, J. R. Ryoo, and M. J. Chung, "Design of a repetitive controller: an application to the track-following servo system of optical disk drives," *IEE Proceedings of Control Theory and Applications*, vol. 153, no. 3, pp. 323–330, 2006.

[4] J. D. Álvarez, L. J. Yebra, and M. Berenguel, "Repetitive control of tubular heat exchangers," *Journal of Process Control*, vol. 17, no. 9, pp. 689–701, 2007.

[5] S. Hara, Y. Yamamoto, T. Omata, and M. Nakano, "Repetitive control system: a new type servo system for periodic exogenous signals," *IEEE Transactions on Automatic Control*, vol. 33, no. 7, pp. 659–668, 1988.

[6] W. Chen and W. Zhang, "Optimality based repetitive controller design for track-following servo system of optical disk drives," *ISA Transactions*, vol. 48, no. 4, pp. 434–438, 2009.

[7] S. Arimoto, "A brief history of iterative learning control," in *Iterative Learning Control Analysis, Design, Integration and Applications*, Kluwer Academic, Boston, Mass, USA, 1998.

[8] J. She, L. Zhou, M. Wu, J. Zhang, and Y. He, "Design of a modified repetitive-control system based on a continuous-discrete 2D model," *Automatica*, vol. 48, no. 5, pp. 844–850, 2012.

[9] L. Zhou, J. She, M. Wu, and J. Zhang, "Design of robust modified repetitive-control system for linear periodic plants," *Journal of Dynamic Systems, Measurement and Control*, vol. 134, no. 1, Article ID 011023, 2012.

[10] Z. Cao and G. Ledwich, "Tracking variable periodic signals with fixed sampling rate-feedforward control," in *Proceedings of the IEEE International Conference on Industrial Technology (ICIT '05)*, pp. 142–145, Hong Kong, China, December 2005.

[11] B.-S. Kim, J. Li, and T.-C. Tsao, "Two-parameter robust repetitive control with application to a novel dual-stage actuator for

noncircular machining," *IEEE/ASME Transactions on Mechatronics*, vol. 9, no. 4, pp. 644–652, 2004.

[12] J. She, M. Wu, Y. Lan, and Y. He, "Simultaneous optimisation of the low-pass filter and state-feedback controller in a robust repetitive-control system," *IET Control Theo y and Applications*, vol. 4, no. 8, pp. 1366–1376, 2010.

[13] L. Zhou, J. She, M. Wu, and Y. He, "Design of a robust observer-based modified repetitive-control system," *ISA Transactions*, vol. 52, no. 3, pp. 375–382, 2013.

[14] K. Zhou, J. C. Doyle, and K. Glover, *Robust and Optimal Control*, Simon and Schuster/A Viacom Company/Prentice Hall, Upper Saddle River, NJ, USA, 1996.

[15] D. W. C. Ho and G. Lu, "Robust stabilization for a class of discrete-time non-linear systems via output feedback: the unified LMI approach," *International Journal of Control*, vol. 76, no. 7, pp. 105–115, 2003.

[16] P. P. Khargonekar, I. R. Petersen, and K. Zhou, "Robust stabilization of uncertain linear systems: quadratic stabilizability and H∞ control theory," *IEEE Transactions on Automatic Control*, vol. 35, no. 3, pp. 356–361, 1990.

[17] U. Mackenroth, *Robust Control Systems—Theory and Case Studies*, Springer, Berlin, Germany, 2004.

[18] D. Z. Zheng, *Linear System Theory*, Tsinghua University Publishing House/Springer, Beijing, China, 2002.

[19] M. Doi, M. Masuko, Y. Ito, and A. Tezuka, "A study on parametric vibration in chuck work," *Bulletion of the Japan Society of Mechanical Engineers*, vol. 28, no. 245, pp. 2774–2780, 1985.

[20] Z.-D. Li and W.-D. Yang, "*H∞* robust repetitive control with output feedback for roll eccentricity compensation," *Control The ry and Applications*, vol. 28, no. 3, pp. 381–388, 2011 (Chinese).

A Study of I-Function of Several Complex Variables

Prathima Jayarama,[1,2] Vasudevan Nambisan Theke Madam,[3] and Shantha Kumari Kurumujji[2,4]

[1] *Department of Mathematics, Manipal Institute of Technology, Manipal, Karnataka 576104, India*
[2] *SCSVMV, Sri Jayendra Saraswathi Street, Enathur, Kanchipuram, Tamil Nadu 631561, India*
[3] *Department of Mathematics, College of Engineering, Trikaripur, Kerala 670307, India*
[4] *Department of Mathematics, P.A. College of Engineering, Mangalore, Karnataka 574153, India*

Correspondence should be addressed to Prathima Jayarama; pamrutharaj@yahoo.co.in

Academic Editor: Alberto Cardona

The aim of this paper is to introduce a natural generalization of the well-known, interesting, and useful Fox H-function into generalized function of several variables, namely, the I-function of "r" variables. For $r = 1$, we get the I-function introduced and studied by Arjun Rathie (1997) and, for $r = 2$, we get I-function of two variables introduced very recently by ShanthaKumari et al. (2012). Convergent conditions, elementary properties, and special cases have also been given. The results presented in this paper generalize the results of H-function of "r" variables available in the literature.

1. Introduction

In 1997, Rathie introduced the generalization of the well-known Fox's H-function [1] which has very recently found interesting applications in wireless communication [2–4]. Motivated by the I-function, very recently Shantha Kumari, Nambisan, and Rathie introduced I-function of two variables [5] which is a natural generalization of the H-function of two variables introduced earlier by Mittal and Gupta [6] and discussed some of its important properties.

In the present paper, we aim to develop I-function of "r" variables which may be regarded as the natural generalization of the H-function of "r" variables introduced earlier by Srivastava and Panda [7]. We also discussed some of the important properties.

The remainder of this paper is organized as follows.

In Section 2, we have defined the I-function of "r" variables by means of multiple Mellin-Barnes type contour integrals. In Section 3, we have given the convergence conditions for this function. In Section 4, we obtained the series representation and behaviour of the function for small values of the variables. In Section 5, we have mentioned special cases of our function giving relations with other functions available in the literature. Finally, in Section 6, we have mentioned a few important properties.

2. The I-Function of Several Variables

The generalized Fox H-function, namely, I-function of "r" variables, is defined and represented in the following manner:

$$I\left[z_1,\ldots,z_r\right] = I^{0,n:m_1,n_1;\ldots;m_r,n_r}_{p,q:p_1,q_1;\ldots;p_r,q_r}\left[\begin{array}{c} z_1 \\ \vdots \\ z_r \end{array}\middle| \begin{array}{l} \left(a_j;\alpha_j^{(1)},\ldots,\alpha_j^{(r)};A_j\right)_{1,p} : \left(c_j^{(1)},\gamma_j^{(1)};C_j^{(1)}\right)_{1,p_1};\ldots;\left(c_j^{(r)},\gamma_j^{(r)};C_j^{(r)}\right)_{1,p_r} \\ \left(b_j;\beta_j^{(1)},\ldots,\beta_j^{(r)};B_j\right)_{1,q} : \left(d_j^{(1)},\delta_j^{(1)};D_j^{(1)}\right)_{1,q_1};\ldots;\left(d_j^{(r)},\delta_j^{(r)};D_j^{(r)}\right)_{1,q_r} \end{array}\right] \quad (1)$$

$$= \frac{1}{(2\pi i)^r}\int_{\mathscr{L}_1}\cdots\int_{\mathscr{L}_r}\phi\left(s_1,\ldots,s_r\right)\theta_1\left(s_1\right)\cdots\theta_r\left(s_r\right)z_1^{s_1}\cdots z_r^{s_r}ds_1\cdots ds_r,$$

where $\phi(s_1,\ldots,s_r)$, $\theta_i(s_i)$, $i=1,\ldots,r$ are given by

$$\phi(s_1,\ldots,s_r)=\prod_{j=1}^{n}\Gamma^{A_j}\left(1-a_j+\sum_{i=1}^{r}\alpha_j^{(i)}s_i\right)\times\left(\prod_{j=n+1}^{p}\Gamma^{A_j}\left(a_j-\sum_{i=1}^{r}\alpha_j^{(i)}s_i\right)\times\prod_{j=1}^{q}\Gamma^{B_j}\left(1-b_j+\sum_{i=1}^{r}\beta_j^{(i)}s_i\right)\right)^{-1}, \tag{2}$$

$$\theta_i(s_i)=\left(\prod_{j=1}^{n_i}\Gamma^{C_j^{(i)}}\left(1-c_j^{(i)}+\gamma_j^{(i)}s_i\right)\times\prod_{j=1}^{m_i}\Gamma^{D_j^{(i)}}\left(d_j^{(i)}-\delta_j^{(i)}s_i\right)\right)\times\left(\prod_{j=n_i+1}^{p_i}\Gamma^{C_j^{(i)}}\left(c_j^{(i)}-\gamma_j^{(i)}s_i\right)\times\prod_{j=m_i+1}^{q_i}\Gamma^{D_j^{(i)}}\left(1-d_j^{(i)}+\delta_j^{(i)}s_i\right)\right)^{-1}, \tag{3}$$

where $i=1,\ldots,r$.

Also,

(i) $z_i\neq 0$, for $i=1,\ldots,r$;

(ii) $i=\sqrt{-1}$;

(iii) an empty product is interpreted as unity;

(iv) the parameters m_j, n_j, p_j, q_j $(j=1,\ldots,r)$, n, p, and q are nonnegative integers such that $0\le n\le p$, $q\ge 0$, $0\le n_j\le p_j$, and $0\le m_j\le q_j$ $(j=1,\ldots,r)$ (not all zero simulataneously);

(v) $\alpha_j^{(i)}$ $(j=1,\ldots,p,\ i=1,\ldots,r)$, $\beta_j^{(i)}$ $(j=1,\ldots,q,\ i=1,\ldots,r)$, $\gamma_j^{(i)}$ $(j=1,\ldots,p_i,\ i=1,\ldots,r)$, and $\delta_j^{(i)}$ $(j=1,\ldots,q_i,\ i=1,\ldots,r)$ are assumed to be positive quantities for standardisation purpose. However, the definition of I-function of "r" variables will have a meaning even if some of the quantities are zero or negative numbers. For these, we may obtain corresponding transformation formulas which will be given in a later section;

(vi) a_j $(j=1,\ldots,p)$, b_j $(j=1,\ldots,q)$, $c_j^{(i)}$ $(j=1,\ldots,p_i,\ i=1,\ldots,r)$, and $d_j^{(i)}$ $(j=1,\ldots,q_i,\ i=1,\ldots,r)$ are complex numbers;

(vii) the exponents A_j $(j=1,\ldots,p)$, B_j $(j=1,\ldots,q)$, $C_j^{(i)}$ $(j=1,\ldots,p_i,\ i=1,\ldots,r)$, and $D_j^{(i)}$ $(j=1,\ldots,q_i,\ i=1,\ldots,r)$ of various gamma functions involved in (2) and (3) may take noninteger values;

(viii) the contour $\mathscr{L}_i$ in the complex s_i-plane is of Mellin-Barnes type which runs from $c-i\infty$ to $c+i\infty$, (c real) with indentation, if necessary, in such a manner that all singularities of $\Gamma^{D_j^{(i)}}(d_j^{(i)}-\delta_j^{(i)}s_i)$, $j=1,\ldots,m_i$ lie to the right and $\Gamma^{C_j^{(i)}}(1-c_j^{(i)}+\gamma_j^{(i)}s_i)$, $j=1,\ldots,n_i$ are to the left of $\mathscr{L}_i$.

Following the results of Braaksma [8] the I-function of "r" variables is analytic if

$$\mu_i=\sum_{j=1}^{p}A_j\alpha_j^{(i)}-\sum_{j=1}^{q}B_j\beta_j^{(i)}+\sum_{j=1}^{p_i}C_j^{(i)}\gamma_j^{(i)}-\sum_{j=1}^{q_i}D_j^{(i)}\delta_j^{(i)}\le 0,\quad i=1,\ldots,r. \tag{4}$$

3. Convergence Conditions

Integral (1) converges absolutely if

$$\left|\arg(z_k)\right|<\frac{1}{2}\Delta_k\pi,\quad k=1,\ldots,r, \tag{5}$$

where

$$\Delta_k=\left[-\sum_{j=n+1}^{p}A_j\alpha_j^{(k)}-\sum_{j=1}^{q}B_j\beta_j^{(k)}+\sum_{1}^{m_k}D_j^{(k)}\delta_j^{(k)}-\sum_{m_k+1}^{q_k}D_j^{(k)}\delta_j^{(k)}+\sum_{j=1}^{n_k}C_j^{(k)}\gamma_j^{(k)}-\sum_{j=n_k+1}^{p_k}C_j^{(k)}\gamma_j^{(k)}\right]>0, \tag{6}$$

and if $|\arg(z_k)|=(1/2)\Delta_k\pi$ and $\Delta_k\ge 0$, $k=1,\ldots,r$, then integral (1) converges absolutely under the following conditions:

(i) $\mu_k=0$, $\Omega_k<-1$, where μ_k is given by (4) and

$$\Omega_k=\sum_{j=1}^{p}\left[\frac{1}{2}-\Re(a_j)\right]A_j-\sum_{j=1}^{q}\left[\frac{1}{2}-\Re(b_j)\right]B_j+\sum_{j=1}^{p_k}\left[\frac{1}{2}-\Re\left(c_j^{(k)}\right)\right]C_j^{(k)}-\sum_{j=1}^{q_k}\left[\frac{1}{2}-\Re\left(d_j^{(k)}\right)\right]D_j^{(k)},\quad k=1,\ldots,r; \tag{7}$$

(ii) $\mu_k\neq 0$ with $s_k=\sigma_k+it_k$, (σ_k and t_k are real, $k=1,\ldots,r$), and σ_k are chosen so that for $|t_k|\to\infty$ we have $\Omega_k+\sigma_k\mu_k<-1$.

Outline of the Proof. The convergence of integral (1) depends on the asymptotic behaviour of the functions $\phi(s_1,\ldots,s_r)$, $\theta_i(s_i)$, $i=1,\ldots,r$ defined by (2) and (3), respectively. Such asymptotic behaviour is based on the following relation for the gamma function $\Gamma(z)$, $z = x + iy$, $x, y \in \mathbb{R}$ [9]:

$$\left|\Gamma(x+iy)\right| \sim \sqrt{2\pi}\,|y|^{x-1/2}\exp\left(-\frac{1}{2}\pi|y|\right), \quad |y| \to \infty. \tag{8}$$

Along the contour $\mathscr{L}_k$, if we put $s_k = \sigma_k + it_k$ and take the limit as $|t_k| \to \infty$ for $k = 1, \ldots, r$, we obtain by virtue of (8) that

$$\begin{aligned} &\left|\Gamma^{A_j}\left(1 - a_j + \sum_{k=1}^{r}\alpha_j^{(k)}s_k\right)\right| \\ &\quad \le (2\pi)^{A_j/2}\left(\alpha_j^{(k)}|t_k|\right)^{[1/2-\Re(a_j)+\alpha_j^{(k)}\sigma_k]A_j} \\ &\quad \times \exp\left[-\frac{\pi}{2}\left(\alpha_j^{(k)}|t_k| + |\Im(a_j)|\right)A_j\right], \end{aligned} \tag{9}$$

$$\begin{aligned} &\prod_{j=1}^{n}\left|\Gamma^{A_j}\left(1 - a_j + \sum_{k=1}^{r}\alpha_j^{(k)}s_k\right)\right| \\ &\quad \le (2\pi)^{\sum_{j=1}^{n}(A_j/2)}\prod_{j=1}^{n}\left(\alpha_j^{(k)}|t_k|\right)^{[1/2-\Re(a_j)+\alpha_j^{(k)}\sigma_k]A_j} \\ &\quad \times \exp\left[-\frac{\Pi}{2}\sum_{j=1}^{n}\left(\alpha_j^{(k)}|t_k| + |\Im(a_j)|\right)A_j\right]. \end{aligned} \tag{10}$$

Similarly, we have

$$\begin{aligned} &\prod_{j=n+1}^{p}\left|\Gamma^{A_j}\left(a_j - \sum_{k=1}^{r}\alpha_j^{(k)}s_k\right)\right| \\ &\quad \ge (2\pi)^{\sum_{j=n+1}^{k}(A_j/2)}\prod_{j=n+1}^{k}\left(\alpha_j^{(k)}|t_k|\right)^{[\Re(a_j)-\alpha_j^{(k)}\sigma_k-1/2]A_j} \\ &\quad \times \exp\left[-\frac{\pi}{2}\prod_{j=n+1}^{p}\left(\alpha_j^{(k)}|t_k| + |\Im(a_j)|\right)A_j\right], \\ &\prod_{j=1}^{q}\left|\Gamma^{B_j}\left(1 - b_j + \sum_{k=1}^{r}\beta_j^{(k)}s_k\right)\right| \\ &\quad \ge (2\pi)^{\sum_{j=1}^{q}(B_j/2)}\prod_{j=1}^{q}\left(\beta_j^{(k)}|t_k|\right)^{[1/2-\Re(b_j)+\beta_j^{(k)}\sigma_k]B_j} \\ &\quad \times \exp\left[-\frac{\Pi}{2}\prod_{j=1}^{q}\left(\beta_j^{(k)}|t_k| + |\Im(a_j)|\right)B_j\right], \\ &\prod_{j=1}^{n_k}\left|\Gamma^{C_j^{(k)}}\left(1 - c_j^{(k)} + \gamma_j^{(k)}s_k\right)\right| \\ &\quad \le (2\pi)^{\sum_{j=1}^{n_k}(C_j^{(k)}/2)}\prod_{j=1}^{n_k}\left(\gamma_j^{(k)}|t_k|\right)^{[1/2-\Re(c_j^{(k)})+\gamma_j^{(k)}\sigma_k]C_j^{(k)}} \\ &\quad \times \exp\left[-\frac{\Pi}{2}\prod_{j=1}^{n_k}\left(\gamma_j^{(k)}|t_k| + |\Im(c_j^{(k)})|\right)C_j^{(k)}\right], \\ &\prod_{j=n_k+1}^{p_k}\left|\Gamma^{C_j^{(k)}}\left(c_j^{(k)} - \gamma_j^{(k)}s_k\right)\right| \\ &\quad \ge (2\pi)^{\sum_{j=n_k+1}^{p_k}(C_j^{(k)}/2)}\prod_{j=n_k+1}^{p_k}\left(\gamma_j^{(k)}|t_k|\right)^{[\Re(c_j^{(k)})-\gamma_j^{(k)}\sigma_k-1/2]C_j^{(k)}} \\ &\quad \times \exp\left[-\frac{\pi}{2}\prod_{j=n_k+1}^{p_k}\left(\gamma_j^{(k)}|t_k| + |\Im(c_j^{(k)})|\right)C_j^{(k)}\right], \\ &\prod_{j=1}^{m_k}\left|\Gamma^{D_j^{(k)}}\left(d_j^{(k)} - \delta_j^{(k)}s_k\right)\right| \\ &\quad \le (2\pi)^{\sum_{j=1}^{m_k}(D_j^{(k)}/2)}\prod_{j=1}^{m_k}\left(\delta_j^{(k)}|t_k|\right)^{[\Re(d_j^{(k)})+\delta_j^{(k)}\sigma_k-1/2]D_j^{(k)}} \\ &\quad \times \exp\left[-\frac{\pi}{2}\prod_{j=1}^{m_k}\left(\delta_j^{(k)}|t_k| + |\Im(d_j^{(k)})|\right)D_j^{(k)}\right], \\ &\prod_{j=m_k+1}^{q_k}\left|\Gamma^{D_j^{(k)}}\left(1 - d_j^{(k)} + \delta_j^{(k)}s_k\right)\right| \\ &\quad \le (2\pi)^{\sum_{j=m_k+1}^{q_k}(D_j^{(k)}/2)}\prod_{j=m_k+1}^{q_k}\left(\delta_j^{(k)}|t_k|\right)^{[1/2-\Re(d_j^{(k)})+\delta_j^{(k)}\sigma_k]D_j^{(k)}} \\ &\quad \times \exp\left[-\frac{\pi}{2}\prod_{j=m_k+1}^{q_k}\left(\delta_j^{(k)}|t_k| + |\Im(d_j^{(k)})|\right)D_j^{(k)}\right]. \end{aligned} \tag{11}$$

Also,

$$\begin{aligned} z_k^{s_k} &= \exp\left[(\sigma_k + it_k)(\log|z_k| + i\arg(z_k))\right] \\ &= \exp\left[\sigma_k\log|z_k| - t_k\arg(z_k)\right] \\ &= |z_k|^{\sigma_k}\exp\left[-t_k\arg(z_k)\right]. \end{aligned} \tag{12}$$

Hence, substituting (10)-(11) in (1) and using (12) we have, after much simplification,

$$\begin{aligned} &\left|\phi(s_1, \ldots, s_r)\,\theta_k(s_k)\,z_k^{s_k}\right| \\ &\quad \sim C_k\,|t_k|^{\Omega_k+\mu_k\sigma_k}\exp\left[-t_k\arg(z_k) - \frac{\pi}{2}|t_k|\,\Delta_k\right], \end{aligned} \tag{13}$$

where C_k is independent of t_k and Δ_k, μ_k, and Ω_k are given by (6), (7), and (8), respectively, for each $k = 1, 2, \ldots, r$.

Hence, the result follows.

where $\phi(s_1,\dots,s_r)$, $\theta_i(s_i)$, $i=1,\dots,r$ are given by

$$\phi(s_1,\dots,s_r)=\prod_{j=1}^{n}\Gamma^{A_j}\left(1-a_j+\sum_{i=1}^{r}\alpha_j^{(i)}s_i\right)\times\left(\prod_{j=n+1}^{p}\Gamma^{A_j}\left(a_j-\sum_{i=1}^{r}\alpha_j^{(i)}s_i\right)\times\prod_{j=1}^{q}\Gamma^{B_j}\left(1-b_j+\sum_{i=1}^{r}\beta_j^{(i)}s_i\right)\right)^{-1}, \quad (2)$$

$$\theta_i(s_i)=\left(\prod_{j=1}^{n_i}\Gamma^{C_j^{(i)}}\left(1-c_j^{(i)}+\gamma_j^{(i)}s_i\right)\times\prod_{j=1}^{m_i}\Gamma^{D_j^{(i)}}\left(d_j^{(i)}-\delta_j^{(i)}s_i\right)\right)\times\left(\prod_{j=n_i+1}^{p_i}\Gamma^{C_j^{(i)}}\left(c_j^{(i)}-\gamma_j^{(i)}s_i\right)\times\prod_{j=m_i+1}^{q_i}\Gamma^{D_j^{(i)}}\left(1-d_j^{(i)}+\delta_j^{(i)}s_i\right)\right)^{-1}, \quad (3)$$

where $i=1,\dots,r$.

Also,

(i) $z_i\neq 0$, for $i=1,\dots,r$;

(ii) $i=\sqrt{-1}$;

(iii) an empty product is interpreted as unity;

(iv) the parameters m_j, n_j, p_j, q_j $(j=1,\dots,r)$, n, p, and q are nonnegative integers such that $0\leq n\leq p$, $q\geq 0$, $0\leq n_j\leq p_j$, and $0\leq m_j\leq q_j$ $(j=1,\dots,r)$ (not all zero simulataneously);

(v) $\alpha_j^{(i)}$ $(j=1,\dots,p,\ i=1,\dots,r)$, $\beta_j^{(i)}$ $(j=1,\dots,q,\ i=1,\dots,r)$, $\gamma_j^{(i)}$ $(j=1,\dots,p_i,\ i=1,\dots,r)$, and $\delta_j^{(i)}$ $(j=1,\dots,q_i,\ i=1,\dots,r)$ are assumed to be positive quantities for standardisation purpose. However, the definition of I-function of "r" variables will have a meaning even if some of the quantities are zero or negative numbers. For these, we may obtain corresponding transformation formulas which will be given in a later section;

(vi) a_j $(j=1,\dots,p)$, b_j $(j=1,\dots,q)$, $c_j^{(i)}$ $(j=1,\dots,p_i,\ i=1,\dots,r)$, and $d_j^{(i)}$ $(j=1,\dots,q_i,\ i=1,\dots,r)$ are complex numbers;

(vii) the exponents A_j $(j=1,\dots,p)$, B_j $(j=1,\dots,q)$, $C_j^{(i)}$ $(j=1,\dots,p_i,\ i=1,\dots,r)$, and $D_j^{(i)}$ $(j=1,\dots,q_i,\ i=1,\dots,r)$ of various gamma functions involved in (2) and (3) may take noninteger values;

(viii) the contour $\mathscr{L}_i$ in the complex s_i-plane is of Mellin-Barnes type which runs from $c-i\infty$ to $c+i\infty$, (c real) with indentation, if necessary, in such a manner that all singularities of $\Gamma^{D_j^{(i)}}(d_j^{(i)}-\delta_j^{(i)}s_i)$, $j=1,\dots,m_i$ lie to the right and $\Gamma^{C_j^{(i)}}(1-c_j^{(i)}+\gamma_j^{(i)}s_i)$, $j=1,\dots,n_i$ are to the left of $\mathscr{L}_i$.

Following the results of Braaksma [8] the I-function of "r" variables is analytic if

$$\mu_i=\sum_{j=1}^{p}A_j\alpha_j^{(i)}-\sum_{j=1}^{q}B_j\beta_j^{(i)}+\sum_{j=1}^{p_i}C_j^{(i)}\gamma_j^{(i)}-\sum_{j=1}^{q_i}D_j^{(i)}\delta_j^{(i)}\leq 0,\quad i=1,\dots,r. \quad (4)$$

3. Convergence Conditions

Integral (1) converges absolutely if

$$\left|\arg(z_k)\right|<\frac{1}{2}\Delta_k\pi,\quad k=1,\dots,r, \quad (5)$$

where

$$\Delta_k=\left[-\sum_{j=n+1}^{p}A_j\alpha_j^{(k)}-\sum_{j=1}^{q}B_j\beta_j^{(k)}+\sum_{1}^{m_k}D_j^{(k)}\delta_j^{(k)}-\sum_{m_k+1}^{q_k}D_j^{(k)}\delta_j^{(k)}+\sum_{j=1}^{n_k}C_j^{(k)}\gamma_j^{(k)}-\sum_{j=n_k+1}^{p_k}C_j^{(k)}\gamma_j^{(k)}\right]>0, \quad (6)$$

and if $|\arg(z_k)|=(1/2)\Delta_k\pi$ and $\Delta_k\geq 0$, $k=1,\dots,r$, then integral (1) converges absolutely under the following conditions:

(i) $\mu_k=0$, $\Omega_k<-1$, where μ_k is given by (4) and

$$\Omega_k=\sum_{j=1}^{p}\left[\frac{1}{2}-\Re\left(a_j\right)\right]A_j-\sum_{j=1}^{q}\left[\frac{1}{2}-\Re\left(b_j\right)\right]B_j+\sum_{j=1}^{p_k}\left[\frac{1}{2}-\Re\left(c_j^{(k)}\right)\right]C_j^{(k)}-\sum_{j=1}^{q_k}\left[\frac{1}{2}-\Re\left(d_j^{(k)}\right)\right]D_j^{(k)},\quad k=1,\dots,r; \quad (7)$$

(ii) $\mu_k\neq 0$ with $s_k=\sigma_k+it_k$, (σ_k and t_k are real, $k=1,\dots,r$), and σ_k are chosen so that for $|t_k|\to\infty$ we have $\Omega_k+\sigma_k\mu_k<-1$.

Outline of the Proof. The convergence of integral (1) depends on the asymptotic behaviour of the functions $\phi(s_1,\dots,s_r)$, $\theta_i(s_i)$, $i=1,\dots,r$ defined by (2) and (3), respectively. Such asymptotic behaviour is based on the following relation for the gamma function $\Gamma(z)$, $z = x + iy$, $x, y \in \mathbb{R}$ [9]:

$$\left|\Gamma(x+iy)\right| \sim \sqrt{2\pi}\,|y|^{x-1/2}\exp\left(-\frac{1}{2}\pi|y|\right), \quad |y| \to \infty. \tag{8}$$

Along the contour $\mathscr{L}_k$, if we put $s_k = \sigma_k + it_k$ and take the limit as $|t_k| \to \infty$ for $k = 1,\ldots,r$, we obtain by virtue of (8) that

$$\begin{aligned}&\left|\Gamma^{A_j}\left(1-a_j+\sum_{k=1}^{r}\alpha_j^{(k)}s_k\right)\right|\\ &\quad\le (2\pi)^{A_j/2}\left(\alpha_j^{(k)}|t_k|\right)^{[1/2-\Re(a_j)+\alpha_j^{(k)}\sigma_k]A_j}\\ &\quad\times\exp\left[-\frac{\pi}{2}\left(\alpha_j^{(k)}|t_k|+|\Im(a_j)|\right)A_j\right],\end{aligned} \tag{9}$$

$$\begin{aligned}&\prod_{j=1}^{n}\left|\Gamma^{A_j}\left(1-a_j+\sum_{k=1}^{r}\alpha_j^{(k)}s_k\right)\right|\\ &\quad\le (2\pi)^{\sum_{j=1}^{n}(A_j/2)}\prod_{j=1}^{n}\left(\alpha_j^{(k)}|t_k|\right)^{[1/2-\Re(a_j)+\alpha_j^{(k)}\sigma_k]A_j}\\ &\quad\times\exp\left[-\frac{\Pi}{2}\sum_{j=1}^{n}\left(\alpha_j^{(k)}|t_k|+|\Im(a_j)|\right)A_j\right].\end{aligned} \tag{10}$$

Similarly, we have

$$\begin{aligned}&\prod_{j=n+1}^{p}\left|\Gamma^{A_j}\left(a_j-\sum_{k=1}^{r}\alpha_j^{(k)}s_k\right)\right|\\ &\quad\ge (2\pi)^{\sum_{j=n+1}^{k}(A_j/2)}\prod_{j=n+1}^{k}\left(\alpha_j^{(k)}|t_k|\right)^{[\Re(a_j)-\alpha_j^{(k)}\sigma_k-1/2]A_j}\\ &\quad\times\exp\left[-\frac{\pi}{2}\prod_{j=n+1}^{p}\left(\alpha_j^{(k)}|t_k|+|\Im(a_j)|\right)A_j\right],\end{aligned}$$

$$\begin{aligned}&\prod_{j=1}^{q}\left|\Gamma^{B_j}\left(1-b_j+\sum_{k=1}^{r}\beta_j^{(k)}s_k\right)\right|\\ &\quad\ge (2\pi)^{\sum_{j=1}^{q}(B_j/2)}\prod_{j=1}^{q}\left(\beta_j^{(k)}|t_k|\right)^{[1/2-\Re(b_j)+\beta_j^{(k)}\sigma_k]B_j}\\ &\quad\times\exp\left[-\frac{\Pi}{2}\prod_{j=1}^{q}\left(\beta_j^{(k)}|t_k|+|\Im(a_j)|\right)B_j\right],\end{aligned}$$

$$\begin{aligned}&\prod_{j=1}^{n_k}\left|\Gamma^{C_j^{(k)}}\left(1-c_j^{(k)}+\gamma_j^{(k)}s_k\right)\right|\\ &\quad\le (2\pi)^{\sum_{j=1}^{n_k}(C_j^{(k)}/2)}\prod_{j=1}^{n_k}\left(\gamma_j^{(k)}|t_k|\right)^{[1/2-\Re(c_j^{(k)})+\gamma_j^{(k)}\sigma_k]C_j^{(k)}}\\ &\quad\times\exp\left[-\frac{\Pi}{2}\prod_{j=1}^{n_k}\left(\gamma_j^{(k)}|t_k|+\left|\Im\left(c_j^{(k)}\right)\right|\right)C_j^{(k)}\right],\end{aligned}$$

$$\begin{aligned}&\prod_{j=n_k+1}^{p_k}\left|\Gamma^{C_j^{(k)}}\left(c_j^{(k)}-\gamma_j^{(k)}s_k\right)\right|\\ &\quad\ge (2\pi)^{\sum_{j=n_k+1}^{p_k}(C_j^{(k)}/2)}\prod_{j=n_k+1}^{p_k}\left(\gamma_j^{(k)}|t_k|\right)^{[\Re(c_j^{(k)})-\gamma_j^{(k)}\sigma_k-1/2]C_j^{(k)}}\\ &\quad\times\exp\left[-\frac{\pi}{2}\prod_{j=n_k+1}^{p_k}\left(\gamma_j^{(k)}|t_k|+\left|\Im\left(c_j^{(k)}\right)\right|\right)C_j^{(k)}\right],\end{aligned}$$

$$\begin{aligned}&\prod_{j=1}^{m_k}\left|\Gamma^{D_j^{(k)}}\left(d_j^{(k)}-\delta_j^{(k)}s_k\right)\right|\\ &\quad\le (2\pi)^{\sum_{j=1}^{m_k}(D_j^{(k)}/2)}\prod_{j=1}^{m_k}\left(\delta_j^{(k)}|t_k|\right)^{[\Re(d_j^{(k)})+\delta_j^{(k)}\sigma_k-1/2]D_j^{(k)}}\\ &\quad\times\exp\left[-\frac{\pi}{2}\prod_{j=1}^{m_k}\left(\delta_j^{(k)}|t_k|+\left|\Im\left(d_j^{(k)}\right)\right|\right)D_j^{(k)}\right],\end{aligned}$$

$$\begin{aligned}&\prod_{j=m_k+1}^{q_k}\left|\Gamma^{D_j^{(k)}}\left(1-d_j^{(k)}+\delta_j^{(k)}s_k\right)\right|\\ &\quad\le (2\pi)^{\sum_{j=m_k+1}^{q_k}(D_j^{(k)}/2)}\prod_{j=m_k+1}^{q_k}\left(\delta_j^{(k)}|t_k|\right)^{[1/2-\Re(d_j^{(k)})+\delta_j^{(k)}\sigma_k]D_j^{(k)}}\\ &\quad\times\exp\left[-\frac{\pi}{2}\prod_{j=m_k+1}^{q_k}\left(\delta_j^{(k)}|t_k|+\left|\Im\left(d_j^{(k)}\right)\right|\right)D_j^{(k)}\right].\end{aligned} \tag{11}$$

Also,

$$\begin{aligned}z_k^{s_k} &= \exp\left[(\sigma_k+it_k)\left(\log|z_k|+i\arg(z_k)\right)\right]\\ &= \exp\left[\sigma_k\log|z_k|-t_k\arg(z_k)\right]\\ &= |z_k|^{\sigma_k}\exp\left[-t_k\arg(z_k)\right].\end{aligned} \tag{12}$$

Hence, substituting (10)-(11) in (1) and using (12) we have, after much simplification,

$$\begin{aligned}&\left|\phi(s_1,\ldots,s_r)\,\theta_k(s_k)\,z_k^{s_k}\right|\\ &\quad\sim C_k\,|t_k|^{\Omega_k+\mu_k\sigma_k}\exp\left[-t_k\arg(z_k)-\frac{\pi}{2}|t_k|\,\Delta_k\right],\end{aligned} \tag{13}$$

where C_k is independent of t_k and Δ_k, μ_k, and Ω_k are given by (6), (7), and (8), respectively, for each $k = 1, 2, \ldots, r$.

Hence, the result follows.

Remark 1. If $D_j^{(i)} = 1$ $(j = 1, \ldots, m_i,\ i = 1, \ldots, r)$ in (1), then the function will be denoted by

$$\bar{\mathrm{I}}\begin{bmatrix} z_1 \\ \vdots \\ z_r \end{bmatrix} = \mathrm{I}^{0,n:m_1,n_1;\ldots;m_r,n_r}_{p,q:p_1,q_1;\ldots;p_r,q_r} \times \left[\begin{matrix} z_1 \\ \vdots \\ z_r \end{matrix}\ \middle|\ \begin{matrix} \left(a_j;\alpha_j^{(1)},\ldots,\alpha_j^{(r)};A_j\right)_{1,p} : \left(c_j^{(1)},\gamma_j^{(1)};C_j^{(1)}\right)_{1,p_1};\ldots;\left(c_j^{(r)},\gamma_j^{(r)};C_j^{(r)}\right)_{1,p_r} \\ \\ \left(b_j;\beta_j^{(1)},\ldots,\beta_j^{(r)};B_j\right)_{1,q} : \left(d_j^{(1)},\delta_j^{(1)};1\right)_{1,m_1},\left(d_j^{(1)},\delta_j^{(1)};D_j^{(1)}\right)_{m_1+1,q_1};\ldots;\left(d_j^{(r)},\delta_j^{(r)};1\right)_{1,m_r},\left(d_j^{(r)},\delta_j^{(r)};D_j^{(r)}\right)_{m_r+1,q_r} \end{matrix}\right]$$
$$= \frac{1}{(2\pi i)^r}\int_{\mathscr{L}_1}\cdots\int_{\mathscr{L}_r}\phi(s_1,\ldots,s_r)\,\overline{\theta_1}(s_1)\cdots\overline{\theta_r}(s_r)\,z_1^{s_1}\cdots z_r^{s_r}\,ds_1\cdots ds_r, \tag{14}$$

where

$$\overline{\theta_i}(s_i) = \prod_{j=1}^{n_i}\Gamma^{C_j^{(i)}}\left(1-c_j^{(i)}+\gamma_j^{(i)}s_i\right)\prod_{j=1}^{m_i}\Gamma\left(d_j^{(i)}-\delta_j^{(i)}s_i\right) \times\left(\prod_{j=n_i+1}^{p_i}\Gamma^{C_j^{(i)}}\left(c_j^{(i)}-\gamma_j^{(i)}s_i\right)\prod_{j=m_i+1}^{q_i}\Gamma^{D_j^{(i)}}\left(1-d_j^{(i)}+\delta_j^{(i)}s_i\right)\right)^{-1}, \tag{15}$$

where $i = 1, \ldots, r$.

Remark 2. If $C_j^{(i)} = 1$ $(j = 1, \ldots, n_i)$, $D_j^{(i)} = 1$ $(j = 1, \ldots, m_i)$, where $i = 1, \ldots, r$ and if $n = 0$ in (1), then the corresponding function will be denoted by

$$\bar{\mathrm{I}}_1\begin{bmatrix} z_1 \\ \vdots \\ z_r \end{bmatrix} = \mathrm{I}^{0,0:m_1,n_1;\ldots;m_r,n_r}_{p,q:p_1,q_1;\ldots;p_r,q_r} \times \left[\begin{matrix} z_1 \\ \vdots \\ z_r \end{matrix}\ \middle|\ \begin{matrix} \left(a_j;\alpha_j^{(1)},\ldots,\alpha_j^{(r)};A_j\right)_{1,p} : \left(c_j^{(1)},\gamma_j^{(1)};1\right)_{1,n_1},\left(c_j^{(1)},\gamma_j^{(1)};C_j^{(1)}\right)_{n_1+1,p_1};\ldots;\left(c_j^{(r)},\gamma_j^{(r)};1\right)_{1,n_r},\left(c_j^{(r)},\gamma_j^{(r)};C_j^{(r)}\right)_{n_r+1,p_r} \\ \\ \left(b_j;\beta_j^{(1)},\ldots,\beta_j^{(r)};B_j\right)_{1,q} : \left(d_j^{(1)},\delta_j^{(1)};1\right)_{1,m_1},\left(d_j^{(1)},\delta_j^{(1)};D_j^{(1)}\right)_{m_1+1,q_1};\ldots;\left(d_j^{(r)},\delta_j^{(r)};1\right)_{1,m_r},\left(d_j^{(r)},\delta_j^{(r)};D_j^{(r)}\right)_{m_r+1,q_r} \end{matrix}\right]$$
$$= \frac{1}{(2\pi i)^r}\int_{\mathscr{L}_1}\cdots\int_{\mathscr{L}_r}\phi_1(s_1,\ldots,s_r)\,\overline{\theta_1}(s_1)\cdots\overline{\theta_r}(s_r)\,z_1^{s_1}\cdots z_r^{s_r}ds_1\cdots ds_r, \tag{16}$$

where

$$\phi_1(s_1,\ldots,s_r) = \frac{1}{\prod_{j-1}^{p}\Gamma^{A_j}\left(a_j-\sum_{i-1}^{r}\alpha_j^{(i)}s_i\right)\prod_{j=1}^{q}\Gamma^{B_j}\left(1-b_j+\sum_{i=1}^{r}\beta_j^{(i)}s_i\right)},$$

$$\overline{\theta_i}(s_i) = \frac{\prod_{j=1}^{n_i}\Gamma\left(1-c_j^{(i)}+\gamma_j^{(i)}s_i\right)\prod_{j=1}^{m_i}\Gamma\left(d_j^{(i)}-\delta_j^{(i)}s_i\right)}{\prod_{j=n_i+1}^{p_i}\Gamma^{C_j^{(i)}}\left(c_j^{(i)}-\gamma_j^{(i)}s_i\right)\prod_{j=m_i+1}^{q_i}\Gamma^{D_j^{(i)}}\left(1-d_j^{(i)}+\delta_j^{(i)}s_i\right)} \quad \forall i = 1,\ldots,r. \tag{17}$$

4. Series Representation

if

(i) $z_i \neq 0$, $(i = 1, \ldots, r)$ and $\mu_i < 0$, where μ_i is given by (4);

(ii) $\delta_{h_i}^{(i)}(d_j^{(i)}+k_i) \neq \delta_j^{(i)}(d_{h_i}^{(i)}+\eta_i)$ for $j \neq h_i$, $j, h_i = 1, \ldots, m_i$, $(i = 1, \ldots, r)$, $k_i, \eta_i = 0, 1, 2, \ldots$ $(i = 1, \ldots, r)$,

then

$$\bar{\mathrm{I}}\begin{bmatrix} z_1 \\ \vdots \\ z_r \end{bmatrix} = \sum_{h_1=1}^{m_1}\cdots\sum_{h_r=1}^{m_r}\sum_{k_1=1}^{\infty}\cdots\sum_{k_r=1}^{\infty} \times\left[\phi_1\left(\frac{dh_1^{(1)}+k_1}{\delta h_1^{(1)}},\ldots,\frac{dh_r^{(r)}+k_r}{\delta h_r^{(r)}}\right) \times\prod_{i=1}^{r}\frac{(-1)^{k_i}}{\delta h_i^{(i)}k_i!}z_i^{(dh_i+k_i)/\delta h_i}\right]_{j\neq h_i}. \tag{18}$$

This result can be proved on computing the residues at the

poles as follows:

$$s_r = \frac{dh_i^{(i)} + k_i}{\delta h_i^{(i)}}, \quad (h_i = 1,\ldots,m_i,\ k_i = 0,1,2,\ldots) \text{ for } i = 1,\ldots,r. \tag{19}$$

The behaviour of the function $\bar{\mathrm{I}}\begin{bmatrix} z_1 \\ \vdots \\ z_r \end{bmatrix}$ is given by

$$\bar{\mathrm{I}}\begin{bmatrix} z_1 \\ \vdots \\ z_r \end{bmatrix} = O\left(\prod_{j=1}^{r} |z_i|^{\phi_j}\right), \quad \max\{|z_1|,\ldots,|z_r|\} \longrightarrow 0, \tag{20}$$

where

$$\phi_j = \min_{1\le j\le m_i}\left[\mathrm{Re}\left(\frac{d_j^{(i)}}{\delta_j^{(i)}}\right)\right], \quad (i = 1,\ldots,r). \tag{21}$$

On the other hand, when $|z_i| \rightarrow \infty$ $(i = 1,\ldots,r)$, the associated function $\bar{\mathrm{I}}_1\begin{bmatrix} z_1 \\ \vdots \\ z_r \end{bmatrix}$ given by (16) has the behaviour

$$\bar{\mathrm{I}}_1\begin{bmatrix} z_1 \\ \vdots \\ z_r \end{bmatrix} = O\left(\prod_{j=1}^{r} |z_j|^{\phi_j}\right), \quad \min\{|z_1|,\ldots,|z_r|\} \longrightarrow 0, \tag{22}$$

where

$$\phi_j = \max_{1\le j\le n_i}\left[\mathrm{Re}\left(\frac{1-c_j^{(i)}}{\gamma_j^{(i)}}\right)\right], \quad (i = 1,\ldots,r). \tag{23}$$

5. Elementary Special Cases

In this section, we mention some interesting and useful special cases of the I-function of "r" variables.

(i) If all the exponents A_j $(j = 1,\ldots,p)$, B_j $(j = 1,\ldots,q)$, $C_j^{(i)}$ $(j = 1,\ldots,p_i,\ i = 1,\ldots,r)$, and $D_j^{(i)}$ $(j = 1,\ldots,q_i,\ i = 1,\ldots,r)$ in (1) are equal to unity, we obtain H-function of "r" variables defined by Srivastava and Panda [7].

(ii) When $p = q = n = 0$, (1) degenerates into the product of r mutually independent I- functions of one variable introduced by Rathie [1].

(iii) When $p = q = n = 0$ and $r = 1$, (1) reduces to the I-function defined by Rathie [1].

(iv) When $n = p$, $m_i = 1$, $n_i = p_i$, $i = 1,\ldots,r$, and $A_j = B_j = C_j = D_j = 1$ and $(d_j^{(i)}, \delta_j^{(i)}; D_j^{(i)})$ is replaced by $(0,1,1)$, $(d_j^{(i)}, \delta_j^{(i)}; D_j^{(i)})$, (1) reduces to the generalized Lauricella function [10].

$$\mathrm{I}^{0,p:1,p_1;\ldots;1,p_r}_{p,q:p_1,q_1+1;\ldots;p_r,q_r+1}\left[\begin{matrix} z_1 \\ \vdots \\ z_r \end{matrix}\ \middle|\ \begin{matrix} \left(a_j;\alpha_j^{(1)},\ldots,\alpha_j^{(r)};1\right)_{1,p} : \left(c_j^{(1)},\gamma_j^{(1)};1\right)_{1,p_1};\ldots;\left(c_j^{(r)},\gamma_j^{(r)};1\right)_{1,p_r} \\ \left(b_j;\beta_j^{(1)},\beta_j^{(1)},\ldots,\beta_j^{(r)};1\right)_{1,q} : (0,1;1),\left(d_j^{(1)},\delta_j^{(1)};1\right)_{1,q_1};\ldots;(0,1;1),\left(d_j^{(r)},\delta_j^{(r)};1\right)_{1,q_r} \end{matrix}\right]$$
$$= \frac{\prod_{j=1}^{p}\Gamma\left(1-a_j\right)\prod_{j=1}^{p_1}\Gamma\left(1-c_j^{(1)}\right)\cdots\prod_{j=1}^{p_r}\Gamma\left(1-c_j^{(r)}\right)}{\prod_{j=1}^{q}\Gamma\left(1-b_j\right)\prod_{j=1}^{q_1}\Gamma\left(1-d_j^{(1)}\right)\cdots\prod_{j=1}^{q_r}\Gamma\left(1-d_j^{(r)}\right)} \tag{24}$$
$$\times F^{p:p_1;\ldots;p_r}_{q:q_1;\ldots;q_r}\left[\begin{matrix} -z_1 \\ \vdots \\ -z_r \end{matrix}\ \middle|\ \begin{matrix} \left(1-a_j;\alpha_j^{(1)},\ldots,\alpha_j^{(r)}\right)_{1,p} : \left(1-c_j^{(1)},\gamma_j^{(1)}\right)_{1,p_1};\ldots;\left(1-c_j^{((r))},\gamma_j^{(r)}\right)_{1,p_r} \\ \left(1-b_j;\beta_j^{(1)},\ldots,\beta_j^{(r)}\right)_{1,q} : \left(1-d_j^{(1)},\delta_j^{(1)}\right)_{1,q_1};\ldots;\left(1-d_j^{(r)},\delta_j^{(r)}\right)_{1,q_r} \end{matrix}\right].$$

(v)
$$\mathrm{I}^{0,0:1,p_1;\ldots 1,p_r}_{0,0:p_1,q_1+1;\ldots;p_r,q_r+1}\left[\begin{matrix} -z_1 \\ \vdots \\ -z_r \end{matrix}\ \middle|\ \begin{matrix} - : \left(1-c_j^{(1)},\gamma_j^{(1)};1\right)_{1,p_1};\ldots;\left(1-c_j^{(r)},\gamma_j^{(r)};1\right)_{1,p_r} \\ - : (0,1;1),\left(1-d_j^{(1)},\delta_j^{(1)};1\right)_{1,q_1};\ldots;(0,1;1),\left(1-d_j^{(r)},\delta_j^{(r)};1\right)_{1,q_r} \end{matrix}\right]$$
$$= {}_{p_1}\Psi_{q_1}\left[\begin{matrix} \left(c_j^{(1)},\gamma_j^{(1)}\right)_{1,p_1} \\ \left(d_j^{(1)},\delta_j^{(1)}\right)_{1,q_1} \end{matrix};z_1\right]\times\cdots\times {}_{p_r}\Psi_{q_r}\left[\begin{matrix} \left(c_j^{(r)},\gamma_j^{(r)}\right)_{1,p_r} \\ \left(d_j^{(r)},\delta_j^{(r)}\right)_{1,q_r} \end{matrix};z_r\right], \tag{25}$$

where the functions ${}_{p_i}\Psi_{q_i}$, $i = 1, \ldots, r$ are Wright's generalized hypergeometric functions [11].

$$\text{(vi)} \quad \mathrm{I}_{0,0:0,2;\ldots;0,2}^{0,0:1,0;\ldots;1,0}\left[\begin{matrix} z_1 \\ \vdots \\ z_r \end{matrix}\,\middle|\, \begin{matrix} -:-;\ldots;- \\ -:(0,1;1),(-\mu_1,\alpha_1;1);\ldots;(0,1;1),(-\mu_r,\alpha_r;1) \end{matrix}\right] = \prod_{i=1}^{r} J_{\mu_i}^{\alpha_i}(z_i), \tag{26}$$

where the functions $J_{\mu_i}^{\alpha_i}(z_i)$ are Wright's generalized Bessel functions [12].

$$\text{(vii)} \quad \mathrm{I}_{0,0:2,2;\ldots;2,2}^{0,0:1,2;\ldots;1,2}\left[\begin{matrix} -z_1 \\ \vdots \\ -z_r \end{matrix}\,\middle|\, \begin{matrix} -:(1,1;1),(1\text{-}\alpha_1,1;\mu_1);\ldots;(1,1;1),(1\text{-}\alpha_r,1;\mu_r) \\ -:(0,1;1),(\text{-}\alpha_1,1;\mu_1);(0,1;1),(\text{-}\alpha_r,1;\mu_r) \end{matrix}\right] = \prod_{i=1}^{r} \Phi(z_i,\mu_i,\alpha_i), \tag{27}$$

where $\Phi(z_i, \mu_i, \alpha_i)$, $i = 1, \ldots, r$ are the generalized Riemann zeta functions [13, page 27, 1.11, (1)], which are the generalizations of Hurwitz zeta functions and Riemann zeta functions [13, page 24, 1.10, (1) and 1.12, (1)].

$$\text{(viii)} \quad \mathrm{I}_{0,0:2,2;\ldots;2,2}^{0,0:1,2;\ldots;1,2}\left[\begin{matrix} -z_1 \\ \vdots \\ -z_r \end{matrix}\,\middle|\, \begin{matrix} -:(1,1;1),(1,1;\mu_1);\ldots;(1,1;1),(1,1;\mu_r) \\ -:(0,1;1),(0,1;\mu_1);\ldots;(0,1;1),(0,1;\mu_r) \end{matrix}\right] = \prod_{i=1}^{r} F(z_i,\mu_i), \tag{28}$$

where $F(z_i, \mu_i)$ are the polylogarithms of order μ_i. For $\mu_i = 2$, $i = 1, \ldots, r$, the R.H.S. of (28) reduces to the product of Euler's dilogarithm [13, page 31, 1.11.1, equation (2)].

6. Elementary Properties and Transformation Formulas

The properties given below are immediate consequence of the definition (1) and hence they are given here without proof:

$$\begin{aligned} &\text{(i)} \quad \mathrm{I}_{p,q:p_1,q_1;\ldots;p_r,q_r}^{0,0:m_1,n_1;\ldots;m_r,n_r}\left[\begin{matrix} z_1 \\ \vdots \\ z_r \end{matrix}\,\middle|\, \begin{matrix} \left(a_j;\alpha_j^{(1)},\ldots,\alpha_j^{(r)};A_j\right)_{1,p} : \left(c_j^{(1)},\gamma_j^{(1)};C_j^{(1)}\right)_{1,p_1};\ldots;\left(c_j^{(r)},\gamma_j^{(r)};C_j^{(r)}\right)_{1,p_r} \\ \left(b_j;\beta_j^{(1)},\ldots,\beta_j^{(r)}B_j\right)_{1,q} : \left(d_j^{(1)},\delta_j^{(1)};D_j^{(1)}\right)_{1,q_1};\ldots;\left(d_j^{(r)},\delta_j^{(r)};D_j^{(r)}\right)_{1,q_r} \end{matrix}\right] \\ &= \mathrm{I}_{q,p:q_1,p_1;\ldots;q_r,p_r}^{0,0:n_1,m_1;\ldots;n_r,m_r}\left[\begin{matrix} z_1^{-1} \\ \vdots \\ z_r^{-1} \end{matrix}\,\middle|\, \begin{matrix} \left(1-b_j;\beta_j^{(1)},\ldots,\beta_j^{(r)};B_j\right)_{1,q} : \left(1-d_j^{(1)},\delta_j^{(1)};D_j^{(1)}\right)_{1,q_1};\ldots;\left(1-d_j^{(r)},\delta_j^{(r)};D_j^{(r)}\right)_{1,q_r} \\ \left(1-a_j;\alpha_j^{(1)},\ldots,\alpha_j^{(r)};A_j\right)_{1,p} : \left(1-c_j^{(1)},\gamma_j^{(1)};C_j^{(1)}\right)_{1,p_1};\left(1-c_j^{(r)},\gamma_j^{(r)};C_j^{(r)}\right)_{1,p_r} \end{matrix}\right], \end{aligned} \tag{29}$$

$$\begin{aligned} &\text{(ii)} \quad z_1^{k_1}\cdots z_r^{k_r}\mathrm{I}\left[z_1\cdots z_r\right] \\ &= \mathrm{I}_{p,q:p_1,q_1;\ldots;p_r,q_r}^{0,n:m_1,n_1;\ldots;m_r,n_r}\left[\begin{matrix} z_1 \\ \vdots \\ z_r \end{matrix}\,\middle|\, \begin{matrix} \left(a_j+\sum_{i=1}^{r}k_i\alpha_j^{(i)};\alpha_j^{(1)},\ldots,\alpha_j^{(r)};A_j\right)_{1,p} : \left(c_j^{(1)}+k_1\gamma_j^{(1)},\gamma_j^{(1)};C_j^{(1)}\right)_{1,p_1};\ldots;\left(c_j^{(r)}+k_r\gamma_j^{(r)},\gamma_j^{(r)};C_j^{(r)}\right)_{1,p_r} \\ \left(b_j+\sum_{i=1}^{r}k_i\beta_j^{(i)};\beta_j^{(1)},\ldots,\beta_j^{(r)};B_j\right)_{1,q} : \left(d_j^{(1)}+k_1\delta_j^{(1)},\delta_j^{(1)};D_j^{(1)}\right)_{1,q_1};\ldots;\left(d_j^{(r)}+k_r\delta_j^{(r)},\delta_j^{(r)};D_j^{(r)}\right)_{1,q_r} \end{matrix}\right], \end{aligned} \tag{30}$$

for $k_i > 0$, $i = 1, \ldots r$,

(iii) $$\frac{1}{k_1}\cdots\frac{1}{k_r}\mathrm{I}\left[z_1,\ldots z_r\right]$$

$$= \mathrm{I}^{0,n:m_1,n_1;\ldots;m_r,n_r}_{p,q:p_1,q_1;\ldots;p_r,q_r}\left[\begin{array}{c|l} z_1^{k_1} & \left(a_j;k_1\alpha_j^{(1)},\ldots,k_r\alpha_j^{(r)};A_j\right)_{1,p}:\left(c_j^{(1)},k_1\gamma_j^{(1)};C_j^{(1)}\right)_{1,p_1};\ldots;\left(c_j^{(r)},k_r\gamma_j^{(r)};C_j^{(r)}\right)_{1,p_r} \\ \vdots & \\ z_r^{k_r} & \left(b_j;k_1\beta_j^{(1)},\ldots,k_r\beta_j^{(r)};B_j\right)_{1,p}:\left(d_j^{(1)},k_1\delta_j^{(1)};D_j^{(1)}\right)_{1,q_1};\ldots;\left(d_j^{(r)},k_r\delta_j^{(r)};D_j^{(r)}\right)_{1,q_r}\end{array}\right], \tag{31}$$

where $k_i > 0, i = 1,\ldots,r$,

(iv) $$\mathrm{I}^{0,n:m_1,n_1;\ldots;m_r,n_r}_{p,q:p_1,q_1;\ldots;p_r,q_r}\left[\begin{array}{c|l} z_1 & (a;\alpha,0,\ldots,0;A),\left(a_j;\alpha_j^{(1)},\ldots,\alpha_j^{(r)};A_j\right)_{2,p}:\left(c_j^{(1)},\gamma_j^{(1)};C_j^{(1)}\right)_{1,p_1};\ldots;\left(c_j^{(r)},\gamma_j^{(r)};C_j^{(r)}\right)_{1,p_r} \\ \vdots & \\ z_r & \left(b_j;\beta_j^{(1)},\ldots,\beta_j^{(r)};B_j\right)_{1,q}:\left(d_j^{(1)},\delta_j^{(1)};D_j^{(1)}\right)_{1,q_1},\ldots,\left(d_j^{(r)},\delta_j^{(r)};D_j^{(r)}\right)_{1,q_r}\end{array}\right]$$

$$= \mathrm{I}^{0,n-1:m_1,n_1+1;\ldots;m_r,n_r}_{p,q:p_1+1,q_1;\ldots;p_r,q_r}\left[\begin{array}{c|l} z_1 & \left(a_j;\alpha_j^{(1)},\ldots,\alpha_j^{(r)};\right)_{2,p}:(a,\alpha;A),\left(c_j^{(1)},\gamma_j^{(1)};C_j^{(1)}\right)_{1,p_1};\ldots;\left(c_j^{(r)},\gamma_j^{(r)};C_j^{(r)}\right)_{1,p_r} \\ \vdots & \\ z_r & \left(b_j;\beta_j^{(1)},\ldots,\beta_j^{(r)};B_j\right)_{1,q}:\left(d_j^{(1)},\delta_j^{(1)};D_j^{(1)}\right)_{1,q_1};\ldots;\left(d_j^{(r)},\delta_j^{(r)};D_j^{(r)}\right)_{1,q_r}\end{array}\right], \tag{32}$$

where $p \geq n \geq 1$,

(v) $$\mathrm{I}^{0,n:m_1,n_1;\ldots;m_r,n_r}_{p,q:p_1,q_1;\ldots;p_r,q_r}\left[\begin{array}{c|l} z_1 & \left(a_j;\alpha_j^{(1)},\ldots\alpha_j^{(r)};A_j\right)_{1,p-1},(a;\alpha,0,\ldots,0;A):\left(c_j^{(1)},\gamma_j^{(1)};C_j^{(1)}\right)_{1,p_1};\ldots;\left(c_j^{(r)},\gamma_j^{(r)};C_j^{(r)}\right)_{1,p_r} \\ \vdots & \\ z_r & \left(b_j;\beta_j^{(1)},\ldots,\beta_j^{(r)};B_j\right)_{1,q}:\left(d_j^{(1)},\delta_j^{(1)};D_j^{(1)}\right)_{1,q_1};\ldots;\left(d_j^{(r)},\delta_j^{(r)};D_j^{(r)}\right)_{1,q_r}\end{array}\right]$$

$$= \mathrm{I}^{0,n:m_1,n_1;\ldots;m_r,n_r}_{p-1,q:p_1+1,q_1;\ldots;p_r,q_r}\left[\begin{array}{c|l} z_1 & \left(a_j;\alpha_j^{(1)},\ldots\alpha_j^{(r)};A_j\right)_{1,p-1}:\left(c_j^{(1)},\gamma_j^{(1)};C_j^{(1)}\right)_{1,p_1},(a,\alpha;A);\ldots;\left(c_j^{(r)},\gamma_j^{(r)};C_j^{(r)}\right)_{1,p_r} \\ \vdots & \\ z_r & \left(b_j;\beta_j^{(1)},\ldots,\beta_j^{(r)};B_j\right)_{1,q}:\left(d_j^{(1)},\delta_j^{(1)};D_j^{(1)}\right)_{1,q_1};\ldots;\left(d_j^{(r)},\delta_j^{(r)};D_j^{(r)}\right)_{1,q_r}\end{array}\right], \tag{33}$$

where $p - 1 \geq n \geq 0$,

(vi) $$\mathrm{I}^{0,n:m_1,n_1;\ldots;m_r,n_r}_{p,q:p_1,q_1;\ldots;p_r,q_r}\left[\begin{array}{c|l} z_1 & \left(a_j;\alpha_j^{(1)},\ldots\alpha_j^{(r)};A_j\right)_{1,p}:\left(c_j^{(1)},\gamma_j^{(1)};C_j^{(1)}\right)_{1,p_1};\ldots;\left(c_j^{(r)},\gamma_j^{(r)};C_j^{(r)}\right)_{1,p_r} \\ \vdots & \\ z_r & \left(b_j;\beta_j^{(1)},\ldots,\beta_j^{(r)};B_j\right)_{1,q-1},(b;\beta,0,\ldots,0;B):\left(d_j^{(1)},\delta_j^{(1)};D_j^{(1)}\right)_{1,q_1};\ldots;\left(d_j^{(r)},\delta_j^{(r)};D_j^{(r)}\right)_{1,q_r}\end{array}\right]$$

$$= \mathrm{I}^{0,n:m_1,n_1;\ldots;m_r,n_r}_{p,q-1:p_1,q_1+1;\ldots;p_r,q_r}\left[\begin{array}{c|l} z_1 & \left(a_j;\alpha_j^{(1)},\ldots\alpha_j^{(r)};A_j\right)_{1,p}:\left(c_j^{(1)},\gamma_j^{(1)};C_j^{(1)}\right)_{1,p_1};\ldots;\left(c_j^{(r)},\gamma_j^{(r)};C_j^{(r)}\right)_{1,p_r} \\ \vdots & \\ z_r & \left(b_j;\beta_j^{(1)},\ldots,\beta_j^{(r)};B_j\right)_{1,q-1}:\left(d_j^{(1)},\delta_j^{(1)};D_j^{(1)}\right)_{1,q_1},(b,\beta;B);\ldots;\left(d_j^{(r)},\delta_j^{(r)};D_j^{(r)}\right)_{1,q_r}\end{array}\right], \tag{34}$$

where $q - 1 \geq 0$,

$$\text{(vii)}\quad I^{0,n:m_1,n_1;\ldots;m_r,n_r}_{p,q:p_1,q_1;\ldots;p_r,q_r}\left[\begin{array}{c} z_1 \\ \vdots \\ z_r \end{array}\middle| \begin{array}{l} (a;0,\ldots,0;A),\left(a_j;\alpha_j^{(1)},\ldots\alpha_j^{(r)};A_j\right)_{2,p}:\left(c_j^{(1)},\gamma_j^{(1)};C_j^{(1)}\right)_{1,p_1};\ldots;\left(c_j^{(r)},\gamma_j^{(r)};C_j^{(r)}\right)_{1,p_r} \\ \left(b_j;\beta_j^{(1)},\ldots,\beta_j^{(r)};B_j\right)_{1,q}:\left(d_j^{(1)},\delta_j^{(1)};D_j^{(1)}\right)_{1,q_1};\ldots;\left(d_j^{(r)},\delta_j^{(r)};D_j^{(r)}\right)_{1,q_r} \end{array}\right]$$
$$= \Gamma^A(1-a) \times I^{0,n-1:m_1,n_1;\ldots;m_r,n_r}_{p-1,q:p_1,q_1;\ldots;p_r,q_r}\left[\begin{array}{c} z_1 \\ \vdots \\ z_r \end{array}\middle| \begin{array}{l} \left(a_j;\alpha_j^{(1)},\ldots\alpha_j^{(r)};A_j\right)_{2,p}:\left(c_j^{(1)},\gamma_j^{(1)};C_j^{(1)}\right)_{1,p_1};\ldots;\left(c_j^{(r)},\gamma_j^{(r)};C_j^{(r)}\right)_{1,p_r} \\ \left(b_j;\beta_j^{(1)},\ldots,\beta_j^{(r)};B_j\right)_{1,q}:\left(d_j^{(1)},\delta_j^{(1)};D_j^{(1)}\right)_{1,q_1};\ldots;\left(d_j^{(r)},\delta_j^{(r)};D_j^{(r)}\right)_{1,q_r} \end{array}\right], \quad (35)$$

where $p \geq n \geq 1$, $\mathfrak{R}(1-a) > 0$,

$$\text{(viii)}\quad I^{0,n:m_1,n_1;\ldots;m_r,n_r}_{p,q:p_1,q_1;\ldots;p_r,q_r}\left[\begin{array}{c} z_1 \\ \vdots \\ z_r \end{array}\middle| \begin{array}{l} \left(a_j;\alpha_j^{(1)},\ldots\alpha_j^{(r)};A_j\right)_{1,p-1},(a;0,\ldots,0;A):\left(c_j^{(1)},\gamma_j^{(1)};C_j^{(1)}\right)_{1,p_1};\ldots;\left(c_j^{(r)},\gamma_j^{(r)};C_j^{(r)}\right)_{1,p_r} \\ \left(b_j;\beta_j^{(1)},\ldots,\beta_j^{(r)};B_j\right)_{1,q}:\left(d_j^{(1)},\delta_j^{(1)};D_j^{(1)}\right)_{1,q_1};\ldots;\left(d_j^{(r)},\delta_j^{(r)};D_j^{(r)}\right)_{1,q_r} \end{array}\right]$$
$$= \frac{1}{\Gamma^A(a)} \times I^{0,n:m_1,n_1;\ldots;m_r,n_r}_{p-1,q:p_1,q_1;\ldots;p_r,q_r}\left[\begin{array}{c} z_1 \\ \vdots \\ z_r \end{array}\middle| \begin{array}{l} \left(a_j;\alpha_j^{(1)},\ldots\alpha_j^{(r)};A_j\right)_{1,p-1}:\left(c_j^{(1)},\gamma_j^{(1)};C_j^{(1)}\right)_{1,p_1};\ldots;\left(c_j^{(r)},\gamma_j^{(r)};C_j^{(r)}\right)_{1,p_r} \\ \left(b_j;\beta_j^{(1)},\ldots,\beta_j^{(r)};B_j\right)_{1,q}:\left(d_j^{(1)},\delta_j^{(1)};D_j^{(1)}\right)_{1,q_1};\ldots;\left(d_j^{(r)},\delta_j^{(r)};D_j^{(r)}\right)_{1,q_r} \end{array}\right], \quad (36)$$

where $p-1 \geq n \geq 0$, $\mathfrak{R}(a) > 0$,

$$\text{(ix)}\quad I^{0,n:m_1,n_1;\ldots;m_r,n_r}_{p,q:p_1,q_1;\ldots;p_r,q_r}\left[\begin{array}{c} z_1 \\ \vdots \\ z_r \end{array}\middle| \begin{array}{l} \left(a_j;\alpha_j^{(1)},\ldots\alpha_j^{(r)};A_j\right)_{1,p}:\left(c_j^{(1)},\gamma_j^{(1)};C_j^{(1)}\right)_{1,p_1};\ldots;\left(c_j^{(r)},\gamma_j^{(r)};C_j^{(r)}\right)_{1,p_r} \\ \left(b_j;\beta_j^{(1)},\ldots,\beta_j^{(r)};B_j\right)_{1,q-1},(b;0,\ldots,0;B):\left(d_j^{(1)},\delta_j^{(1)};D_j^{(1)}\right)_{1,q_1};\ldots;\left(d_j^{(r)},\delta_j^{(r)};D_j^{(r)}\right)_{1,q_r} \end{array}\right]$$
$$= \frac{1}{\Gamma^B(1-b)} \times I^{0,n:m_1,n_1;\ldots;m_r,n_r}_{p,q-1:p_1,q_1;\ldots;p_r,q_r}\left[\begin{array}{c} z_1 \\ \vdots \\ z_r \end{array}\middle| \begin{array}{l} \left(a_j;\alpha_j^{(1)},\ldots\alpha_j^{(r)};A_j\right)_{1,p}:\left(c_j^{(1)},\gamma_j^{(1)};C_j^{(1)}\right)_{1,p_1};\ldots;\left(c_j^{(r)},\gamma_j^{(r)};C_j^{(r)}\right)_{1,p_r} \\ \left(b_j;\beta_j^{(1)},\ldots,\beta_j^{(r)};B_j\right)_{1,q-1}:\left(d_j^{(1)},\delta_j^{(1)};D_j^{(1)}\right)_{1,q_1};\ldots;\left(d_j^{(r)},\delta_j^{(r)};D_j^{(r)}\right)_{1,q_r} \end{array}\right], \quad (37)$$

where $q-1 \geq 0$, $\mathfrak{R}(1-b) > 0$,

$$\text{(x)}\quad I^{0,n:m_1,n_1;\ldots;m_r,n_r}_{p,q:p_1,q_1;\ldots;p_r,q_r}\left[\begin{array}{c} z_1 \\ \vdots \\ z_r \end{array}\middle| \begin{array}{l} \left(a_j;\alpha_j^{(1)},\ldots\alpha_j^{(r)};A_j\right)_{1,p}:(c;0;C),\left(c_j^{(1)},\gamma_j^{(1)};C_j^{(1)}\right)_{2,p_1};\ldots;\left(c_j^{(r)},\gamma_j^{(r)};C_j^{(r)}\right)_{1,p_r} \\ \left(b_j;\beta_j^{(1)},\ldots,\beta_j^{(r)};B_j\right)_{1,q}:\left(d_j^{(1)},\delta_j^{(1)};D_j^{(1)}\right)_{1,q_1};\ldots;\left(d_j^{(r)},\delta_j^{(r)};D_j^{(r)}\right)_{1,q_r} \end{array}\right]$$
$$= \Gamma^C(1-c) \times I^{0,n:m_1,n_1-1;\ldots;m_r,n_r}_{p,q:p_1-1,q_1;\ldots;p_r,q_r}\left[\begin{array}{c} z_1 \\ \vdots \\ z_r \end{array}\middle| \begin{array}{l} \left(a_j;\alpha_j^{(1)},\ldots\alpha_j^{(r)};A_j\right)_{1,p}:\left(c_j^{(1)},\gamma_j^{(1)};C_j^{(1)}\right)_{2,p_1};\ldots;\left(c_j^{(r)},\gamma_j^{(r)};C_j^{(r)}\right)_{1,p_r} \\ \left(b_j;\beta_j^{(1)},\ldots,\beta_j^{(r)};B_j\right)_{1,q}:\left(d_j^{(1)},\delta_j^{(1)};D_j^{(1)}\right)_{1,q_1};\ldots;\left(d_j^{(r)},\delta_j^{(r)};D_j^{(r)}\right)_{1,q_r} \end{array}\right], \quad (38)$$

where $p_1 \geq n_1 \geq 1$, $\mathfrak{R}(1-c) > 0$,

$$\text{(xi)} \quad I^{0,n:m_1,n_1;\dots;m_r,n_r}_{p,q:p_1,q_1;\dots;p_r,q_r}\left[\begin{array}{c|l} z_1 & \left(a_j;\alpha_j^{(1)},\dots\alpha_j^{(r)};A_j\right)_{1,p}:\left(c_j^{(1)},\gamma_j^{(1)};C_j^{(1)}\right)_{1,p_1-1},(c;0;C);\dots;\left(c_j^{(r)},\gamma_j^{(r)};C_j^{(r)}\right)_{1,p_r} \\ \vdots & \\ z_r & \left(b_j;\beta_j^{(1)},\dots,\beta_j^{(r)};B_j\right)_{1,q}:\left(d_j^{(1)},\delta_j^{(1)};D_j^{(1)}\right)_{1,q_1};\dots;\left(d_j^{(r)},\delta_j^{(r)};D_j^{(r)}\right)_{1,q_r} \end{array}\right]$$
$$= \frac{1}{\Gamma^C(c)} \times I^{0,n:m_1,n_1;\dots;m_r,n_r}_{p,q:p_1-1,q_1;\dots;p_r,q_r}\left[\begin{array}{c|l} z_1 & \left(a_j;\alpha_j^{(1)},\dots\alpha_j^{(r)};A_j\right)_{1,p}:\left(c_j^{(1)},\gamma_j^{(1)};C_j^{(1)}\right)_{1,p_1-1};\dots;\left(c_j^{(r)},\gamma_j^{(r)};C_j^{(r)}\right)_{1,p_r} \\ \vdots & \\ z_r & \left(b_j;\beta_j^{(1)},\dots,\beta_j^{(r)};B_j\right)_{1,q}:\left(d_j^{(1)},\delta_j^{(1)};D_j^{(1)}\right)_{1,q_1};\dots;\left(d_j^{(r)},\delta_j^{(r)};D_j^{(r)}\right)_{1,q_r} \end{array}\right], \tag{39}$$

where $p_1 - 1 \geq n_1 \geq 0$, $\mathfrak{R}(c) > 0$,

$$\text{(xii)} \quad I^{0,n:m_1,n_1;\dots;m_r,n_r}_{p,q:p_1,q_1;\dots;p_r,q_r}\left[\begin{array}{c|l} z_1 & \left(a_j;\alpha_j^{(1)},\dots\alpha_j^{(r)};A_j\right)_{1,p}:\left(c_j^{(1)},\gamma_j^{(1)};C_j^{(1)}\right)_{1,p_1};\dots;\left(c_j^{(r)},\gamma_j^{(r)};C_j^{(r)}\right)_{1,p_r} \\ \vdots & \\ z_r & \left(b_j;\beta_j^{(1)},\dots,\beta_j^{(r)};B_j\right)_{1,q}:(d;0;D),\left(d_j^{(1)},\delta_j^{(1)};D_j^{(1)}\right)_{2,q_1};\dots;\left(d_j^{(r)},\delta_j^{(r)};D_j^{(r)}\right)_{1,q_r} \end{array}\right]$$
$$= \Gamma^D(d) \times I^{0,n:m_1-1,n_1;\dots;m_r,n_r}_{p,q:p_1,q_1-1;\dots;p_r,q_r}\left[\begin{array}{c|l} z_1 & \left(a_j;\alpha_j^{(1)},\dots\alpha_j^{(r)};A_j\right)_{1,p}:\left(c_j^{(1)},\gamma_j^{(1)};C_j^{(1)}\right)_{1,p_1};\dots;\left(c_j^{(r)},\gamma_j^{(r)};C_j^{(r)}\right)_{1,p_r} \\ \vdots & \\ z_r & \left(b_j;\beta_j^{(1)},\dots,\beta_j^{(r)};B_j\right)_{1,q}:\left(d_j^{(1)},\delta_j^{(1)};D_j^{(1)}\right)_{2,q_1};\dots;\left(d_j^{(r)},\delta_j^{(r)};D_j^{(r)}\right)_{1,q_r} \end{array}\right], \tag{40}$$

where $q_1 \geq m_1 \geq 1$, $\mathfrak{R}(d) > 0$,

$$\text{(xiii)} \quad I^{0,n:m_1,n_1;\dots;m_r,n_r}_{p,q:p_1,q_1;\dots;p_r,q_r}\left[\begin{array}{c|l} z_1 & \left(a_j;\alpha_j^{(1)},\dots\alpha_j^{(r)};A_j\right)_{1,p}:\left(c_j^{(1)},\gamma_j^{(1)};C_j^{(1)}\right)_{1,p_1};\dots;\left(c_j^{(r)},\gamma_j^{(r)};C_j^{(r)}\right)_{1,p_r} \\ \vdots & \\ z_r & \left(b_j;\beta_j^{(1)},\dots,\beta_j^{(r)};B_j\right)_{1,q}:\left(d_j^{(1)},\delta_j^{(1)};D_j^{(1)}\right)_{1,q_1-1},(d;0;D);\dots;\left(d_j^{(r)},\delta_j^{(r)};D_j^{(r)}\right)_{1,q_r} \end{array}\right]$$
$$= \frac{1}{\Gamma^D(1-d)} \times I^{0,n:m_1,n_1;\dots;m_r,n_r}_{p,q:p_1,q_1-1;\dots;p_r,q_r}\left[\begin{array}{c|l} z_1 & \left(a_j;\alpha_j^{(1)},\dots\alpha_j^{(r)};A_j\right)_{1,p}:\left(c_j^{(1)},\gamma_j^{(1)};C_j^{(1)}\right)_{1,p_1};\dots;\left(c_j^{(r)},\gamma_j^{(r)};C_j^{(r)}\right)_{1,p_r} \\ \vdots & \\ z_r & \left(b_j;\beta_j^{(1)},\dots,\beta_j^{(r)};B_j\right)_{1,q}:\left(d_j^{(1)},\delta_j^{(1)};D_j^{(1)}\right)_{1,q_1-1};\dots;\left(d_j^{(r)},\delta_j^{(r)};D_j^{(r)}\right)_{1,q_r} \end{array}\right], \tag{41}$$

provided that $q_1 - 1 \geq m_1 \geq 0$, $\mathfrak{R}(1-d) \geq 0$,

$$\text{(xiv)} \quad I^{0,n:m_1,n_1;\dots;m_r,n_r}_{p,q:p_1,q_1;\dots;p_r,q_r}$$
$$\left[\begin{array}{c|l} z_1 & \left(a_j;\alpha_j^{(1)},\alpha_j^{(r)};A_j\right)_{1,p}:\left(c_j^{(1)},\gamma_j^{(1)};C_j^{(1)}\right)_{1,p_1};\dots;\left(c_j^{(r)},\gamma_j^{(r)};C_j^{(r)}\right)_{1,p_r} \\ \vdots & \\ z_r & \left(b_j;\beta_j^{(1)},\dots,\beta_j^{(r)};B_j\right)_{1,q-1},\left(a_1;\alpha_1^{(1)},\dots,\alpha_1^{(r)};A_1\right):\left(d_j^{1},\delta_j^{(1)};D_j^{(1)}\right)_{1,q_1-1},\left(c_1^{(1)},\gamma_1^{(1)};C_1^{(1)}\right);\dots;\left(d_j^{(r)},\delta_j^{(r)};D_j^{(r)}\right)_{1,q_r-1},\left(c_1^{(r)},\gamma_1^{(r)};C_1^{(r)}\right) \end{array}\right]$$
$$= I^{0,n-1:m_1,n_1-1;\dots;m_r,n_r-1}_{p-1,q-1:p_1-1,q_1-1;\dots;p_r-1,q_r-1}\left[\begin{array}{c|l} z_1 & \left(a_j;\alpha_j^{(1)},\dots,\alpha_j^{(r)};A_j\right)_{2,p}:\left(c_j^{(1)},\gamma_j^{(1)};C_j^{(1)}\right)_{2,p_1};\dots;\left(c_j^{(r)},\gamma_j^{(r)};C_j^{(r)}\right)_{2,p_r} \\ \vdots & \\ z_r & \left(b_j;\beta_j^{(1)},\beta_j^{(r)};B_j\right)_{1,q-1}:\left(d_j^{(1)},\delta_j^{(1)};D_j^{(1)}\right)_{1,q_1-1};\dots;\left(d_j^{(r)},\delta_j^{(r)};D_j^{(r)}\right)_{1,q_r-1} \end{array}\right], \tag{42}$$

provided that $p \geq n \geq 1$, $p_i \geq n_i \geq 1$, $i = 1,\dots,r$, and $q \geq 1$, $q_i \geq m_i + 1$, $i = 1,\dots,r$,

(xv) $$I^{0,n:m_1,n_1;\dots;m_r,n_r}_{p,q:p_1,q_1;\dots;p_r,q_r}\left[\begin{array}{c|l} z_1 & \left(a_j;\alpha_j^{(1)},\dots,\alpha_j^{(r)};A_j\right)_{1,p}:\left(c_j^{(1)},\gamma_j^{(1)};C_j^{(1)}\right)_{1,p_1-1},\left(d_1^{(1)},\delta_1^{(1)};D_1^{(1)}\right);\dots;\left(c_j^{(r)},\gamma_j^{(r)};C_j^{(r)}\right)_{1,p_r-1},\left(d_1^{(r)},\delta_1^{(r)};D_1^{(r)}\right) \\ \vdots & \\ z_r & \left(b_j;\beta_j^{(1)},\beta_j^{(r)};B_j\right)_{1,q}:\left(d_j^{(1)},\delta_j^{(1)};D_j^{(1)}\right)_{1,q_1};\dots;\left(d_j^{(r)},\delta_j^{(r)};D_j^{(r)}\right)_{1,q_r} \end{array}\right]$$

$$= I^{0,n:m_1-1,n_1;\dots;m_r-1,n_r}_{p,q:p_1-1,q_1-1;\dots;p_r-1,q_r-1}\left[\begin{array}{c|l} z_1 & \left(a_j;\alpha_j^{(1)},\dots,\alpha_j^{(r)};A_j\right)_{1,p}:\left(c_j^{(1)},\gamma_j^{(1)};C_j^{(1)}\right)_{1,p_1-1};\dots;\left(c_j^{(r)},\gamma_j^{(r)};C_j^{(r)}\right)_{1,p_r-1} \\ \vdots & \\ z_r & \left(b_j;\beta_j^{(1)},\beta_j^{(r)};B_j\right)_{1,q}:\left(d_j^{(1)},\delta_j^{(1)};D_j^{(1)}\right)_{2,q_1};\dots;\left(d_j^{(r)},\delta_j^{(r)};D_j^{(r)}\right)_{2,q_r} \end{array}\right], \quad (43)$$

provided that $p \geq n+1$, $q \geq 1$, $p_i \geq n_i + 1$, and $q_i \geq m_i \geq 1$, $i = 1, \dots, r$,

(xvi) $$I^{0,n:m_1,n_1;\dots;m_r,n_r}_{p,q:p_1,q_1;\dots;p_r,q_r}\left[\begin{array}{c|l} z_1 & \left(a_j;\alpha_j^{(1)},\dots,\alpha_j^{(r)};A_j\right)_{1,p}:\left(c_j^{(1)},\gamma_j^{(1)};C_j^{(1)}\right)_{1,p_1};\dots;\left(c_j^{(r)},\gamma_j^{(r)};C_j^{(r)}\right)_{1,p_r} \\ \vdots & \\ z_r & \left(b_j;\beta_j^{(1)},\dots,\beta_j^{(r)};B_j\right)_{1,q}:\left(c_{p_1}^{(1)},\gamma_{p_1}^{(1)};C_{p_1}^{(1)}\right),\left(d_j^{(1)},\delta_j^{(1)};D_j^{(1)}\right)_{2,q_1};\dots;\left(c_{p_r}^{(r)},\gamma_{p_r}^{(r)};C_{p_r}^{(r)}\right),\left(d_j^{(r)},\delta_j^{(r)};D_j^{(r)}\right)_{2,q_r} \end{array}\right] \quad (44)$$

$$= I^{0,n:\; m_1-1,n_1;\dots;\; m_r-1,n_r}_{p,q:\; p_1-1,q_1-1;\dots;\; p_r-1,q_r-1}\left[\begin{array}{c|l} z_1 & \left(a_j;\alpha_j^{(1)},\dots,\alpha_j^{(r)};A_j\right)_{1,p}:\left(c_j^{(1)},\gamma_j^{(1)};C_j^{(1)}\right)_{1,p_1-1};\dots;\left(c_j^{(r)},\gamma_j^{(r)};C_j^{(r)}\right)_{1,p_r-1} \\ \vdots & \\ z_r & \left(b_j;\beta_j^{(1)},\dots,\beta_j^{(r)};B_j\right)_{1,q}:\left(d_j^{(1)},\delta_j^{(1)};D_j^{(1)}\right)_{2,q_1};\dots;\left(d_j^{(r)},\delta_j^{(r)};D_j^{(r)}\right)_{2,q_r} \end{array}\right],$$

provided that $p_i \geq n_i + 1$, $q_i \geq m_i$, $i = 1, \dots, r$,

(xvii) $$I^{0,n:m_1,n_1;\dots;m_r,n_r}_{pq:p_1,q_1;\dots;p_r,q_r}\left[\begin{array}{c|l} z_1 & \left(a_j;\alpha_j^{(1)},\dots,\alpha_j^{(r)};A_j\right)_{1,p}:\left(d_{q_1}^{(1)},\delta_{q_1}^{(1)},D_{q_1}^{(1)}\right),\left(c_j^{(1)},\gamma_j^{(1)};C_j^{(1)}\right)_{2,p_1};\dots;\left(d_{q_r}^{(r)},\delta_{q_r}^{(r)},D_{q_r}^{(r)}\right),\left(c_j^{(r)},\gamma_j^{(r)};C_j^{(r)}\right)_{2,p_r} \\ \vdots & \\ z_r & \left(b_j;\beta_j^{(1)},\beta_j^{(r)};B_j\right)_{1,q}:\left(d_j^{(1)},\delta_j^{(1)};D_j^{(1)}\right)_{1,q_1};\dots;\left(d_j^{(r)},\delta_j^{(r)};D_j^{(r)}\right)_{1,q_r} \end{array}\right] \quad (45)$$

$$= I^{0,n:m_1,n_1-1;\dots;m_r,n_r-1}_{p,q:p_1-1,q_1-1;\dots;p_r-1,q_r-1}\left[\begin{array}{c|l} z_1 & \left(a_j;\alpha_j^{(1)},\dots,\alpha_j^{(r)};A_j\right)_{1,p}:\left(c_j^{(1)},\gamma_j^{(1)};C_j^{(1)}\right)_{2,p_1};\dots;\left(c_j^{(r)},\gamma_j^{(r)};C_j^{(r)}\right)_{2,p_r} \\ \vdots & \\ z_r & \left(b_j;\beta_j^{(1)},\beta_j^{(r)};B_j\right)_{1,q}:\left(d_j^{(1)},\delta_j^{(1)};D_j^{(1)}\right)_{1,q_1-1};\dots;\left(d_j^{(r)},\delta_j^{(r)};D_j^{(r)}\right)_{1,q_r-1} \end{array}\right],$$

where $p \geq n$, $q \geq 1$, $p_i > n_i \geq 1$, $q_i - 1 \geq m_i$, $i = 1, \dots, r$.

7. Special Cases

When $r = 2$ and all the exponents A_j $(j = 1, \dots, p)$, B_j $(j = 1, \dots, q)$, $C_j^{(i)}$ $(j = 1, \dots, p_i,\ i = 1 \dots, r)$, and $D_j^{(i)}$ $(1, \dots, q_i,\ i = 1 \dots, r)$ the I-function of "r" variables reduces to H-function of two variables and therefore we obtain the corresponding results in H-function of two variables [14].

Conflict of Interests

The authors declare that there is no conflict of interests regarding the publication of this paper.

Acknowledgment

The authors are immensely grateful to the worthy referee for some useful and valuable suggestions for the improvement of this paper which led to a better presentation.

References

[1] A. K. Rathie, "A new generalization of generalized hypergeometric functions," *Le Matematiche*, vol. 52, no. 2, pp. 297–310, 1997.

[2] I. S. Ansari, F. Yilmaz, M. S. Alouni, and O. Kucur, "New results on the sum of Gamma random variates with application to the performance of wireless communication systems over nakagami-m fading channels," http://arxiv-web3.library.cornell.edu/abs/1202.2576v4.

[3] I. S. Ansari and Yilmaz, "On the sum of squared n-Random variates with application to the performance of wireless communication systems," http://arxiv-web3.library.cornell.edu/abs/1210.0100v1.

[4] X. Minghua, W. Yik-Chung, and A. Sonia, "Exact outage probability of dual-hop CSI-assisted AF relaying over nakagami-m fading channels," *IEEE Transactions on Signal Processing*, vol. 60, no. 10, pp. 5578–5583, 2012.

[5] K. ShanthaKumari, T. M. Vasudevan Nambisan, and A. K. Rathie, "A study of the I-function of two variables," http://arxiv.org/abs/1212.6717 .

[6] P. K. Mittal and K. C. Gupta, "An integral involving generalized function of two variables," *Proceedings of the Indian Academy of Sciences A*, vol. 75, no. 3, pp. 117–123, 1972.

[7] H. M. Srivastava and R. Panda, "Some bilateral generating functions for a class of generalkized hypergeometric polynomials," *Journal Für die Reine und Angewandte Mathematik*, vol. 17, no. 288, pp. 265–274, 1976.

[8] B. L. J. Braaksma, "Asymptotic expansions and analytic continuations for a class of Barnesintegrals," *Compositio Mathematicah*, vol. 15, pp. 239–341, 1964.

[9] Y. L. Luke, *Th Special Functions and Their Approximations*, vol. 1, Academic Press, New York, NY, USA, 1969.

[10] H. M. Srivastava and M. C. Daoust, "On Eulerian integrals associated with Kampé de Fériet's function," *Publications De L'Institut Mathématique*, vol. 9, no. 23, pp. 199–202, 1969.

[11] H. M. Srivastava and H. L. Manocha, *A Treatise on Generating Functions*, Halsted Press, Chichester, UK.

[12] E. M. Wright, "The asymptotic expansion of the generalized Bessel function," *Proceedings of London Mathematical Society*, vol. 38, pp. 257–270, 1935.

[13] A. Erdelyi, *Higher Transcendental Functions*, vol. 1, McGraw-Hill, New York, NY, USA, 1953.

[14] H. M. Srivastava, K. C. Gupta, and S. P. Goyal, *Th H—Functions of One and Two Variables With Applications*, South Asian Publishers, New Delhi, India, 1982.

19

On Third-Order Nonlinearity of Biquadratic Monomial Boolean Functions

Brajesh Kumar Singh

Department of Mathematics, School of Allied Sciences, Graphic Era Hill University, Dehradun, Uttarakhand 248002, India

Correspondence should be addressed to Brajesh Kumar Singh; bksingh0584@gmail.com

Academic Editor: J. A. Tenreiro Machado

The rth-order nonlinearity of Boolean function plays a central role against several known attacks on stream and block ciphers. Because of the fact that its maximum equals the covering radius of the rth-order Reed-Muller code, it also plays an important role in coding theory. The computation of exact value or high lower bound on the rth-order nonlinearity of a Boolean function is very complicated problem, especially when $r > 1$. This paper is concerned with the computation of the lower bounds for third-order nonlinearities of two classes of Boolean functions of the form $\mathrm{Tr}_1^n(\lambda x^d)$ for all $x \in \mathbb{F}_{2^n}$, $\lambda \in \mathbb{F}_{2^n}^*$, where (a) $d = 2^i + 2^j + 2^k + 1$, where i, j, and k are integers such that $i > j > k \geq 1$ and $n > 2i$, and (b) $d = 2^{3\ell} + 2^{2\ell} + 2^{\ell} + 1$, where ℓ is a positive integer such that $\gcd(\ell, n) = 1$ and $n > 6$.

1. Introduction

Boolean functions are the building blocks for the design and the security of symmetric cryptographic systems and for the definition of some kinds of error correcting codes, sequences, and designs. The rth-order nonlinearity, $\mathrm{nl}_r(f)$, of a Boolean function $f \in \mathcal{B}_n$ is defined by the minimum Hamming distance of f to $\mathrm{RM}(r, n)$-*Reed-Muller code* of length 2^n and order r ($\mathrm{RM}(r,n) := \{f \in \mathcal{B}_n : \deg(f) \leq r\}$). The nonlinearity of f is given by $\mathrm{nl}(f) = \mathrm{nl}_1(f)$ and is related to the immunity of f against *best affine approximation attacks* [1] and *fast correlation attacks* [2], when f is used as a combiner function or a filter function in a stream cipher. The rth-order nonlinearity is an important parameter, which measures the resistance of the function against various *low-order approximation attacks* [1, 3, 4]. In cryptographic framework, within a trade-off with the other important criteria, the rth-order nonlinearity must be as large as possible; see [5–9]. Since, the maximal rth-order nonlinearity of all Boolean functions equals the *covering radius* of $\mathrm{RM}(r, n)$, it also has an application in coding theory. Besides these applications, an interesting connection between the rth-order nonlinearity and the *fast algebraic attacks* has been introduced, recently in [9], which claims that a cryptographic Boolean function should have high rth-order nonlinearity to resist the *fast algebraic attack*.

Unlike nonlinearity there is no efficient algorithm to compute second-order nonlinearities for $n > 11$. The most efficient algorithm is introduced by Fourquet and Tavernier [10] which works for $n \leq 11$ and up to $n = 13$ for some special functions. Thus, to identify a class of Boolean function with high rth-order nonlinearity, even for $r = 2$, is a very relevant area of research. In 2008, Carlet has devolved a technique to compute rth-order nonlinearity recursively in [11], and using this technique he has obtained the lower bounds of nonlinearity profiles for functions belonging to several classes of functions: Kasami functions, Welch functions, inverse functions, and so forth. Based on this technique, the lower bound for rth-order nonlinearity, for $r \geq 2$, is obtained for some specific classes of Boolean functions, in many articles; see, for example, [11–14] and the references therein. The best known asymptotic upper bound for $\mathrm{nl}_3(f)$ given by Carlet and Mesnager [15] is as follows:

$$\mathrm{nl}_3(f) \leq 2^{n-1} - \sqrt{15} \cdot \left(1 + \sqrt{2}\right) \cdot 2^{n/2-1} + O(n). \qquad (1)$$

The classes of Boolean functions for which the lower bound of third nonlinearity is known are inverse functions [11], Dillon functions [16], and Kasami functions, $f(x) = \mathrm{Tr}_1^n(\lambda x^{57})$ [12]. In this paper, we deduce the theoretical lower bounds on third-order nonlinearities of two classes of biquadratic monomial Boolean functions $\mathrm{Tr}_1^n(\lambda x^d)$ for all $x \in \mathbb{F}_{2^n}$, where $\lambda \in \mathbb{F}_{2^n}^*$ and (a) $d = 2^i+2^j+2^k+1$, where i, j, and k are integers such that $i > j > k \geq 1$ and $n > 2i$, and (b) $d = 2^{3\ell}+2^{2\ell}+2^{\ell}+1$, where ℓ is a positive integer such that $\gcd(\ell, n) = 1$ and $n > 6$.

Remainder of the paper is organized as follows. In Section 2 some basic definitions and notations required for the subsequent sections are reviewed. The main results on lower bounds of third-order nonlinearities are presented in Section 3. The numerical compression of our bounds with the previous known results is provided in Section 4. Section 5 is conclusion.

2. Preliminaries

Let $\mathbb{F}_{2^n}$ be the finite field consisting of 2^n elements. The group of units of $\mathbb{F}_{2^n}$, denoted by $\mathbb{F}_{2^n}^*$, is a cyclic group consisting of $2^n - 1$ elements. An element $\alpha \in \mathbb{F}_{2^n}$ is said to be a primitive element if it is a generator of the multiplicative group $\mathbb{F}_{2^n}^*$. A function from $\mathbb{F}_{2^n}$ to $\mathbb{F}_2$ is said to be a Boolean function on n variables; the set of such functions is denoted by $\mathcal{B}_n$. Let $\mathbb{Z}$ and $\mathbb{Z}_q$, where q is a positive integer, denote the ring of integers and integers modulo q, respectively. A cyclotomic coset modulo $2^n - 1$ of $s \in \mathbb{Z}$ is defined as

$$C_s = \left\{s, s2, s2^2, \ldots, s2^{n_s-1}\right\}, \tag{2}$$

where n_s is the smallest positive integer such that $s \equiv s2^{n_s} \pmod{2^n - 1}$ [17, page 104]. It is a convention to choose the subscript s to be the smallest integer in C_s and refer to it as the coset leader of C_s and n_s denotes the size of C_s. The trace function $\mathrm{Tr}_1^n : \mathbb{F}_{2^n} \to \mathbb{F}_2$ is defined by $\mathrm{Tr}_1^n(x) = \sum_{i=0}^{n-1} x^{2^i}$ for all $x \in \mathbb{F}_{2^n}$. The trace representation [18] of a function $f \in \mathcal{B}_n$ is

$$f(x) = \sum_{k\in\Gamma(n)} \mathrm{Tr}_1^{n_k}\left(A_k x^k\right) + A_{2^n-1}x^{2^n-1}, \quad \forall x \in \mathbb{F}_{2^n}, \tag{3}$$

where $\Gamma(n)$ is the set of all coset leaders modulo $2^n - 1$ and $A_k \in \mathbb{F}_{2^{n_k}}$, $A_{2^n-1} \in \mathbb{F}_2$, for all $k \in \Gamma(n)$. A Boolean function is said to be a *monomial trace function* if its trace representation consists of single trace term. The *binary representation* of an integer $d \in \mathbb{Z}$ is

$$d = d_{m-1}2^{m-1} + d_{m-2}2^{m-2} + \cdots + d_1 2 + d_0, \tag{4}$$

where $d_0, d_1, \ldots, d_{m-1} \in \{0, 1\}$. The Hamming weight of d is $w_H(d) = \sum_{i=0}^{m-1} d_i$, where the sum is over $\mathbb{Z}$. The algebraic degree, denoted by $\deg(f)$, of $f \in \mathcal{B}_n$, as represented in (3), is the largest positive integer w for which $w_H(k) = w$ and $A_k \neq 0$. The support of $f \in \mathcal{B}_n$ is $\mathrm{supp}(f) = \{x \in \mathbb{F}_{2^n} : f(x) \neq 0\}$. The weight of f is $w_H(f) = |\{x \in \mathbb{F}_{2^n} : f(x) \neq 0\}|$, where $|S|$ is the cardinality of any set S. The Hamming distance between two functions f, $g \in \mathcal{B}_n$ is defined by $d_H(f, g) = |\{x \in \mathbb{F}_{2^n} : f(x) \neq g(x)\}|$.

The *Walsh-Hadamard transform* (WHT) of a Boolean function $f \in \mathcal{B}_n$ at $\lambda \in \mathbb{F}_{2^n}$ is defined by $W_f(\lambda) := \sum_{x\in\mathbb{F}_{2^n}} (-1)^{f(x)+\mathrm{Tr}_1^n(\lambda x)}$. The nonlinearity of $f \in \mathcal{B}_n$ in terms of its *Walsh-Hadamard spectrum* (WHS) is given by

$$\mathrm{nl}(f) = 2^{n-1} - \frac{1}{2}\max_{\lambda\in\mathbb{F}_{2^n}} \left|W_f(\lambda)\right|. \tag{5}$$

The set $\{W_f(\lambda) : \lambda \in \mathbb{F}_{2^n}\}$ is referred to as the WHS of $f \in B_n$ which satisfies the Parseval's identity: $\sum_{\lambda\in\mathbb{F}_2^n} W_f(\lambda)^2 = 2^{2n}$ which implies that $\max\{|W_f(\lambda)| : \lambda \in \mathbb{F}_{2^n}\} \geq 2^{n/2}$, and so $\mathrm{nl}(f) \leq 2^{n-1} - 2^{(n/2)-1}$. The function $f \in \mathcal{B}_n$ achieving maximum possible nonlinearity $2^{n-1} - 2^{n/2-1}$ are said to be bent functions (exists only for even n), were introduced by Rothaus [19].

The derivative of $f \in \mathcal{B}_n$ with respect to $a \in \mathbb{F}_{2^n}$ is defined by $D_a f(x) = f(x) + f(x + a)$ for all $x \in \mathbb{F}_{2^n}$. The second-order derivatives of $f \in \mathcal{B}_n$ with respect to $V = \langle a, b\rangle$ is the Boolean function $D_V f \in \mathcal{B}_n$ which is defined by $D_V f(x) = D_b D_a f(x) = f(x) + f(x + a) + f(x + b) + f(x + a + b)$, where V is two-dimensional subspace of $\mathbb{F}_{2^n}$ generated by a and b; for details on higher derivatives, see [5, 11]. The rth-order nonlinearity of $f \in \mathcal{B}_n$ is defined as

$$\begin{aligned}\mathrm{nl}_r(f) &= \min_{h\in\mathrm{RM}(r,n)} d_H(f, h)\\ &= 2^{n-1} - \frac{1}{2}\max_{h\in\mathrm{RM}(r,n)} \left|\sum_{x\in\mathbb{F}_{2^n}} (-1)^{f(x)+h(x)}\right|.\end{aligned} \tag{6}$$

The sequence $\{\mathrm{nl}_r(f)\}_{r=1}^{n-1}$ is called the nonlinearity profile of f. Also, $\mathrm{nl}_r(f) \leq \mathrm{nl}_{r-1}(f)$ because $\mathrm{RM}(r - 1, n) \subset \mathrm{RM}(r, n)$. The notion of rth-order bent functions was introduced by Iwata and Kurosawa [4]. A function $f \in \mathcal{B}_n$ is said to be rth-order bent (for $r \leq n - 3$) if and only if $\mathrm{nl}_r(f) \geq 2^{n-r-3}(r + 4)$, for even r, and $\mathrm{nl}_r(f) \geq 2^{n-r-3}(r + 5)$, for odd r.

Carlet's [11] recursive lower bounds for third-order nonlinearities which we use to compute our bounds, are given below in Propositions 1 and 2.

Proposition 1 (see [11, Proposition 2]). *Let* $f \in \mathcal{B}_n$*; then* $nl_3(f) \geq (1/4)\max\{nl(D_b D_a f) : a, b \in \mathbb{F}_{2^n}\}$.

Proposition 2 (see [11, Equation (1)]). *Let* $f \in \mathcal{B}_n$*. Then*

$$nl_3(f) \geq 2^{n-1} - \frac{1}{2}\sqrt{\sum_{a\in\mathbb{F}_{2^n}} \sqrt{2^{2n} - 2\sum_{b\in\mathbb{F}_{2^n}} nl(D_a D_b f)}}. \tag{7}$$

Proposition 3 (see [17, Chapter 15, Corollary 13] (*McEliece's theorem*)). *Th* r*th-order nonlinearities of a Boolean function* $f \in \mathcal{B}_n$ *with algebraic degree* d *are divisible by* $2^{\lceil n/d\rceil-1}$*, where* $\lceil u\rceil$ *denotes the ceiling of* u *(the smallest integer greater than or equal to* u*).*

Proposition 4 (see [20, Corollary 1]). *Let* $L(x) = \sum_{i=0}^{v} c_i x^{2^{ik}}$ *be a linearized polynomial over* $\mathbb{F}_{2^n}$*, where* v*,* k *are positive integers such that* $\gcd(n, k) = 1$*. Then zeroes of the linearized polynomial* $L(x)$ *in* $\mathbb{F}_{2^n}$ *are at most* 2^v.

The result in Proposition 4 above was introduced by Bracken et al. [20]. The bilinear form [17] associated with a quadratic Boolean function $f \in \mathcal{B}_n$ is defined by $B(x, y) := f(0) + f(x) + f(y) + f(x + y)$ and the kernel, $\mathcal{E}_f$ of $B(x, y)$ is the subspace of $\mathbb{F}_{2^n}$ defined by

$$\mathcal{E}_f = \{x \in \mathbb{F}_{2^n} : B(x, y) = 0\ \forall y \in \mathbb{F}_{2^n}\}. \tag{8}$$

An element $c \in \mathcal{E}_f$ is called a linear structure of f. Next, if V is a vector space over a field $\mathbb{F}_q$ of characteristic 2 and $Q : V \to \mathbb{F}_q$ a quadratic form, then $\dim(V)$ and $\dim(\mathcal{E}_Q)$ have the same parity [21]. The distribution of the WHT values of a quadratic Boolean function $f \in \mathcal{B}_n$ is given in the following theorem which claims that the weight distribution of the values in the WHS of f depends only on the dimension k of $\mathcal{E}_f$.

Theorem 5 (see [17, 21]). *Let $f \in \mathcal{B}_n$ be a quadratic Boolean function and $k = \dim(\mathcal{E}_f)$, where $\mathcal{E}_f$ is defined in (8); then the weight distribution of the WHT values of f is given by*

$$W_f(\lambda) = \begin{cases} 0, & 2^n - 2^{n-k} \text{ times}, \\ 2^{(n+k)/2}, & 2^{n-k-1} + (-1)^{f(0)} 2^{(n-k-2)/2} \text{ times}, \\ -2^{(n+k)/2}, & 2^{n-k-1} - (-1)^{f(0)} 2^{(n-k-2)/2} \text{ times}. \end{cases} \quad (9)$$

3. Main Results

In this section, using Carlet's recursive technique [11], the theoretical lower bounds for third-order nonlinearities of two general classes of monomial Boolean functions of degree 4 are obtained.

Theorem 6. *Let $f_\lambda(x) = Tr_1^n(\lambda x^{2^i+2^j+2^k+1})$, for all $x \in \mathbb{F}_{2^n}$, where $\lambda \in \mathbb{F}_{2^n}^*$ and i, j, and k are integers such that $i > j > k \geq 1$ and $n > 2i$. Then*

$$nl_3(f_\lambda) \geq \begin{cases} 2^{n-3} - 2^{(n+2i-6)/2}, & \text{if } n = 0 \mod 2, \\ 2^{n-3} - 2^{(n+2i-7)/2}, & \text{if } n = 1 \mod 2. \end{cases} \quad (10)$$

In particular, if $\gcd(j-k, n) = 1$, then

$$nl_3(f_\lambda) \geq \begin{cases} 2^{n-1} - \frac{1}{2}\sqrt{(2^n-1)\sqrt{2^{(3n+2i)/2} + 2^{n+1} - 2^{(n+2i+2)/2}}}, & \text{if } n = 0 \mod 2, \\ 2^{n-1} - \frac{1}{2}\sqrt{(2^n-1)\sqrt{2^{(3n+2i-1)/2} + 2^{n+1} - 2^{(n+2i+1)/2}}}, & \text{if } n = 1 \mod 2. \end{cases} \quad (11)$$

Proof. Derivative of f_λ with respect to $a \in \mathbb{F}_{2^n}^*$ is

$$\begin{aligned} D_a f_\lambda(x) &= f_\lambda(x+a) + f_\lambda(x) \\ &= \mathrm{Tr}_1^n\left(\lambda (x+a)^{2^i+2^j+2^k+1}\right) + \mathrm{Tr}_1^n\left(\lambda x^{2^i+2^j+2^k+1}\right) \\ &= \mathrm{Tr}_1^n\left(\lambda\left(a x^{2^i+2^j+2^k} + a^{2^i} x^{2^j+2^k+1} \right.\right. \\ &\quad \left.\left. + a^{2^j} x^{2^i+2^k+1} + a^{2^k} x^{2^i+2^j+1}\right)\right) + q(x), \end{aligned} \quad (12)$$

where q is quadratic. The second derivative $D_b D_a f_\lambda$ with respect to $a, b \in \mathbb{F}_{2^n}^*$, where $a \neq b$, is

$$\begin{aligned} &D_b D_a f_\lambda(x) \\ &= f_\lambda(x+a+b) + f_\lambda(x+a) + f_\lambda(x+b) + f_\lambda(x) \\ &= \mathrm{Tr}_1^n\left(\lambda(x+a+b)^{2^i+2^j+2^k+1}\right) + \mathrm{Tr}_1^n\left(\lambda(x+b)^{2^i+2^j+2^k+1}\right) \\ &\quad + \mathrm{Tr}_1^n\left(\lambda(x+a)^{2^i+2^j+2^k+1}\right) + \mathrm{Tr}_1^n\left(\lambda x^{2^i+2^j+2^k+1}\right) \\ &= l(x) + \mathrm{Tr}_1^n\left(\lambda\left(\left(ab^{2^k} + a^{2^k}b\right)x^{2^i+2^j}\right.\right. \\ &\quad + \left(ab^{2^j} + a^{2^j}b\right)x^{2^i+2^k} \\ &\quad + \left(ab^{2^i} + a^{2^i}b\right)x^{2^j+2^k} \\ &\quad + \left(a^{2^j}b^{2^k} + a^{2^k}b^{2^j}\right)x^{2^i+1} \\ &\quad + \left(a^{2^i}b^{2^k} + a^{2^k}b^{2^i}\right)x^{2^j+1} \\ &\quad \left.\left. + \left(a^{2^i}b^{2^j} + a^{2^j}b^{2^i}\right)x^{2^k+1}\right)\right), \end{aligned} \quad (13)$$

where l is an affine function. If $D_b D_a f_\lambda$ is quadratic, then the WHS of $D_b D_a f_\lambda$ is equivalent to the WHS of the function h_λ obtained by removing l from $D_b D_a f_\lambda$:

$$\begin{aligned} &h_\lambda(x) \\ &= \mathrm{Tr}_1^n\left(\lambda\left(\left(ab^{2^k} + a^{2^k}b\right)x^{2^i+2^j} + \left(ab^{2^j} + a^{2^j}b\right)x^{2^i+2^k}\right.\right. \\ &\quad + \left(a^{2^j}b^{2^k} + a^{2^k}b^{2^j}\right)x^{2^i+1} + \left(ab^{2^i} + a^{2^i}b\right)x^{2^j+2^k} \\ &\quad + \left(a^{2^i}b^{2^k} + a^{2^k}b^{2^i}\right)x^{2^j+1} \\ &\quad \left.\left. + \left(a^{2^i}b^{2^j} + a^{2^j}b^{2^i}\right)x^{2^k+1}\right)\right). \end{aligned} \quad (14)$$

Further, $\mathcal{E}_{h_\lambda} = \{x \in \mathbb{F}_{2^n} : B(x,y) = 0 \text{ for all } y \in \mathbb{F}_{2^n}\}$, where $B(x,y)$ is the bilinear form associated with h_λ. Now, using $x^{2^n} = x$, $y^{2^n} = y$, and $\mathrm{Tr}_1^n(x^{2^i}) = \mathrm{Tr}_1^n(x)$, for all $x, y \in \mathbb{F}_{2^n}$, we compute $B(x,y)$ as follows

$$\begin{aligned} B(x,y) &= h_\lambda(0) + h_\lambda(x) + h_\lambda(y) + h_\lambda(x+y) \\ &= \mathrm{Tr}_1^n\left(\lambda\left(y^{2^i}\left(\left(ab^{2^k} + a^{2^k}b\right)x^{2^j} + \left(ab^{2^j} + a^{2^j}b\right)x^{2^k}\right.\right.\right. \\ &\quad \left. + \left(a^{2^j}b^{2^k} + a^{2^k}b^{2^j}\right)x\right) \\ &\quad + y^{2^j}\left(\left(ab^{2^k} + a^{2^k}b\right)x^{2^i} + \left(ab^{2^i} + a^{2^i}b\right)x^{2^k}\right. \\ &\quad \left. + \left(a^{2^i}b^{2^k} + a^{2^k}b^{2^i}\right)x\right) \end{aligned}$$

$$
\begin{aligned}
&+ y^{2^k}\Big(\big(ab^{2^j} + a^{2^j}b\big)x^{2^i} + \big(ab^{2^i} + a^{2^i}b\big)x^{2^j} \\
&\qquad + \big(a^{2^i}b^{2^j} + a^{2^j}b^{2^i}\big)x\Big) \\
&+ y\Big(\big(a^{2^j}b^{2^k} + a^{2^k}b^{2^j}\big)x^{2^i} \\
&\qquad + \big(a^{2^i}b^{2^k} + a^{2^k}b^{2^i}\big)x^{2^j} \\
&\qquad + \big(a^{2^i}b^{2^j} + a^{2^j}b^{2^i}\big)x^{2^k}\Big)\Big)\Big) \\
&= \mathrm{Tr}_1^n\left(yP\left(x\right)\right),
\end{aligned} \tag{15}
$$

where

$$
\begin{aligned}
P(x) = &\Big(\lambda\big(ab^{2^k} + a^{2^k}b\big)x^{2^j} + \lambda\big(ab^{2^j} + a^{2^j}b\big)x^{2^k} \\
&+\lambda\big(a^{2^j}b^{2^k} + a^{2^k}b^{2^j}\big)x\Big)^{2^{n-i}} \\
&+ \Big(\lambda\big(ab^{2^j} + a^{2^j}b\big)x^{2^i} \\
&\quad + \lambda\big(ab^{2^i} + a^{2^i}b\big)x^{2^j} + \lambda\big(a^{2^i}b^{2^j} + a^{2^j}b^{2^i}\big)x\Big)^{2^{n-j}} \\
&+ \Big(\lambda\big(ab^{2^j} + a^{2^j}b\big)x^{2^i} + \lambda\big(ab^{2^i} + a^{2^i}b\big)x^{2^j} \\
&\quad + \lambda\big(a^{2^i}b^{2^j} + a^{2^j}b^{2^i}\big)x\Big)^{2^{n-k}} \\
&+ \lambda\big(a^{2^j}b^{2^k} + a^{2^k}b^{2^j}\big)x^{2^i} \\
&+ \lambda\big(a^{2^i}b^{2^k} + a^{2^k}b^{2^i}\big)x^{2^j} + \lambda\big(a^{2^i}b^{2^j} + a^{2^j}b^{2^i}\big)x^{2^k}.
\end{aligned} \tag{16}
$$

Therefore,

$$
\mathscr{E}_{h_\lambda} = \left\{x \in \mathbb{F}_{2^n} : P(x) = 0 = P(x)^{2^i}\right\}. \tag{17}
$$

Let $L_{(\lambda,a,b)}(x) = P(x)^{2^i}$. Using $x^{2^n} = x$, $y^{2^n} = y$, $a^{2^n} = a$, $b^{2^n} = b$, and $\lambda^{2^n} = \lambda$, for all $x, y, a, b, \lambda \in \mathbb{F}_{2^n}$, we have

$$
\begin{aligned}
L_{(\lambda,a,b)}(x) &= \left(P\left(x\right)\right)^{2^i} \\
&= \lambda\Big(\big(ab^{2^j} + a^{2^j}b\big)x^{2^k} \\
&\qquad + \big(ab^{2^k} + a^{2^k}b\big)x^{2^j} \\
&\qquad + \big(a^{2^j}b^{2^k} + a^{2^k}b^{2^j}\big)x\Big) \\
&\quad + \lambda^{2^i}\Big(\big(a^{2^{i+j}}b^{2^{i+k}} + a^{2^{i+k}}b^{2^{i+j}}\big)x^{2^{2i}} \\
&\qquad + \big(a^{2^{i+k}}b^{2^{2i}} + a^{2^{2i}}b^{2^{i+k}}\big)x^{2^{i+j}} \\
&\qquad + \big(a^{2^{2i}}b^{2^{i+j}} + a^{2^{i+j}}b^{2^{2i}}\big)x^{2^{i+k}}\Big) \\
&\quad + \lambda^{2^{i-j}}\Big(\big(a^{2^{i-j}}b^{2^i} + a^{2^i}b^{2^{i-j}}\big)x^{2^{2i-j}} \\
&\qquad + \big(a^{2^{i-j}}b^{2^{2i-j}} + a^{2^{2i-j}}b^{2^{i-j}}\big)x^{2^i} \\
&\qquad + \big(a^{2^{2i-j}}b^{2^i} + a^{2^i}b^{2^{2i-j}}\big)x^{2^{i-j}}\Big) \\
&\quad + \lambda^{2^{i-k}}\Big(\big(a^{2^{i-k}}b^{2^{i+j-k}} + a^{2^{i+j-k}}b^{2^{i-k}}\big)x^{2^{2i-k}} \\
&\qquad + \big(a^{2^{i-k}}b^{2^{2i-k}} + a^{2^{2i-k}}b^{2^{i-k}}\big)x^{2^{i+j-k}} \\
&\qquad + \big(a^{2^{2i-k}}b^{2^{i+j-k}} + a^{2^{i+j-k}}b^{2^{2i-k}}\big)x^{2^{i-k}}\Big).
\end{aligned} \tag{18}
$$

The coefficient of x in $L_{(\lambda,a,b)}(x)$ is zero if and only if $a^{2^j}b^{2^k} + a^{2^k}b^{2^j} = 0$; that is, $a^{2^{j-k}}b + ab^{2^{j-k}} = 0$ which implies that $b \in a\mathbb{F}_{2^{j-k}}$. Therefore, for every $0 \neq a$, $b \in \mathbb{F}_{2^n}$ such that $b \notin a\mathbb{F}_{2^{j-k}}$, the degree of linearized polynomial, $L_{(\lambda,a,b)}$, in x is at most 2^{2i}; this implies that the dimension of the kernel $\mathscr{E}_{D_bD_af_\lambda}$ associated with $D_bD_af_\lambda$ is $k(a,b) \leq 2i$ if n is even; otherwise $k(a,b) \leq 2i - 1$. The WHT of $D_bD_af_\lambda$ at $\mu \in \mathbb{F}_{2^n}$ is

$$
W_{D_bD_af_\lambda}(\mu) \leq \begin{cases} 2^{(n+2i)/2}, & \text{if } n = 0 \mod 2, \\ 2^{(n+2i-1)/2}, & \text{if } n = 1 \mod 2. \end{cases} \tag{19}
$$

Therefore,

$$
\mathrm{nl}\left(D_bD_af_\lambda\right) = \begin{cases} 2^{n-1} - 2^{(n+2i-2)/2}, & \text{if } n = 0 \mod 2, \\ 2^{n-1} - 2^{(n+2i-3)/2}, & \text{if } n = 1 \mod 2. \end{cases} \tag{20}
$$

Using Proposition 1, we have

$$
\mathrm{nl}_3\left(f_\lambda\right) \geq \begin{cases} 2^{n-3} - 2^{(n+2i-6)/2}, & \text{if } n = 0 \mod 2, \\ 2^{n-3} - 2^{(n+2i-7)/2}, & \text{if } n = 1 \mod 2. \end{cases} \tag{21}
$$

In particular, if $\gcd(j-k, n) = 1$, we have $k(a,b) \leq 2i$ if n is even; otherwise $k(a,b) \leq 2i - 1$ for all $a, b \in \mathbb{F}_{2^n}$ such that $a \neq 0$ and $b \notin a\mathbb{F}_2$. Therefore, (20) holds for all $a, b \in \mathbb{F}_{2^n}$ such that $a \neq 0$ and $b \notin a\mathbb{F}_2$.

Using Proposition 2, we have the following.

(i) When $n = 0 \mod 2$,

$$
\begin{aligned}
\mathrm{nl}_3\left(f_\lambda\right) &\geq 2^{n-1} \\
&\quad - \frac{1}{2}\sqrt{\left(2^n - 1\right)\sqrt{2^{2n} - 2\left(2^n - 2\right)\left(2^{n-1} - 2^{(n+2i-2)/2}\right)}} \\
&= 2^{n-1} - \frac{1}{2}\sqrt{\left(2^n - 1\right)\sqrt{2^{(3n+2i)/2} + 2^{n+1} - 2^{(n+2i+2)/2}}}.
\end{aligned} \tag{22}
$$

(ii) When $n = 1 \mod 2$,

$$\mathrm{nl}_3(f_\lambda) \geq 2^{n-1} - \frac{1}{2}\sqrt{(2^n-1)\sqrt{2^{2n}-2(2^n-2)(2^{n-1}-2^{(n+2i-3)/2})}} = 2^{n-1} - \frac{1}{2}\sqrt{(2^n-1)\sqrt{2^{(3n+2i-1)/2}+2^{n+1}-2^{(n+2i+1)/2}}}. \quad (23)$$

Theorem 7. *Let $g_\lambda(x) = Tr_1^n(\lambda x^{2^{3\ell}+2^{2\ell}+2^\ell+1})$, for all $x \in \mathbb{F}_{2^n}$ and $\lambda \in \mathbb{F}_{2^n}^*$, where ℓ is a positive integer such that $\gcd(\ell, n) = 1$ and $n > 6$. Then*

$$nl_3(g_\lambda) \geq \begin{cases} 2^{n-1} - \frac{1}{2}\sqrt{(2^n-1)\sqrt{2^{(3n+6)/2}+2^{n+1}-2^{(n+8)/2}}}, & \text{if } n = 0 \mod 2, \\ 2^{n-1} - \frac{1}{2}\sqrt{(2^n-1)\sqrt{2^{(3n+5)/2}+2^{n+1}-2^{(n+7)/2}}}, & \text{if } n = 1 \mod 2. \end{cases} \quad (24)$$

Proof. The proof is similar to that of Theorem 6 up to (18). Here the kernel of $B(x,y)$ associated with $D_bD_ag_\lambda$ is $\mathscr{E} = \{x \in \mathbb{F}_{2^n} : P(x) = 0 = L_{(\lambda,a,b)}(x)\}$, where $L_{(\lambda,a,b)}(x)$ is obtained by replacing i, j, and k in (18) by 3ℓ, 2ℓ, and ℓ, respectively:

$$\begin{aligned} L_{(\lambda,a,b)}(x) = P(x)^{2^{3\ell}} &= \lambda^{2^{3\ell}}\Big(\big(a^{2^{5\ell}}b^{2^{4\ell}} + a^{2^{4\ell}}b^{2^{5\ell}}\big)x^{2^{6\ell}} + \big(a^{2^{4\ell}}b^{2^{6\ell}} + a^{2^{6\ell}}b^{2^{4\ell}}\big)x^{2^{5\ell}} + \big(a^{2^{6\ell}}b^{2^{5\ell}} + a^{2^{5\ell}}b^{2^{6\ell}}\big)x^{2^{4\ell}}\Big) \\ &+ \lambda^{2^{\ell}}\Big(\big(a^{2^{2\ell}}b^{2^{3\ell}} + a^{2^{3\ell}}b^{2^{2\ell}}\big)x^{2^{4\ell}} + \big(a^{2^{2\ell}}b^{2^{4\ell}} + a^{2^{4\ell}}b^{2^{2\ell}}\big)x^{2^{3\ell}} + \big(a^{2^{4\ell}}b^{2^{3\ell}} + a^{2^{3\ell}}b^{2^{4\ell}}\big)x^{2^{2\ell}}\Big) \\ &+ \lambda^{2^{2\ell}}\Big(\big(a^{2^{2\ell}}b^{2^{4\ell}} + a^{2^{4\ell}}b^{2^{2\ell}}\big)x^{2^{5\ell}} + \big(a^{2^{2\ell}}b^{2^{5\ell}} + a^{2^{5\ell}}b^{2^{2\ell}}\big)x^{2^{4\ell}} + \big(a^{2^{5\ell}}b^{2^{4\ell}} + a^{2^{4\ell}}b^{2^{5\ell}}\big)x^{2^{2\ell}}\Big) \\ &+ \lambda\big(ab^{2^{2\ell}} + a^{2^{2\ell}}b\big)x^{2^\ell} + \lambda\big(ab^{2^\ell} + a^{2^\ell}b\big)x^{2^{2\ell}} + \lambda\big(a^{2^{2\ell}}b^{2^\ell} + a^{2^\ell}b^{2^{2\ell}}\big)x. \end{aligned} \quad (25)$$

The coefficient of x in $L_{(\lambda,a,b)}(x)$ is zero if and only if $a^{2^{2\ell}}b^{2^\ell} + a^{2^\ell}b^{2^{2\ell}} = 0$; that is, $a^{2^\ell}b + ab^{2^\ell} = 0$. Moreover, $\gcd(\ell, n) = 1$ and so, by Proposition 4, $b \in a\mathbb{F}_2$. The polynomial $L_{(\lambda,a,b)}(x)$ as represented in (25) is of the form $\sum_{i=0}^{6} c_i x^{2^{i\ell}}$ and so, again by Proposition 4, the equation $L_{(\lambda,a,b)}(x) = 0$ has at most 2^6 roots for all $a, b \in \mathbb{F}_{2^n}$ such that $a \neq 0$ and $b \notin a\mathbb{F}_2$. This implies that $k(a,b) \leq 6$ if n is even; otherwise $k(a,b) \leq 5$. The WHT of $D_bD_ag_\lambda$ at $\mu \in \mathbb{F}_{2^n}$ is

$$W_{D_bD_ag_\lambda}(\mu) \leq \begin{cases} 2^{(n+6)/2}, & \text{if } n = 0 \mod 2, \\ 2^{(n+5)/2}, & \text{if } n = 1 \mod 2. \end{cases} \quad (26)$$

Therefore,

$$\mathrm{nl}(D_bD_ag_\lambda) \geq \begin{cases} 2^{n-1} - 2^{(n+4)/2}, & \text{if } n = 0 \mod 2, \\ 2^{n-1} - 2^{(n+3)/2}, & \text{if } n = 1 \mod 2. \end{cases} \quad (27)$$

Using Proposition 1, we have

$$\mathrm{nl}_3(g_\lambda) \geq \begin{cases} 2^{n-3} - 2^{n/2}, & \text{if } n = 0 \mod 2, \\ 2^{n-3} - 2^{(n-1)/2}, & \text{if } n = 1 \mod 2. \end{cases} \quad (28)$$

Using Proposition 2, we have the following.

(i) When $n = 0 \mod 2$,

$$\mathrm{nl}_3(g_\lambda) \geq 2^{n-1} - \frac{1}{2}\sqrt{(2^n-1)\sqrt{2^{2n}-2(2^n-2)(2^{n-1}-2^{(n+4)/2})}} = 2^{n-1} - \frac{1}{2}\sqrt{(2^n-1)\sqrt{2^{(3n+6)/2}+2^{n+1}-2^{(n+8)/2}}}. \quad (29)$$

(ii) When $n = 1 \mod 2$,

$$\mathrm{nl}_3(g_\lambda) \geq 2^{n-1} - \frac{1}{2}\sqrt{(2^n-1)\sqrt{2^{2n}-2(2^n-2)(2^{n-1}-2^{(n+3)/2})}} = 2^{n-1} - \frac{1}{2}\sqrt{(2^n-1)\sqrt{2^{(3n+5)/2}+2^{n+1}-2^{(n+7)/2}}}. \quad (30)$$

Remark 8. Let $f \in \mathscr{B}_n$ be a biquadratic Boolean function. If there exists at least elements $a, b \in \mathbb{F}_2^n$ such that D_bD_af is quadratic, then $\mathrm{nl}_3(f) \geq 2^{n-4}$. This result follows from Proposition 1 and the fact that the nonlinearity of any quadratic function in $\mathscr{B}_n$ is at least 2^{n-2} [11, 22].

4. Comparison

The theoretical lower bounds for third-order nonlinearities obtained by using Theorem 6 for $i = 3, 4, 5$ and j, k are taken in such a way that $\gcd(j-k, n) = 1$ and reported in Tables 1 and 2. The bounds are compared with the general bounds for third-order nonlinearity: $\mathrm{nl}_3(f) \geq 2^{n-4}$, for any biquadratic

Table 1: The lower bounds on the third-order nonlinearities obtained by Theorem 6 for odd n and $i = 3, 4, 5$.

	n						
	7	9	11	13	15	17	19
$i = 3$	11	75	415	2047	9493	42361	184199
$i = 4$	—	41	330	1660	8191	37979	169457
$i = 5$	—	—	163	1200	6642	32767	151923
General bounds	8	32	128	512	2048	8192	32768

Table 2: The lower bounds on the third-order nonlinearities obtained by Theorem 6 for even n and $i = 3, 4, 5$.

	n						
	8	10	12	14	16	18	20
$i = 3$	21	150	830	4094	18988	84726	368407
$i = 4$	—	82	560	3321	16283	75960	338919
$i = 5$	—	—	326	2400	13284	65535	303849
General bounds	16	64	256	1024	4096	16384	65536

Table 3: Comparison of the value of lower bounds on third-order nonlinearities obtained by Theorem 6 with the bound obtained in [4, 11, 12] for odd n.

n	Theorem 6	[12]	[4]	[11]
7	12	8	16	6
9	76	—	64	60
11	416	240	256	360
13	2048	992	1024	1864
15	9496	—	4096	8872
17	42368	16256	16384	40272
19	184208	65280	65536	177168

Table 4: Comparison of the value of lower bounds on third-order nonlinearities obtained by Theorem 6 with the bound obtained in [4, 11, 12] for even n.

n	Theorem 6	[12]	[4]	[11]
8	22	28	32	20
10	152	120	128	152
12	832	—	512	828
14	4096	2016	2048	4096
16	18992	—	8192	18992
18	84736	—	32768	84736
20	368416	130816	131072	368416

Boolean function. It is evident that the bounds for $i = 3, 4$ are efficiently large and decrease with increasing the value of i. It is to be noted that Class (a) is the more general class of biquadratic monomial Boolean functions containing several classes of highly nonlinear Boolean functions. In particular, for $i = 5$, $j = 4$, and $k = 3$ Class (a) coincides with Kasami functions of algebraic degree 4.

The theoretical bounds for third-order nonlinearities obtained by using Theorem 7 and Proposition 3 are compared with known classes of functions [4, 11, 12] and reported in Tables 3 and 4. It is to be noted that the lower bounds for third-order nonlinearities of the inverse functions ($\mathrm{nl}_3(f_{\mathrm{inv}}) \geq 2^{n-1} - 2^{(7n-2)/8}$) are larger than that of the Dillon functions ($\mathrm{nl}_3(f_{\mathrm{dillon}}) \geq 2^{n-1} - 2^{7n/8}$) for all n. Thus, it is demonstrated that the lower bound obtained by Theorem 7 is better than the bounds obtained by Gode and Gangopadhyay [12] for Kasami functions: $\mathrm{Tr}(\lambda x^{57})$, Iwata and Kurosawa's general bound [4] for all $n > 8$. Also these bounds are improved upon Carlet's [11] bound for inverse function when n is odd, or $n = 8, 12$, and equal for the rest of values of even n.

5. Conclusion

In this paper, using recursive approach introduced in [11], we have computed the lower bounds of third-order nonlinearities of two general classes of biquadratic monomial Boolean functions. It is demonstrated that in some cases our bounds are better than the bounds obtained previously.

Conflict of Interests

The author declares that there is no conflict of interests regarding the publication of this paper.

Acknowledgments

The author would like to thank the anonymous referees for their time, effort, and extensive comments on the revision of the paper which improve the quality of the presentation of the paper. The work is supported by Council of Scientific and Industrial Research, New Delhi, India.

References

[1] J. Golic, "Fast low order approximation of cryptographic functions," in *Advances in Cryptology—EUROCRYPT '96*, vol. 1070 of *Lecture Notes in Computer Science*, pp. 268–282, Springer, 1996.

[2] W. Meier and O. Staffelbach, "Nonlinearity criteria for cryptographic functions," in *Advanced in Cryptology—EUROCRYPT*

'89, vol. 434 of *Lecture Notes in Computer Science*, pp. 549–562, Springer, 1990.

[3] N. Courtois, "Higher order correlation attacks, XL algorithm and cryptanalysis of Toyocrypt," in *Information Security and Cryptology—ICISC, 2002*, vol. 2587 of *Lecture Notes in Computer Science*, pp. 182–199, Springer, 2002.

[4] T. Iwata and K. Kurosawa, "Probabilistic higher order differential attack and higher order bent functions," in *Advances in Cryptology—ASIACRYPT '99*, vol. 1716 of *Lecture Notes in Computer Science*, pp. 62–74, Springer, 1999.

[5] C. Carlet, "Boolean functions for cryptography and error correcting codes," in *Boolean Models and Methods in Mathematics, Computer Science and Engineering*, Y. Crama and P. Hammer, Eds., chapter of the monograph, pp. 257–397, Cambridge University Press, 2010.

[6] C. Ding, G. Xiao, and W. Shan, *Th Stability Theory of Stream Ciphers*, vol. 561 of *Lecture Notes in Computer Science*, Springer, 1991.

[7] L. Knudsen and M. Robshaw, "Non-linear approximations in linear cryptanalysis," in *Advances in Cryptology—EUROCRYPT '96*, vol. 1070 of *Lecture Notes in Computer Science*, pp. 224–236, Springer, 1996.

[8] T. Shimoyama and T. Kaneko, "Quadratic relation of S-box and its application to the linear attack of full round DES," in *Advanced in Cryptology—CRYPTO '98*, vol. 1462 of *Lecture Notes in Computer Science*, pp. 200–211, Springer, 1998.

[9] Q. Wang and T. Johansson, "A note on fast algebraic attacks and higher order nonlinearities," *Lecture Notes in Computer Science (including subseries Lecture Notes in Artificial Intelligence and Lecture Notes in Bioinformatics)*, vol. 6584, pp. 404–414, 2011.

[10] R. Fourquet and C. Tavernier, "An improved list decoding algorithm for the second order Reed-Muller codes and its applications," *Designs, Codes, and Cryptography*, vol. 49, no. 1–3, pp. 323–340, 2008.

[11] C. Carlet, "Recursive lower bounds on the nonlinearity profile of Boolean functions and their applications," *IEEE Transactions on Information Theory*, vol. 54, no. 3, pp. 1262–1272, 2008.

[12] R. Gode and S. Gangopadhyay, "Third-order nonlinearities of a subclass of Kasami functions," *Cryptography and Communications*, vol. 2, no. 1, pp. 69–83, 2010.

[13] S. Gangopadhyay and B. K. Singh, "On second-order nonlinearities of some $\mathscr{D}_0$ type bent functions," *Fundamenta Informaticae*, vol. 114, no. 3-4, pp. 271–285, 2012.

[14] B. K. Singh, "On second-order nonlinearity and maximum algebraic immunity of some bent functions in $\mathscr{PS}^+$," *Journal of Applied Mathematics and Computing*, 2014.

[15] C. Carlet and S. Mesnager, "Improving the upper bounds on the covering radii of binary Reed-Muller codes," *IEEE Transactions on Information Theory*, vol. 53, no. 1, pp. 162–173, 2007.

[16] C. Carlet, "More vectorial Boolean functions with unbounded nonlinearity profile," *International Journal of Foundations of Computer Science*, vol. 22, no. 6, pp. 1259–1269, 2011.

[17] F. J. MacWilliams and N. J. A. Sloane, *Th Theo y of Error Correcting Codes*, North-Holland, Amsterdam, The Netherlands, 1977.

[18] S. W. Golomb and G. Gong, *Signal Design for Good Correlation: For Wireless Communication, Cryptography and Radar*, Cambridge University Press, 2005.

[19] O. S. Rothaus, "On "bent" functions," *Journal of Combinatorial Theo y A*, vol. 20, no. 3, pp. 300–305, 1976.

[20] C. Bracken, E. Byrne, N. Markin, and G. McGuire, "Determining the nonlinearity of a new family of APN functions," *Applied Algebra, Algebraic Algorithms and Error-Correcting Codes*, vol. 4851, pp. 72–79, 2007.

[21] A. Canteaut, P. Charpin, and G. M. Kyureghyan, "A new class of monomial bent functions," *Finite Fields and Their Applications*, vol. 14, no. 1, pp. 221–241, 2008.

[22] J. Seberry, X. M. Zhang, and Y. Zheng, "Relationships among nonlinearity criteria," in *Advanced in Cryptology—EUROCRYPT '94*, vol. 950 of *Lecture Notes in Computer Science*, pp. 376–388, Springer, 1995.

Permissions

The contributors of this book come from diverse backgrounds, making this book a truly international effort. This book will bring forth new frontiers with its revolutionizing research information and detailed analysis of the nascent developments around the world.

We would like to thank all the contributing authors for lending their expertise to make the book truly unique. They have played a crucial role in the development of this book. Without their invaluable contributions this book wouldn't have been possible. They have made vital efforts to compile up to date information on the varied aspects of this subject to make this book a valuable addition to the collection of many professionals and students.

This book was conceptualized with the vision of imparting up-to-date information and advanced data in this field. To ensure the same, a matchless editorial board was set up. Every individual on the board went through rigorous rounds of assessment to prove their worth. After which they invested a large part of their time researching and compiling the most relevant data for our readers. Conferences and sessions were held from time to time between the editorial board and the contributing authors to present the data in the most comprehensible form. The editorial team has worked tirelessly to provide valuable and valid information to help people across the globe.

Every chapter published in this book has been scrutinized by our experts. Their significance has been extensively debated. The topics covered herein carry significant findings which will fuel the growth of the discipline. They may even be implemented as practical applications or may be referred to as a beginning point for another development. Chapters in this book were first published by Hindawi Publishing Corporation; hereby published with permission under the Creative Commons Attribution License or equivalent.

The editorial board has been involved in producing this book since its inception. They have spent rigorous hours researching and exploring the diverse topics which have resulted in the successful publishing of this book. They have passed on their knowledge of decades through this book. To expedite this challenging task, the publisher supported the team at every step. A small team of assistant editors was also appointed to further simplify the editing procedure and attain best results for the readers.

Our editorial team has been hand-picked from every corner of the world. Their multi-ethnicity adds dynamic inputs to the discussions which result in innovative outcomes. These outcomes are then further discussed with the researchers and contributors who give their valuable feedback and opinion regarding the same. The feedback is then collaborated with the researches and they are edited in a comprehensive manner to aid the understanding of the subject.

Apart from the editorial board, the designing team has also invested a significant amount of their time in understanding the subject and creating the most relevant covers. They scrutinized every image to scout for the most suitable representation of the subject and create an appropriate cover for the book.

The publishing team has been involved in this book since its early stages. They were actively engaged in every process, be it collecting the data, connecting with the contributors or procuring relevant information. The team has been an ardent support to the editorial, designing and production team. Their endless efforts to recruit the best for this project, has resulted in the accomplishment of this book. They are a veteran in the field of academics and their pool of knowledge is as vast as their experience in printing. Their expertise and guidance has proved useful at every step. Their uncompromising quality standards have made this book an exceptional effort. Their encouragement from time to time has been an inspiration for everyone.

The publisher and the editorial board hope that this book will prove to be a valuable piece of knowledge for researchers, students, practitioners and scholars across the globe.

List of Contributors

Priyanka Kokil, V. Krishna Rao Kandanvli and Haranath Kar
Department of Electronics and Communication Engineering, Motilal Nehru National Institute of Technology Allahabad, Allahabad 211004, India

Dulal Pal
Department of Mathematics, Visva-Bharati University, Santiniketan, West Bengal 731235, India

Babulal Talukdar
Department of Mathematics, Gobindapur High School, Kalabagh, Murshidabad, West Bengal 742213, India

D. R. V. S. R. K. Sastry
Department of Mathematics, Aditya Engineering College, Surampalem, Ardhra Pradesh 533437, India

A. S. N. Murti and T. Poorna Kantha
Department of Engineering Mathematics, GITAM University, Visakhapatnam, Ardhra Pradesh 530023, India

M. Qasim
Department of Mathematics, COMSATS Institute of Information Technology, Park Road, Chak Shahzad, Islamabad 44000, Pakistan

S. Noreen
Department of Mathematics, COMSATS Institute of Information Technology, Attock 43600, Pakistan

Mohammad Najafi, Maliheh Najafi and Somayeh Arbabi
Department of Mathematics, Anar Branch, Islamic Azad University, Anar, Iran

Ramakanta Meher and Srikanta K. Meher
Department of Mathematics, Sardar Vallabhbhai National Institute of Technology, Surat 395007, India
Department of Mathematics, Anand Agriculture University, Dahod, Gujarat, India

Benito A. Stradi-Granados
Department of Materials Science and Engineering, Institute of Technology of Costa Rica, Cartago 07050, Costa Rica

Zhenguo Luo and Jianhua Huang
Department of Mathematics, National University of Defense Technology, Changsha 410073, China

Zhenguo Luo and Liping Luo
Department of Mathematics, Hengyang Normal University, Hengyang, Hunan 421008, China

Binxiang Dai
School of Mathematical Sciences and Statistics, Central South University, Changsha, Hunan 410075, China

D. Srinivasacharya and O. Surender
Department of Mathematics, National Institute of Technology, Warangal Andhra Pradesh 506 004, India

Xiaofeng Wang and Dongyang Shi
School of Mathematics and Statistics, Zhengzhou University, Zhengzhou 450001, China

Xiaofeng Wang
School of Mathematical Sciences, Henan Institute of Science and Technology, Xinxiang 453003, China

Ishfaq Ahmad Ganaie and V. K. Kukreja
Department of Mathematics, SLIET, Longowal, Punjab 148106, India

Shelly Arora
Department of Mathematics, Punjabi University, Patiala, Punjab 147002, India

Rajitha Gurijala and Malla Reddy Perati
Department of Mathematics, Kakatiya University, Andhra Pradesh 506009, Warangal, India

Mario Lefebvre and Fatima Bensalma
Département de Mathématiques et de Génie Industriel, École Polytechnique, C.P. 6079, Succursale Centre-ville, Montréal, QC, Canada H3C 3A7

Nirmal C. Sacheti and Pallath Chandran
Department of Mathematics & Statistics, College of Science, Sultan Qaboos University, Al Khod, 123 Muscat, Oman

Ashok K. Singh and Beer S. Bhadauria
Department of Mathematics, Banaras Hindu University, Varanasi 221005, India

Beer S. Bhadauria
Department of Applied Mathematics, School of Physical Sciences, Babasaheb Bhimrao Ambedkar University, Lucknow 226025, India

Nassar H. S. Haidar
Center for Research in Applied Mathematics & Statistics (CRAMS), AUL, Lebanon

Anuradha Singh and J. P. Jaiswal
Department of Mathematics, Maulana Azad National Institute of Technology, Bhopal 462051, India

Lan Zhou, Shaowu Zhou and Qiwei Chen
School of Information and Electrical Engineering, Hunan University of Science and Technology, Xiangtan 411201, China

Jinhua She
School of Computer Science, Tokyo University of Technology, Tokyo 192-0982, Japan

Prathima Jayarama
Department of Mathematics, Manipal Institute of Technology, Manipal, Karnataka 576104, India

Prathima Jayarama and Shantha Kumari Kurumujji
SCSVMV, Sri Jayendra Saraswathi Street, Enathur, Kanchipuram, Tamil Nadu 631561, India

Vasudevan Nambisan Theke Madam
Department of Mathematics, College of Engineering, Trikaripur, Kerala 670307, India

Shantha Kumari Kurumujji
Department of Mathematics, P.A. College of Engineering, Mangalore, Karnataka 574153, India

Brajesh Kumar Singh
Department of Mathematics, School of Allied Sciences, Graphic Era Hill University, Dehradun, Uttarakhand 248002, India

www.ingramcontent.com/pod-product-compliance
Lightning Source LLC
Chambersburg PA
CBHW070155240326
41458CB00126B/5248